PROPERTIES OF Plastics

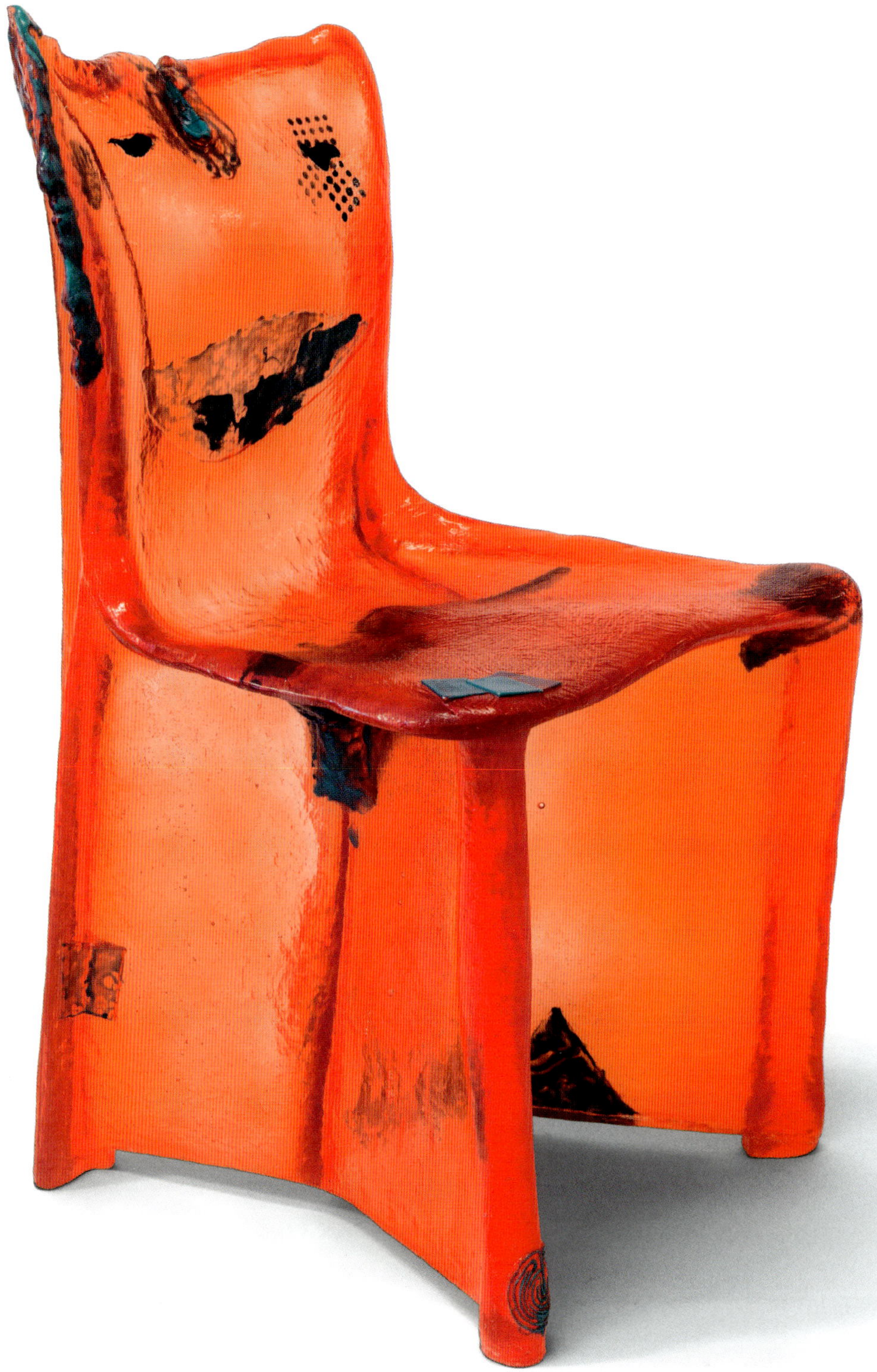

PROPERTIES OF

Plastics

A GUIDE FOR CONSERVATORS

THEA B. VAN OOSTEN

With contributions by

Lydia Beerkens

Ana Cudell

Anna Laganà

Rita Veiga

Getty Conservation Institute Los Angeles

Getty Conservation Institute
Timothy P. Whalen, *John E. and Louise Bryson Director*
Jeanne Marie Teutonico, *Associate Director, Strategic Initiatives and Publications*

The Getty Conservation Institute (GCI) works internationally to advance conservation practice in the visual arts—broadly interpreted to include objects, collections, architecture, and sites. The Institute serves the conservation community through scientific research, education and training, field projects, and the dissemination of information. In all its endeavors, the GCI creates and delivers knowledge that contributes to the conservation of the world's cultural heritage.

Published by the Getty Conservation Institute, Los Angeles
Getty Publications
1200 Getty Center Drive, Suite 500
Los Angeles, California 90049-1682
getty.edu/publications

Laura diZerega, *Project Editor*
Leslie Tilley, *Manuscript Editor*
Kurt Hauser, *Designer*
Victoria Gallina, *Production*
Nancy Rivera and Leslie Rollins, *Image and Rights Acquisition*

Distributed in the United States and Canada by the
University of Chicago Press

Distributed outside the United States and Canada by
Yale University Press, London

Printed in China

Library of Congress Cataloging-in-Publication Data
Names: Oosten, Thea van, author. | Beerkens, Lydia, contributor. | Cudell, Ana, contributor. | Laganà, Anna, contributor. | Veiga, Rita, contributor. | Getty Conservation Institute, issuing body.
Title: Properties of plastics : a guide for conservators / Thea B. van Oosten ; with contributions by Lydia Beerkens, Ana Cudell, Anna Laganà, Rita Veiga.
Description: Los Angeles : Getty Conservation Institute, [2022] | Includes bibliographical references and index. | Summary: "Intended as a highly practical guide for the conservation community, this book offers information essential to understanding plastics, polymers, and rubber/elastomers and their behaviors in the cultural heritage context"— Provided by publisher.
Identifiers: LCCN 2021058482 (print) | LCCN 2021058483 (ebook) | ISBN 9781606066935 (paperback) | ISBN 9781606067680 (adobe pdf)
Subjects: LCSH: Plastics—Conservation. | Art—Conservation. | Art, Modern—20th century. | Art, Modern—21st century. | Museum conservation methods. | Plastics as art material. | Plastics.
Classification: LCC N8560 .O59 2022 (print) | LCC N8560 (ebook) | DDC 702.8/8—dc23/eng/20220210
LC record available at https://lccn.loc.gov/2021058482
LC ebook record available at https://lccn.loc.gov/2021058483

Front cover: Gaetano Pesce (Italian, b. 1939), *Vase, Fish Design Collection*, 1995 (detail, fig. 5.19)

Back cover: Jean Shin (South Korean, b. 1971), *Chemical Balance II*, 2005 (fig. 2.23)

Page 2: Gaetano Pesce (Italian, b. 1939), *Pratt Chair #7*, 1984/2018. Polyurethane resin, Smooth-On pigment, 94 x 48 x 50 cm. Courtesy of the artist and Salon 94 Design, New York. © Gaetano Pesce

Page 14: Heringa/van Kalsbeek (Dutch, b. 1966/b. 1962), *Untitled*, 2007 (detail, fig. 9.1)

Page 16: David Harman Powell (British, b. 1931), red "Melaware" cup and saucer, designed 1958, manufactured by Ranton & Company, Ltd., United Kingdom. Compression-molded melamine-formaldehyde, 12.3 x 6 cm. Bournemouth, Arts University, Museum of Design in Plastics, AIBDC: 007869.1

Page 122: Cellulose acetate sheet from Lourdes Castro (Portuguese, 1930–2022), *Ombres Transparentes*, 1967 (detail, fig. 8.6a)

Page 168: Jean Shin (South Korean, b. 1971), *Chemical Balance II*, 2005 (detail, fig. 2.23)

Illustration Credits

Contents

Foreword

Plastics have become one of the most ubiquitous materials used in manufacturing and design. Since methods of mass producing plastics have multiplied in the last century, objects made of this versatile class of material—ranging from works of art to furniture and household goods—have been finding their way into collections around the world. And as curators, conservators, and collectors have learned, despite the stability and longevity of many types of plastic, others are inherently unstable, and many objects have begun to deteriorate as a result, often very quickly, by yellowing, crazing, cracking, warping, powdering, or becoming sticky.

Conservators therefore face significant challenges in caring for objects made from this relatively new type of material. Since 2007, the Getty Conservation Institute (GCI) has been dedicated to advancing the conservation of plastics as part of its Modern and Contemporary Art Research Initiative, which aims to study a wide range of plastics to understand how to prevent or slow their degradation and to establish appropriate means of preservation and conservation treatments. A valuable partner in this work has been Thea van Oosten, a senior conservation scientist and author of this guide.

With her extensive knowledge of the chemical properties of plastics, Thea has been at the forefront of plastics conservation since 1989, when she began her first research project on plastics at the Cultural Heritage Agency of the Netherlands (RCE). Throughout her career at the RCE, she has spearheaded numerous international plastics research projects, which have helped establish standards in practice. Among these was the POPART (Preservation of Plastic Artefacts in Museum Collections) project, a major international research effort running from 2008 to 2012 that involved twelve research institutions, including the RCE and the GCI, and that remains the largest, most coordinated research effort in this area to date. In 2012, Thea was knighted by Queen Beatrix of the Netherlands in recognition of her outstanding contribution to the field.

In addition to her research, she has taught countless courses and workshops, and has published widely on the subject of conserving plastics. It is no surprise, then, that her proposal to publish this guide was enthusiastically supported by the GCI and her colleagues. In 2015–16, she was a GCI guest scholar, and it was here that she began working on this book in earnest.

Properties of Plastics: A Guide for Conservators is an accessible resource that offers conservators and students the most essential information for understanding each of the main classes of plastics. Beginning with an explanation of polymer structure and the manufacture of plastic and elastomer objects, the guide then summarizes the mechanical, physical, thermal, and optical properties of plastics that determine not only their use and appearance but also their durability and deterioration. Practitioners will find the carefully compiled fact sheets, summarizing the properties of fifty-six plastics and elastomers most commonly found in collections, a valuable tool in conservation labs.

We're grateful to Thea for her many years of research and teaching, her ongoing commitment to the preservation of plastics in collections, and for entrusting us with this important book. Her extensive knowledge, enthusiasm, and commitment to sharing this knowledge are reflected in the pages of this volume.

Despite recent advances in the conservation of plastics, there is still a need to better understand the aging and deterioration of these materials and to improve conservation strategies to preserve them in cultural heritage collections. We believe that this book will contribute greatly to spreading that knowledge and will be a valuable resource for curators, conservators, and collectors.

Timothy P. Whalen
John E. and Louise Bryson Director
Getty Conservation Institute

Acknowledgments

The inspiration to write this book came while I was teaching workshops on plastics for conservators. During these many workshops, participants often asked the same questions: How does a specific plastic react to temperature, to moisture, and to light? What cleaning agent is best to use, and what adhesive should I apply? All these questions can be answered when one knows the properties of plastics. Although information on properties can be found in handbooks on polymer technology and in manufacturers' catalogs, I realized that a book that presented all properties of plastics in one place would be very useful for conservators in helping them understand the behavior of plastics and in their conservation of plastic objects. It was this need for a single source of information that motivated me to write this book.

I would like to thank the Cultural Heritage Agency of the Netherlands (RCE), which made my research into the conservation of plastics possible and gave me the opportunity to travel around the world to gain and exchange plastics knowledge with other conservation scientists and conservators—Tim Bechthold, Christine Frohnert, Roger Griffith, Brenda Keneghan, Kathrin Kessler, Yvonne Shashoua, Friederike Waentig, and many others. The RCE also made it possible for me to give national and international workshops and participate in international research projects on plastics, such as the AXA Art project and the POPART project. I am grateful to all my dear colleagues, but especially to Agnes Brokerhof for her inspiring discussions about conservation, Suzan de Groot for providing FTIR spectra and reference materials, and Henk van Keulen for identifying rubber objects—they made working in the same laboratory more than pleasant.

I would also like to thank the Foundation for the Conservation of Modern and Contemporary Art (SBMK) for letting me organize the content of "plastics days" for conservators and curators of the Dutch modern and contemporary art museums.

Without the help of the Getty Conservation Institute (GCI), this book would not have seen the light of day. First, I thank Tim Whalen, Tom Learner, and Jeanne Marie Teutonico for giving me the opportunity to be a scholar at the GCI, where I met fellow scholars David Saunders and Colm Murray. They have become special friends. Together with Nikki van Basten, we shared many coffee breaks and discussions that bonded us even more closely. Also at the GCI, thanks to Anna Duer for providing extensive literature searches, to Michael Doutre for making the drawings of the chemical structures, and to Joy Mazurek for her helpful comments. To the team at Getty Publications, thanks to Laura diZerega and Leslie Tilley for their meticulous editing of this rather technical manuscript, and to Kurt Hauser, Victoria Gallina, Leslie Rollins, and Nancy Rivera for making the book look so great. Thanks also to Janet Bridgland, from whom I have learned to write clear sentences. I have to thank all of the abovementioned people at the Getty for their faith in my ability to write such technical material in a non-native language. But mostly my thanks go to Cynthia Godlewski, who always gave me encouragement and energy and never gave up on me and this book.

Without the practical case studies, for which data on the properties of plastics were used to determine conservation treatments, this would have remained solely a technical book. I am therefore most grateful to Lydia Beerkens, Ana Cudell, Anna Laganà, and Rita Veiga for their contributions and discussions about the most important properties of plastics in terms of conservation.

Thanks also go to my friend the art historian Annemiek Ouwerkerk, who helped me with the provenance of the photos of artworks in the book, and to my fellow conservation scientists from Italy (Costanza Cucci), Portugal (Ana Ramos), France (Natalie Balcar), and Spain/Germany (Alberto de Tagle) for the translation of the list of terms from English into their respective native languages.

Finally, a very special thank-you to two dear friends. The first, Colin Williamson, was my first invited speaker from abroad in 1993, when I organized the initial plastics day for conservators. Since that time, he has provided me with a great deal of knowledge about plastics, as well as vintage plastics and technical books from his personal collection. The second, Anna Laganà, arrived at my laboratory in 2008 to study plastics conservation research and has since become both a very dear friend and a plastics conservation specialist, always pleasant to work with and to whom I can always turn for excellent advice whenever faced with plastics conservation issues.

Introduction

Almost every museum in the world is confronted with plastics in its collection, and conservators, curators, and collection managers are aware that plastics do not last forever. Since 1960, articles have been published dealing with concerns about the conservation of modern art and modern materials (Althöfer 1980). Further problems with plastics have been noted, such as those affecting cellulose acetate sculptures by the artist Naum Gabo (Pullen and Heuman 1988), cellulose nitrate objects by Antoine Pevsner (Derrick, Stulik, and Ordonez 1993), and polyurethane foam artworks by Piero Gilardi (Hummelen, Sillé, and Zijlmans 1999).

Initially, research into plastics conservation focused on inventories of problems in collections in which an abundance of plastics was normally found, such as visual art, science, textile, and design collections. Studies showed that cellulose esters (nitrate, acetate, acetate-butyrate), polyurethane foams, natural rubber, and plasticized polyvinyl chloride were the most vulnerable to decay over time (Keneghan 1996; Quye 1993; Shashoua and Ward 1995; Then and Oakley 1993; van Oosten et al. 2002; van Oosten, Laganà, and Wagenaar 2012).

Concerns with plastics in works of art and design were also highlighted at a conference devoted to plastics that had the striking title "Saving the Twentieth Century" (Grattan 1993). In the 1990s, museum and conservation staff were often not equipped with the knowledge to properly care for works of art and design made of plastics, rubbers, or elastomers. Acknowledgment of the need for education and training about modern materials and plastics was one of the outcomes of the conference "Modern Art: Who Cares?" (Hummelen, Sillé, and Zijlmans 1999, 297). Over the last twenty-five years, conferences and research initiatives—ranging from the development of methodologies for conservation to intervention on plastics—have increased, and dissemination of this new knowledge has begun to address the gaps in the field (see appendix B).

As of 1998, only a few education facilities, such as the Limburg Conservation Institute—and later the University of Amsterdam—and the Danish Konservatorskolen, in Copenhagen, offered special curricula in the conservation of plastics in modern and contemporary art. Since 1999, when the definition of the profession of conservator/restorer specified that practitioners should have a degree at a master's or doctoral research level, several universities have expanded their programs to include modern and contemporary art, time-based media, and materials, leading to an increase in research projects in other universities and institutes.[1]

Today, programs in conservation centers at universities all over the world, such as the Institute of Fine Arts at New York University, offer a specialty in modern and contemporary

1. ENCoRE, the European Network for Conservation-Restoration Education, provides a list of institutions that offer a specialization or full program in modern and contemporary art conservation. See http://www.encore-edu.org/specialisations.htm.

art. New doctoral programs have been established as well, such as New Approaches in the Conservation of Contemporary Art (NACCA) at Maastricht University, in the Netherlands, and Science and Engineering in Arts, Heritage and Archaeology (SEAHA), in London.

Since 1996, "plastics" workshops, intended to disseminate knowledge about plastics and provide advanced professional development, have been held around the globe. Today, workshops and master classes dealing with plastics conservation have evolved from a theoretical approach to a more hands-on practice—for example, the Conservation of Plastics in Museum Collections workshop series, organized by the Getty Conservation Institute, and the Canadian Conservation Institute's 2018 advanced professional development workshop Conservation of Plastics.

Most books on polymers (large molecules composed of many repeating smaller molecules) and plastics (materials based on composed polymers) are meant for chemistry students or intended to acquaint industry polymer chemists with new applications and manufacturing processes. These references contain a wealth of information about the physical, mechanical, thermal, and optical properties of plastics, although different standard units are used across texts, depending on the country of publication, making the comparison of data sheets on the various plastics and their properties difficult.

What was needed for the conservation community was an easily accessible resource that would clearly outline the complex behaviors of plastics to help in conserving cultural heritage objects made of plastic. This book aims to provide conservators with the essential information to understand polymers, plastics, and elastomers and their behaviors, and to compare the various properties of those materials during conservation. To understand the behavior of plastics and elastomers and their various properties, an explanation of polymer structure and the manufacture of plastic and elastomer objects will be of help.

This book is divided into three parts. Part I provides basic information about polymers and polymerization, plastics, and elastomers. Chapters 1 and 2 provide a general introduction to polymers and plastics. Chapter 3 describes the history and types of elastomers: natural and synthetic rubbers and thermoplastic elastomers. Chapter 4 presents graphs as well as text explanations of the mechanical, physical, thermal, and optical properties of plastics and elastomers. Chapter 5 goes into how polymers are compounded, processed, and manufactured into objects. All of the information in these chapters has been kept simple; suggestions for further reading are given for those who desire more depth.

Part II, chapters 6 through 11, consists of six case studies that illustrate the application of the properties of plastics and elastomers to conservation.

Part III comprises information sheets, here called fact sheets, in which general information and specific data on individual materials' properties have been compiled in a compact format that conservators can utilize in their day-to-day work. Plastics and elastomers were selected for this treatment because they are often found alongside each other in modern and contemporary works of art and are commonly found in cultural heritage objects dating from the beginning of the development of plastics to the newest biobased and biodegradable plastics.[2]

Appendix A is a schema of the plastics and elastomers mostly found in cultural heritage collections. Appendix B lists conferences, research projects, and related publications. Appendix C provides translations of relevant terminology into Dutch, French, German, Italian, Spanish, and Portuguese. Appendix D provides a comparison of the naming conventions for plastics and elastomers used in this book alongside those of the International Organization for Standardization and ASTM International for the plastics found in Part III. Finally, a glossary of technical terms is included at the end of the book.

Our hope is that this information will assist practitioners in choosing the appropriate materials for cleaning, adhering, and consolidating plastic and elastomer objects, with the result that collections will attain longer life spans.

2. Coatings and fibers are not included in this book, nor are fluorinated and high-performance plastics, as the latter are not often found in cultural heritage museums.

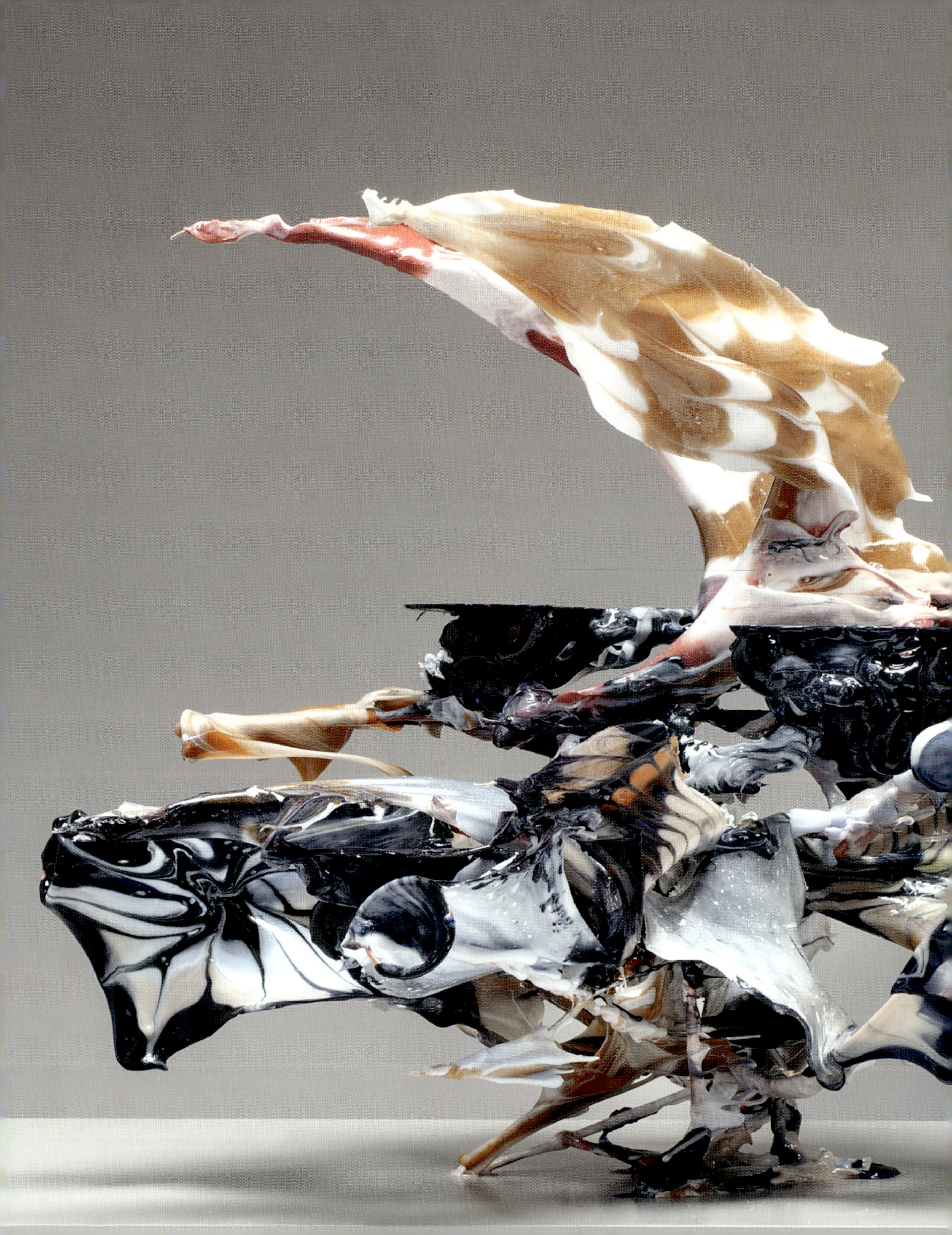

Part I

Polymers, Plastics, and Elastomers

The word *plastic* refers to the property of plasticity: the ability to deform without breaking. Plastics are polymeric materials produced primarily from petroleum and other fossil fuels—and more recently also from natural ("green"), renewable materials made from plant sources such as soy, corn, and wood fibers. Plastics can be molded or shaped by the application of heat and pressure.

Plastics and polymers are not synonymous; all plastics are polymers, but not all polymers are plastics. Polymers are a group of organic chemical substances containing large polymer molecules that serve as the raw materials for plastics. As described in chapter 1, these molecules are formed by linking together small molecules called monomers through a process called polymerization. Plastics are polymers that have been compounded with modifiers, fillers, and additives—such as colorants, plasticizers, stabilizers, and reinforcements—as we will see in chapter 5.

Plastics are versatile materials. Whether derived from petroleum feedstock, biobased feedstock, or biodegradable feedstock, plastics have countless advantages over other materials: They are inexpensive, impermeable, and insulating. They have a low density and thus can be shaped into thin films while maintaining good strength and flexibility, which is important when using a material in packaging to save weight, space, and associated transportation costs. In addition, plastics have lower melting temperatures than glass and metals; therefore, less energy is needed to shape them for a given purpose. Plastics' range of use has been further extended through the industry practice of tailoring properties by mixing different polymers (copolymers and blends) and additives (fibers and fillers) in the same compound.

Nowadays plastics are used in every facet of human activity and industry—from packaging and household items, furniture, fashion, and jewelry to items essential for health care to parts for the automotive, electronics, construction, architecture, and aerospace industries. In fact, plastics are so ubiquitous that they have even found their way into works of art and design (Campos 2007; Sparke 1990).

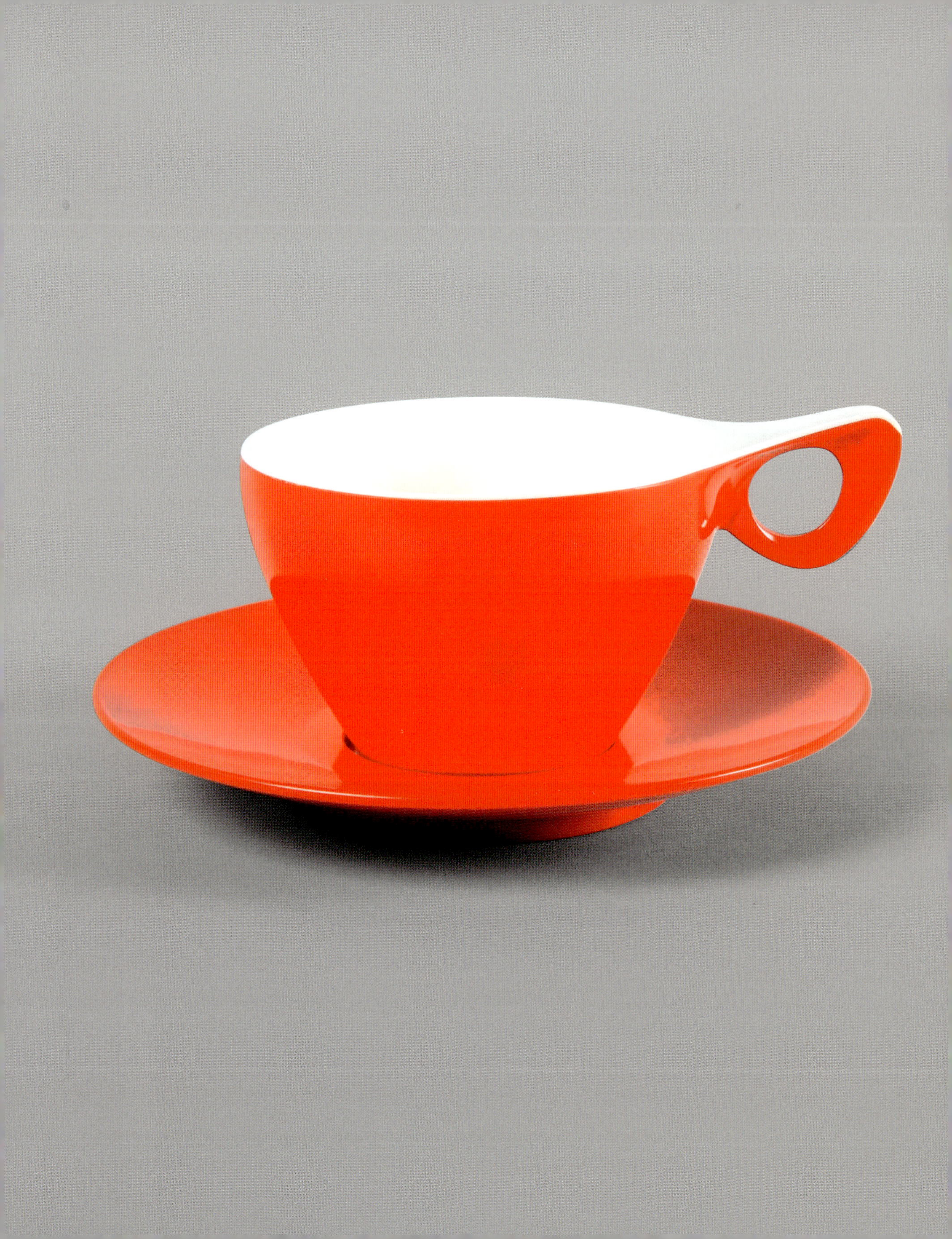

Chapter 1 | An Introduction to Polymers

Many people can easily recognize the difference between types of wood, such as oak and walnut. Similarly, properties of metals are commonly understood; for example, we know that lead is heavier than aluminum. In contrast, names, types, and properties of plastics and elastomers[1] are largely unfamiliar (Lokensgard 2004).

To understand the various types of plastics and elastomers and their characteristics and properties, knowledge of their basic chemistry is helpful. Therefore, this chapter discusses polymers, how their general structure is created through polymerization, and how that structure determines their fundamental properties. This information provides context for the fact sheets on the individual plastics and elastomers found in part III of this book.

Polymers are natural, semisynthetic, or synthetic chemical compounds or mixtures of compounds consisting of long chains of molecules with the repeating links formed from monomers. These terms are derived from the Greek words *polys*, for "many," and *meros*, for "part." Likewise, *monomer* comes from *mono*, meaning "one." Monomers usually have one or more double bonds, or reactive sites, in their chemical structure; they bind to one another to form a repeating chain molecule via the polymerization process.

Polymers have many different physical properties and behaviors, and can be grouped in various ways. One is to classify first on the basis of origin: natural, semisynthetic, or synthetic. Other ways of grouping polymers are by the polymerization mechanism employed, by type of monomers, by molecular structure, or by molecular forces. We will consider each of these in this chapter.

1.1 POLYMERS BY ORIGIN

Polymers are classified into three main categories: natural polymers, semisynthetic polymers, and synthetic polymers.

Natural Polymers

Natural polymers are found in organic sources—plants, trees, and animals. Examples are the cellulose and starch found in plants, rubber from *Hevea brasiliensis* trees, and shellac found in the resinous material of *Kerria lacca* insects. Some natural polymers, including horn, tortoiseshell, and shellac, were the first materials to be plasticized and molded into objects.

Semisynthetic Polymers

Semisynthetic polymers are produced by the artificial modification of natural polymers. These polymers are formed by a chemical reaction of an organic compound and a natural polymer. For example, adding acetic anhydride to cellulose and using sulfuric acid as a catalyst forms cellulose acetate, and adding formaldehyde to casein (milk protein) yields

1. Elastomers are best described as cross-linked rubbery polymers. Originally, *rubber* referred to natural rubber. Later, *elastomer* (elastic polymer) came into use to denote synthetic rubbers. Most rubbery materials are now considered to be a type of elastomeric material, thus *elastomers* in this book should be understood to include rubbers unless otherwise noted. Elastomers are discussed in chapter 3.

casein-formaldehyde. Another example is the vulcanization of rubber by reacting sulfur with rubber (isoprene) molecules.

Synthetic Polymers

Synthetic polymers are industrially produced polymers made by linking monomers derived from fossil feedstock (petroleum, coal, natural gas). A common example is polyvinyl chloride, which is polymerized from vinyl chloride monomers.

1.2 POLYMERIZATION PROCESSES

Polymerization is the process by which monomers are made to react together to form linear polymer chains or a three-dimensional network of polymer chains. Polymerization is caused by the application of light, heat, or pressure and the addition of an organometallic compound as a catalyst and an organic peroxide, such as methyl ethyl ketone peroxide or benzoyl peroxide, as an initiator. Any of these factors can cause the reactive side (double bonds) in monomers to break, and the resulting reactive monomers can generate breakage in other molecules to form long chains of polymers (Brydson 1999).

A monomer with two covalent bonds (a functionality of 2) will lead to a linear polymer. A monomer with a functionality greater than 2 will lead to a network: a cross-linked structure. Most polymerization reactions will fall into one of three classifications: addition polymerization, condensation polymerization, or ring-opening polymerization.

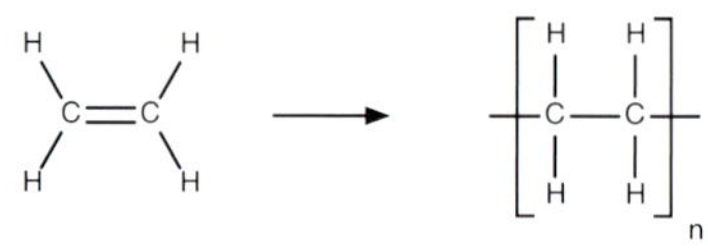

FIGURE 1.1
Addition polymerization of polyethylene from the monomer ethylene.

Addition Polymerization

Addition, or chain-growth, polymerization is the reaction that occurs when at least two reactive groups react with each other without the formation of by-products. Addition polymerization consists of three steps: initiation, propagation, and termination.

Initiation may require a catalyst to begin or speed up the reaction. Propagation involves the continued addition of monomers to the chains. Termination is the final step; at that point all the monomers are used up, causing the reaction to cease (Brydson 1999, 20–27).

The simplest plastic formed by addition polymerization is polyethylene—$(C_2H_4)_n$ (where n is the number of repeat monomer units)—which is polymerized from ethylene monomers (C_2H_4) (fig. 1.1). Since the 1950s, organometallic compounds called Ziegler-Natta catalysts[2] have been used at polymerization. In addition to general cost savings, among the advantages of using these catalysts are that linear polyethylene can be polymerized without branching (stereospecific) and that polypropylene can be polymerized into useful stereoregular products.

Condensation Polymerization

Condensation, or step-growth, polymerization is the reaction that occurs when at least two reactive groups react with each other and a small molecule (usually water) is released

2. These catalysts are named for Karl Ziegler and Giulio Natta, who, respectively, discovered them and developed the process for using them to create polymers. The chemists were awarded a Nobel Prize for their work in 1963.

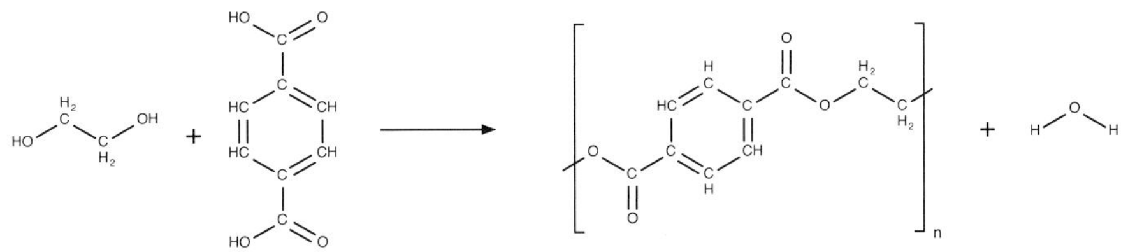

FIGURE 1.2
Condensation polymerization of 1,2-ethanediol and terephthalic acid.

as a by-product. A number of important polymers are made using condensation polymerization, including polyamides (nylons), unsaturated polyesters, and polyurethanes.

One downside of this process is that the hydrolytic stability of some polymers formed by condensation reactions is not very good, and moisture absorption is often a practical problem in the use of this class of polymers.

An example of condensation polymerization with hydrolytic stability is the reaction of 1,2-ethanediol (a bifunctional alcohol) and terephthalic acid (a bifunctional acid) in the formation of polyethylene terephthalate, a linear thermoplastic polyester. This polymer formation is a two-step process that begins when one diacid molecule reacts with one dialcohol molecule, forming a polyester and a water molecule (fig. 1.2).

The ester unit has an alcohol on one end and an acid on the other, which is available for further addition polymerization in which no by-product (water) is released. The eventual result is a saturated polyester (a linear thermoplastic) called polyethylene terephthalate (PET) that has good hydrolytic stability (Brydson 1999; Elias 2003) (fig. 1.3).

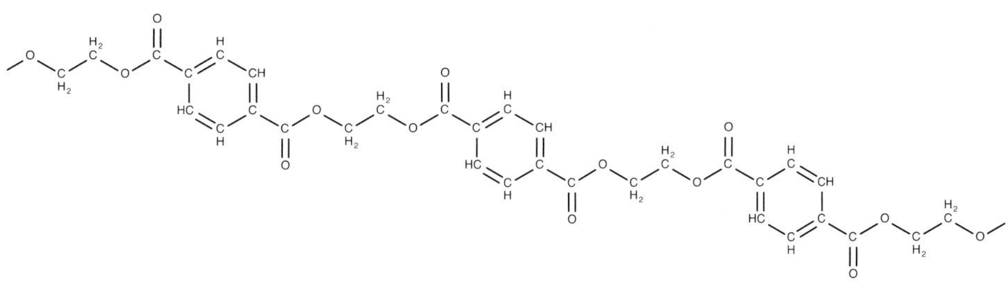

FIGURE 1.3
Polyethylene terephthalate (saturated polyester).

Ring-Opening Polymerization

Ring-opening, or rearrangement, polymerization exhibits the general characteristics of addition polymerization but with an additional (ring-opening) step that occurs at about 260°C in an inert atmosphere without releasing a by-product. The polymerization of ε-caprolactam into nylon-6 proceeds via this ring-opening mechanism (fig. 1.4a). The resulting polyamides (nylon-6) are much less influenced by environmental moisture (water) than nylon-6,6, which is polymerized via condensation polymerization whereby water is released as a by-product (fig. 1.4b).

FIGURE 1.4
Ring-opening polymerization of ε-caprolactam, forming nylon-6 (a); condensation polymerization of adipic acid and hexamethylene diamine, forming nylon-6,6 (b).

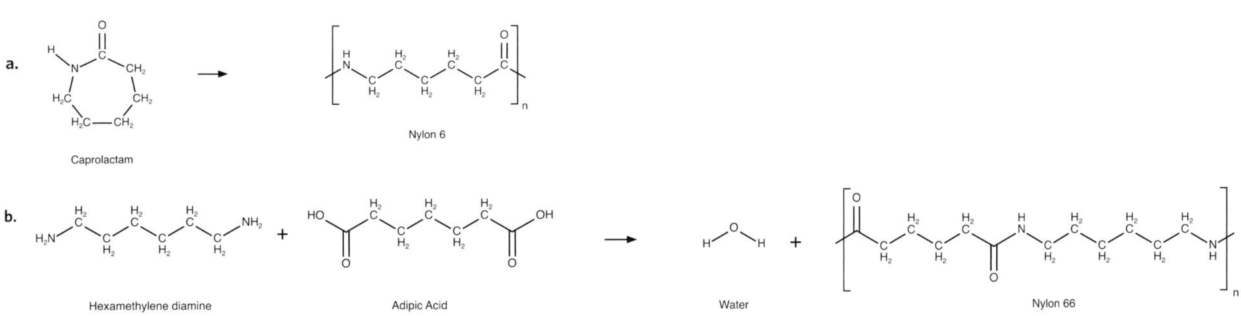

Polymerization and Molecular Weight

Polymers can be distinguished from other types of organic compounds by their unusually large molecules, named macromolecules, the result of polymerization. Molecule size is measured by molecular weight, the sum of the atomic weights of all the atoms in that molecule. A polymeric molecule can consist of several thousand to millions of repeating units (monomers) linked together by covalent or molecular bonds—chemical bonds created by sharing electron pairs between atoms (see "Molecular Forces," later in this chapter). Thus, the molecular weights of polymers are very large.

Because polymerization produces chains of different lengths and molecule size, molecular weights for polymers are average values, calculated by molecular weight distribution. Molecular weight distribution is a key parameter influencing the physical properties of polymers. In data sheets, molecular weight is usually characterized by either the number average molecular weight (Mn) or the weight average molecular weight (Mw). The number of repeat units (n) in a chain is called the degree of polymerization or chain length (Brydson 1999, 40–41).

Some examples of natural, semisynthetic, and synthetic polymers are listed in table 1.1 along with their weight average molecular weights.

Table 1.1. Examples of polymers with their weight average molecular weights (Mw)

Type of polymer	Monomer	Material	Mw (g/mol)
Natural polymer	Rosin	Shellac	10,000–20,000
	Cellobiose	Cellulose	50,000
	Isoprene	Natural rubber	100,000–1,000,000
Semisynthetic polymer	Cellobiose	Cellulose acetate	30,000–60,000
		Cellulose nitrate	30,000–60,000
Synthetic polymer	Ethylene	Polyethylene	3,000,000–6,000,000

1.3 POLYMERS BY MONOMER CONFIGURATION

The terms *homopolymer, heteropolymer, copolymer, terpolymer,* and *polyblends* are frequently used in polymer chemistry. They refer to how many types of monomers repeat in the polymer chain and how they are formed (Young and Lovell 1991, 3–10).

Homopolymers

In some cases, a homopolymer is a polymer consisting of only one type of monomer, and thus one type of repeat unit. Polyvinyl chloride and polyethylene are two common examples of the many existing "poly-plastics." However, in condensation polymerization, often two different monomers are required to synthesize the repeat unit, as is the case in nylon-6,6 and polyethylene terephthalate. These materials are also called homopolymers, since they also have only one type of repeat unit.

Polyvinyl chloride is a linear homopolymer polymerized from the monomer vinyl chloride (fig. 1.5).

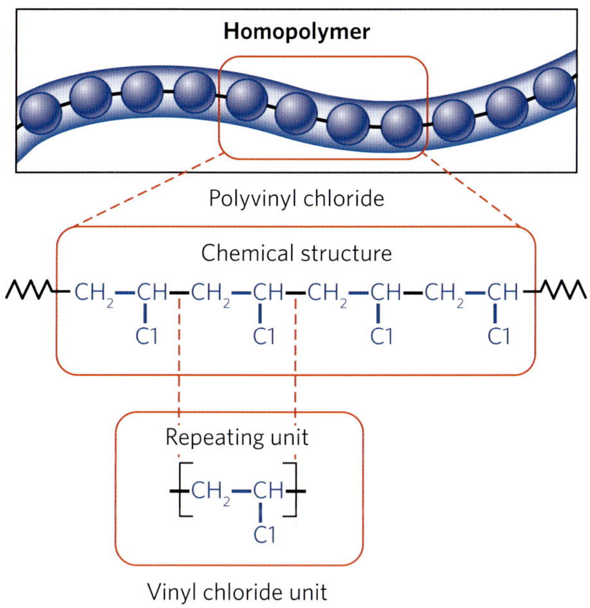

FIGURE 1.5
Polyvinyl chloride, a homopolymer with repeating monomers.

Copolymers

In contrast to a homopolymer, a copolymer has two types of repeat units (referred to generically as A and B). According to this definition, the polyester synthesized from ethylene glycol and terephthalic acid is a homopolymer. However, if the terephthalic acid is replaced by a mixture of terephthalic acid and isophthalic acid, the resulting product is a copolymer.

Copolymers are subdivided by their chain topology. When two different monomers are involved, four possible combinations arise, and their names are based on the monomer sequence distribution on the polymer chain:

- In random copolymers, the monomer sequence distribution is essentially by chance, such as AABBABAB.

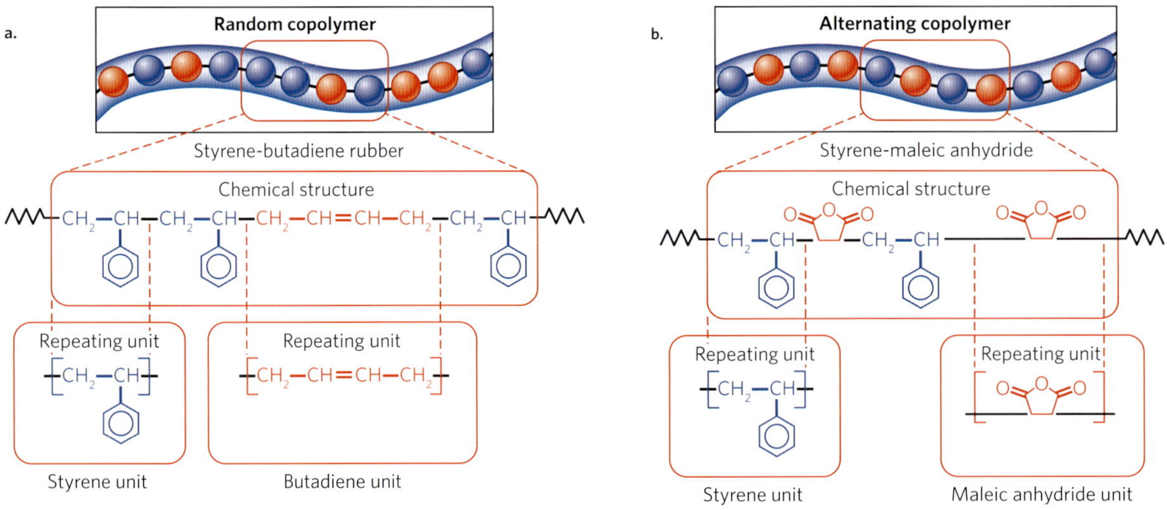

FIGURE 1.6

Styrene-butadiene rubber, a random copolymer (a), and styrene-maleic anhydride, an alternating copolymer (b).

- In alternating copolymers, monomers are built into the polymer chain in pairs: ABABAB.
- In block copolymers, long sequences of the same monomer occur: AAAABBBBAAAABBBB.
- Finally, there are graft copolymers, which have a more complex molecular structure, as shown in figure 1.7b.

The order of the repeating units in a polymer chain is responsible for the properties of a polymer. The repeat units derived from the different monomers may be present in random order, as in styrene-butadiene rubber, which is used to make tires, among other things (fig. 1.6a). Or repeat units of one kind may alternate with repeat units of another, such as in styrene-maleic anhydride, which is used as a coating (fig. 1.6b). Clearly, these two copolymers have very different properties.

Certain homopolymers may have properties that are suited to an intended application yet deficient in some respect. For example, polystyrene is a brittle plastic that can be improved by adding butadiene at polymerization; doing so results in a copolymer with improved impact strength.

When copolymerized, the various polymers can be arranged in a kind of block formation, as in styrene-isoprene rubber, a thermoplastic elastomer (fig. 1.7a), or grafted onto a polymer backbone, as in ethylene-acrylonitrile, for enhancing thermal resistance (fig. 1.7b).

Terpolymers

If three separate monomers (repeat units) are polymerized, the result is called a terpolymer. Examples such as acrylonitrile-butadiene-styrene and acrylonitrile-styrene-acrylate are formed from three different monomers. This combination gives the plastic its good properties, which are not present in the homopolymers of the separate monomers.

Polymer Blends

Polymer blends, or polyblends, are mixtures of polymers that are not chemically bonded to each other. They typically are used in the manufacture of a plastic to achieve a specific property, to improve certain properties, or to lower costs. For example, the blend of polyvinyl chloride and polymethyl methacrylate has greater impact strength (Elias 2003, 323).

FIGURE 1.7
Styrene-isoprene rubber, a block copolymer (a), and ethylene-acrylonitrile, a graft copolymer (b).

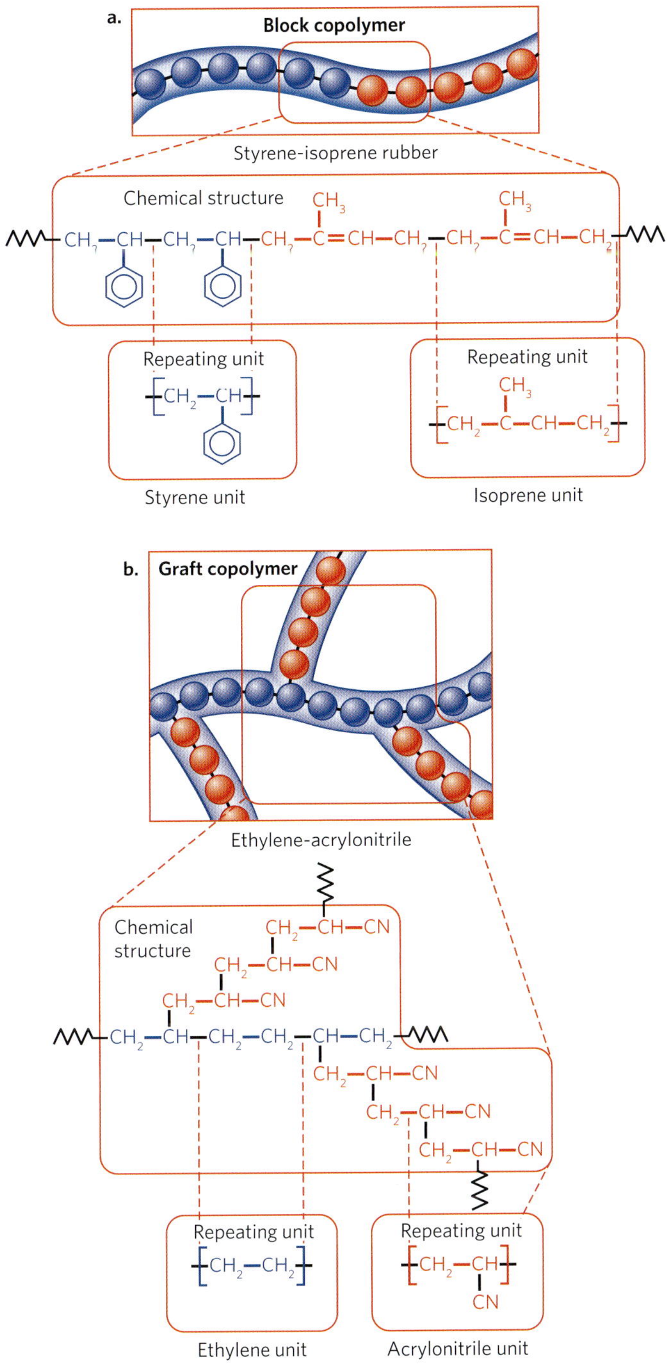

FIGURE 1.8
Properties of copolymers/terpolymers of polystyrene. HIPS = high-impact polystyrene, ABS = acrylonitrile-butadiene-styrene, GPPS = general-purpose polystyrene, SAN = styrene-acrylonitrile.

FIGURE 1.9
Disposable cups made of general-purpose polystyrene.

FIGURE 1.10
Style 20 cooler, designed ca. 1965–69, manufactured by Quattro Elle, Italy. Injection-molded high-impact polystyrene, 40 cm x 36 cm x 22 cm. Bournemouth, Arts University, Museum of Design in Plastics, AIBDC: 005533.

FIGURE 1.11
David Harman Powell (British, b. 1931), Ecko Nova tableware, designed ca. 1966, manufactured by Ekco Plastics, United Kingdom. Injection-molded styrene-acrylonitrile, H: 60 mm; Diam: 82 mm. Bournemouth, Arts University, Museum of Design in Plastics, AIBDC: 001487.

FIGURE 1.12
Joe Colombo (Italian, 1930–1971), Universale chair, designed 1965–67, manufactured by Kartell. Injection-molded acrylonitrile-butadiene-styrene, 72.5 × 43.5 × 43.5 × 43.5 cm (seat height). London, Victoria and Albert Museum. CIRC.887-1968, purchased from Conran Contracts Ltd., London.

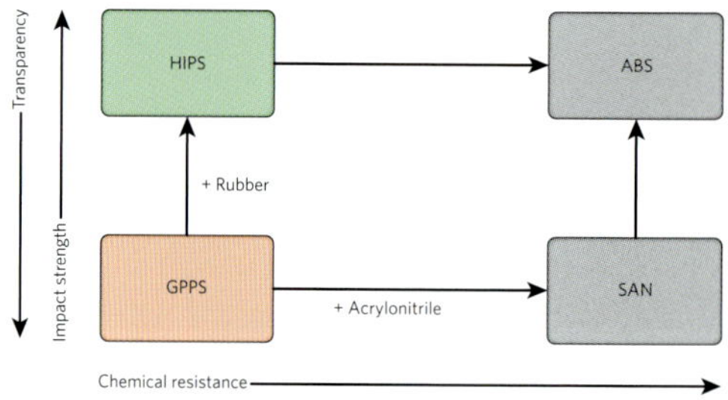

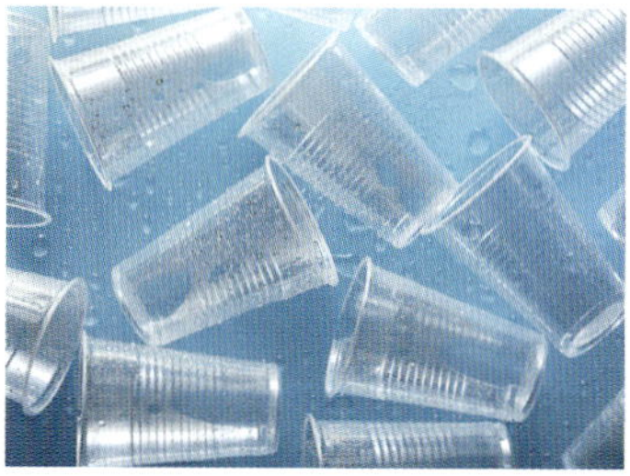

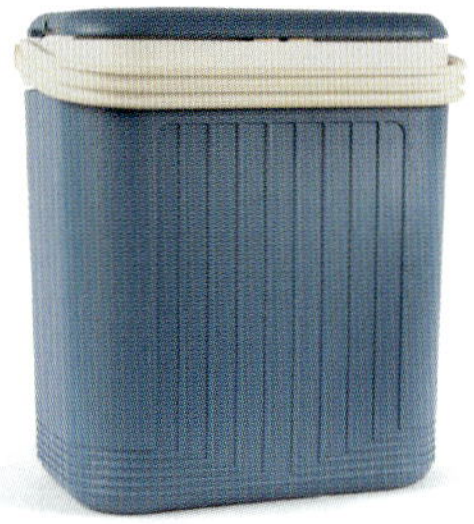

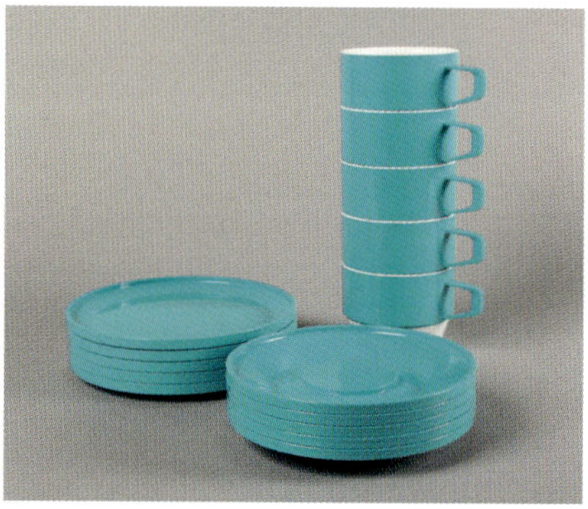

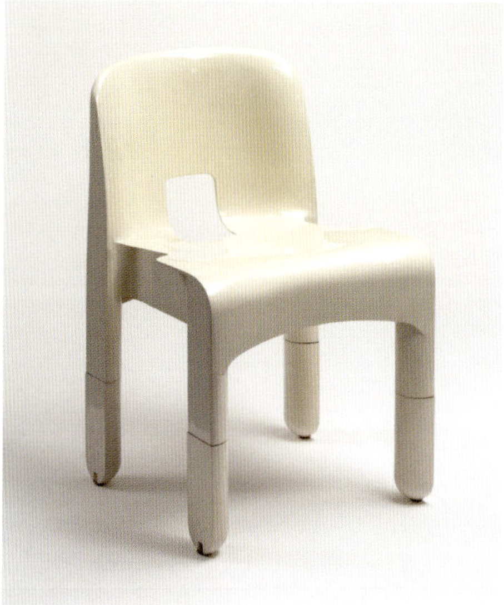

Examples of copolymers, terpolymers, and polyblends are illustrated in the schematic in figure 1.8, where chemical resistance and impact strength increases and transparency decreases according to the composition of the copolymer, terpolymer, or polyblend.

The typical brittle polystyrene known as general-purpose polystyrene is an amorphous, colorless homopolymer in which styrene is the monomer (fig. 1.9). Styrene can be combined with a rubbery polymer to enhance toughness (shock absorption). The arrangement of rubbery particles in the brittle component matrix is responsible for the improvement in mechanical properties. These rubbery chains can absorb energy, which makes the polymer blend (high-impact polystyrene) stronger, less brittle, and capable of taking harder impacts without breaking, compared to regular polystyrene (fig. 1.10).

Styrene-acrylonitrile is a copolymer of styrene and acrylonitrile that provides better chemical resistance than when polystyrene is the sole polymer (fig. 1.11).

Acrylonitrile-butadiene-styrene is a terpolymer that has both impact strength, derived from polybutadiene rubber, and chemical resistance, derived from acrylonitrile (fig. 1.12).

1.4 MOLECULAR STRUCTURE

For polymerization to occur, monomers must be able to link to two or more other monomer molecules. Polymer molecules may exhibit linear, branched, or network (cross-linked) structures, depending on the structure of the monomer(s). Linear or branched polymers are polymerized from monomers that have two active sites (bifunctional monomers), while polymers with a network structure are polymerized from monomers with at least three active sites (trifunctional monomers).

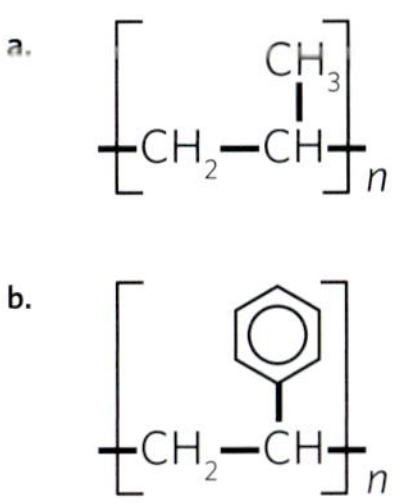

FIGURE 1.13
Polypropylene's branching structure (a) and polystyrene's side-chain structure (b).

Pendant Groups

A group that is part of a repeating unit and attached to the polymer backbone is called a pendant group. Pendant groups are divided into branching and side-chain categories. Branching groups consist of the same molecules as the polymer backbone chain; side chains are made up of molecules different from those of the polymer backbone. For example, polypropylene has a branching methyl (CH_3) group (fig. 1.13a), and polyvinyl chloride and polystyrene have side chains (benzene rings) (fig. 1.13b).

As we will see in chapter 4, the presence and the size of pendant groups modifies the properties of a linear polymer: the flexibility of a polymer chain depends on the groups in the polymer backbone. Polypropylene is much more flexible than polystyrene.

Linear and Branched

As mentioned, synthetic polymers are very large molecules with unusual physical properties that depend on the interactions between their chains. Their structures can take either linear or branched forms. In linear polymers, such as high-density polyethylene (PE-HD) and ultra-high molecular weight polyethylene (PE-UHMW), the molecules can pack closely together. As the intermolecular interactions increase, the plastics become stronger (fig. 1.14a).

Branched polymers, such as low-density polyethylene (PE-LD), have branches of different sizes irregularly spaced along the chain (fig. 1.14b). The branches prevent the nonlinear molecules from packing as closely as they do in the linear polymers, reducing their density and strength (Kaufman 1968, 64).

FIGURE 1.14
Linear and branched polymer chains in high-density (a) and low-density (b) polyethylene, respectively.

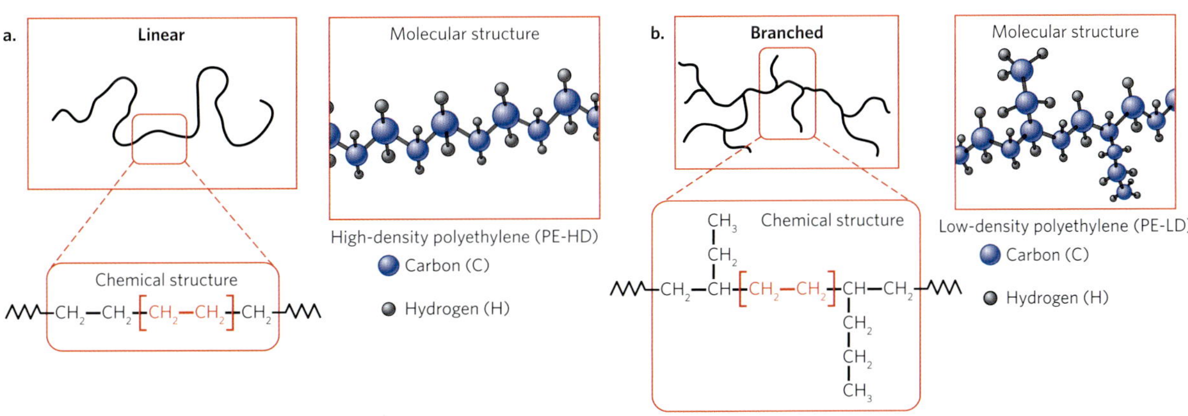

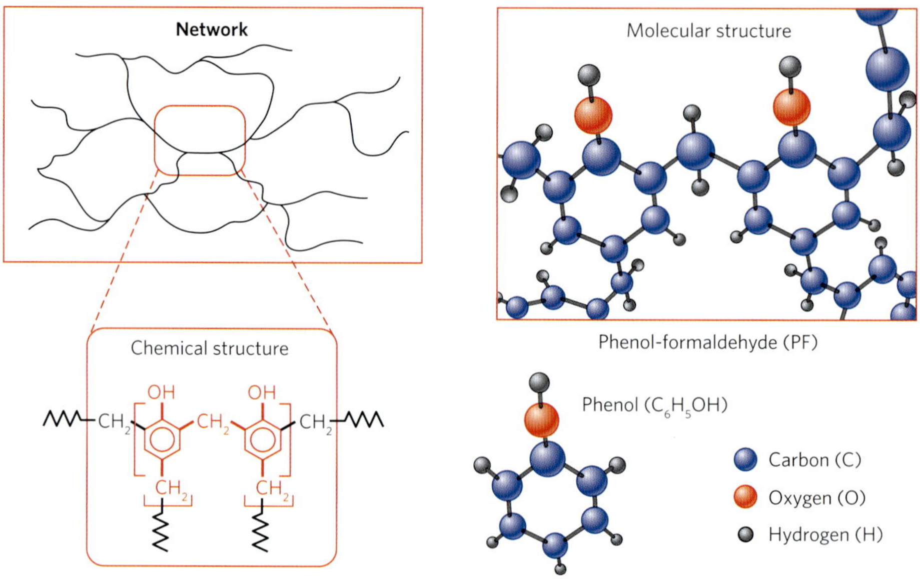

FIGURE 1.15
Phenol-formaldehyde, a three-dimensional network polymer.

Network

When condensation polymerization occurs and bifunctional or trifunctional groups (or both) are present in the monomers, a three-dimensional (cross-linked) network is formed. One example of a three-dimensional network polymer is phenol-formaldehyde, better known under the trade name Bakelite (fig. 1.15).

1.5 MOLECULAR FORCES

Molecular forces determine the chemical and physical properties of the resulting polymer and thus play a part in the behavior of a polymer. There are two types of molecular forces: intramolecular and intermolecular. The first holds together the component atoms of a molecule, while the second bonds molecules together. The bonding by intermolecular forces at crossing points of polymer chains—due to the coiling of long polymer chains (chain entanglement)—affects many properties of plastics.

Intramolecular Forces

The atoms that make up polymers are held together by shared electrons; this is referred to as covalent bonding. This is the most important type of bond for polymers and can involve single, double, or triple bonds between atoms. Chemical covalent bonds between polymer molecules are called cross-links.

Intermolecular Forces

Intermolecular interactions are attractions between molecules; they are much weaker than covalent bonds. Intermolecular forces are called van der Waals forces and comprise three types: dispersive forces (also called London dispersion forces), dipole-dipole forces, and hydrogen bonding. Hydrogen bonds are the strongest intermolecular forces, and dispersion forces are the weakest.

Dispersion forces occur between all molecules of polymer chains. As the size of polymer chains increases, these forces also increase, and this explains why polymers are such strong materials: the molecular weights of polymers range from the tens of thousands up to several million atomic mass units (Brydson 1999, 76–81).

Dipole-dipole forces result from the attraction between the difference in electronegativity of polar groups, such as those in polyesters (C=O) and vinyl polymers (C–Cl).

Hydrogen bonding is a special type of dipole interaction that occurs when the polymer molecule contains C–OH or C–NH groups, such as those in polyamides and polyvinyl alcohol (Kaufman 1968). Hydrogen bonds are the strongest of the intermolecular forces, due to the strong attraction force between hydrogen and oxygen atoms and hydrogen and nitrogen atoms.

Chain Entanglement

As discussed earlier, the linked units of a polymer are typically called a chain. The atoms linked by chemical bonds are said to compose the backbone of the chain. For example, polymers have carbon backbones—chains of carbon (C) linked by a single (C–C) or double (C=C) bond. This structure makes them highly flexible because rotation around single carbon-carbon bonds enables the molecules to assume many different configurations. In other words, the polymer chains can coil up.

Polymer molecules are always long or very long chains, and they can become entangled with one another (fig. 1.16). The intermolecular forces at the crossing points of the coiled-up polymer chains mean that chain entanglement contributes to the unique physical properties of polymers. Because their chains are so difficult to untangle, polymers are extremely strong and resilient (Brydson 1999).

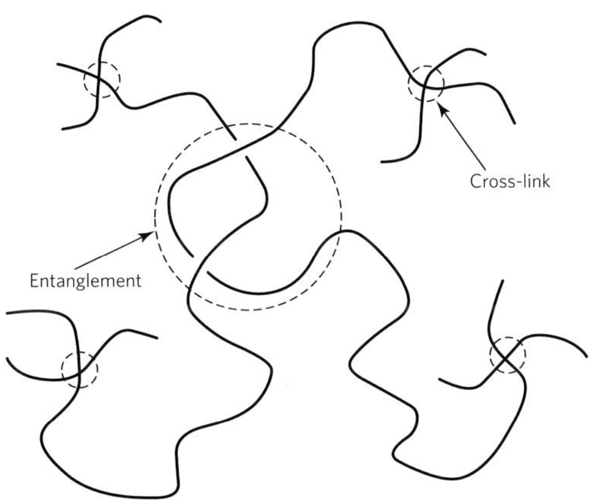

FIGURE 1.16
Cross-linking and entanglement in polymer chains.

> If polymer chains can pack very closely due to their linear structure, the intermolecular forces are stronger. If the molecular weight is also greater (more monomers connected to each other by covalent (intramolecular) bonds, then the resulting plastic is very strong. Polyethylene again provides a good example: PE-UHMW, with a molecular weight of about 3,000,000–6,000,000, is used to make bulletproof vests, while PE-LD, with a molecular weight of 200,000–500,000, is used to make packaging films, grocery bags, household containers, and the like (Rodriguez et al. 2016).

1.6 POLYMER CHAINS IN THERMOPLASTICS, THERMOSETS, AND ELASTOMERS

As mentioned at the beginning of the chapter, polymers consist of long chains of molecules. For the purposes of this book, we are considering three main polymer categories: thermoplastics, thermosets, and elastomers.

Thermoplastic polymer chains are long, unbranched or only slightly branched chains (fig. 1.17a). This structure gives thermoplastics the ability to be remelted and reformed.

Thermosets are densely cross-linked polymer chains (fig. 1.17b). They can be formed only once.

Elastomers' polymer chains fall between thermoplastics and thermosets in terms of structure. They are moderately cross-linked (fig. 1.17c), which in combination with other properties typically gives them good elasticity and allows them to be processed like rubber. They are often termed *millable* or *malleable*.

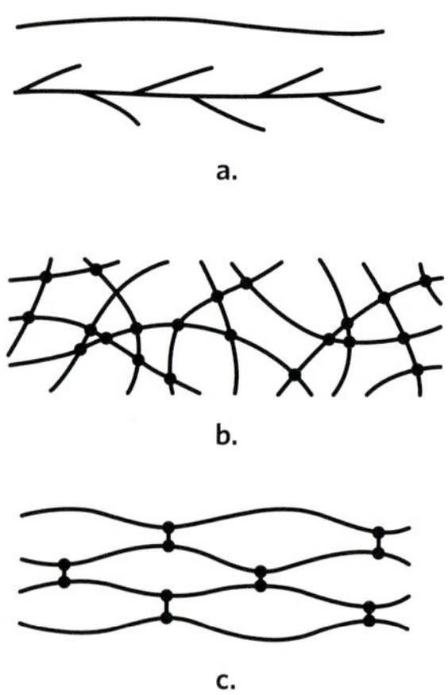

FIGURE 1.17
Thermoplastic (a), thermosetting (b), and elastomeric (c) polymer chains.

Thermoplastics

As mentioned, thermoplastics' mostly linear polymer chains enable them to soften when heated, so they require cooling to solidify them once they are shaped. Thermoplastics are usually created through a process called addition polymerization, in which no cross-linking takes place; therefore, thermoplastics can be repeatedly melted and solidified without changing the chemistry of their mechanical properties. This process is similar to the repeated melting and hardening of chocolate. Examples of thermoplastics are polyethylene, polypropylene, polyamide, and polyvinyl chloride.

Amorphous versus semicrystalline thermoplastics | Thermoplastics can be divided into two types: amorphous and semicrystalline (no polymer is entirely crystalline). Semicrystalline polymers have both crystalline and amorphous regions (fig. 1.18). The classification of a polymer as semicrystalline or amorphous has to do with polymer morphology—whether polymer chains can align and form crystalline regions.

Amorphous regions of a polymer are made up of randomly coiled and entangled chains that resemble cooked spaghetti on a plate. Amorphous thermoplastics are softer, have lower melting points, and are more vulnerable to solvents than semicrystalline thermoplastics. For example, polystyrene and polymethyl methacrylate have side groups that

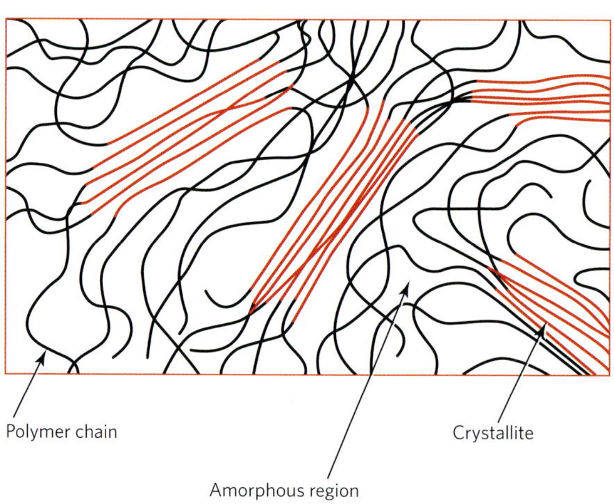

FIGURE 1.18
Amorphous regions and crystallites in semicrystalline polymers.

prevent the chains from coiling up in an orderly way; those polymers are therefore amorphous. In use, amorphous polymers tend to be rather stiff, brittle, and optically transparent.

The less the polymer chains are branched, the higher the crystallinity of the polymer. For instance, the crystallinity of polyethylene can range from 35% in PE-LD to 80% in PE-HD.

Semicrystallinity is an advantageous property in most plastics because it combines the strength found in crystalline polymers with the flexibility of amorphous polymers, allowing the material to bend without breaking. Typical semicrystalline thermoplastics include polyamides (nylons), polypropylene, and polyacetal (polyoxymethylene). Such polymers are usually opaque and tough.

Highly semicrystalline polymers such as nylon have high melting points (due to their hydrogen bonding), generally are rigid, show greater tensile strength than low crystalline polymers, and are less affected by solvent penetration. However, crystallinity makes polymers less impact resistant (Young and Lovell 1991, 252–53).

Crystallinity can be improved by drawing, or stretching, the bulk material. This is performed during synthesis or processing, most commonly during the manufacture of films. When a film is formed, the small crystallites are usually positioned randomly. Drawing the film causes individual chains to be pulled into a somewhat parallel organization (fig. 1.19). These films can be oriented in one direction (uniaxially) or in two directions (biaxially).

Transparent plastics have randomly coiled-up polymer chains with none or very few crystalline-aligned regions. Opaque plastics have regularly aligned polymer chains and a high degree of crystallinity. Light that passes through a plastic is scattered by crystalline areas, so the more ordered the crystalline areas, the more the light is scattered, and the more opaque the plastic (Brydson 1999, 121).

Stereoregularity | Molecular weight, linearity, branching, pendant groups, and stereoregularity are factors that affect the way polymer chains are arranged.

Polymers with side chains and branching can be divided into two classes: stereoregular, in which there is a recurring pattern of pendant groups, and atactic, which has a random structure. When a polymer molecule contains a chiral center (an asymmetrical carbon atom), different configurations, or optical isomers, are possible.

Polypropylene is an example of stereoregularity. Polypropylene has asymmetrical carbon atoms, which allow it to be polymerized to form three basic chain structures. Two are

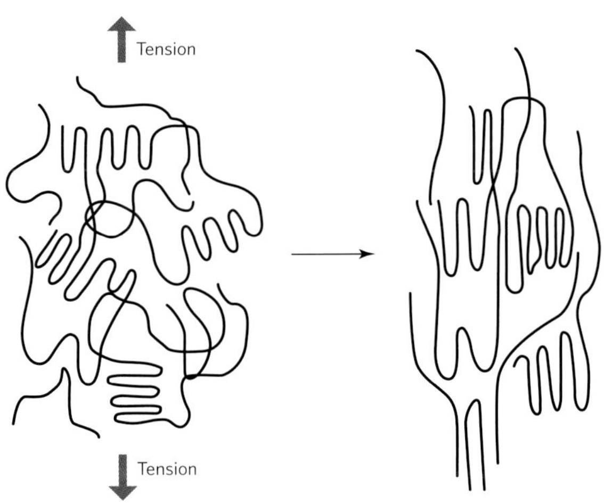

FIGURE 1.19
Arrangement of polymer chains under tension.

stereoregular: isotactic, which has a repeating pendant group on the same side of the polymer backbone (fig. 1.20a), and syndiotactic, which has a regular repeating pendant group on both sides of the polymer backbone (fig. 1.20b). The third structure is atactic; it has a randomly irregular repeating pendant group on both sides of the polymer backbone (fig. 1.20c).

Stereoregularity affects the arrangements of polymer chains. Regular and close packing of polymer chains, such as in isotactic polypropylene, provides a highly crystalline material with a melting point of 160°C. Syndiotactic polypropylene, with slightly differently arranged polymer chains, is also crystalline; it melts at 130°C. The randomly ordered polymer chains in atactic polypropylene provide a soft, noncrystalline polymer with rubberlike properties that melts at only 75°C.

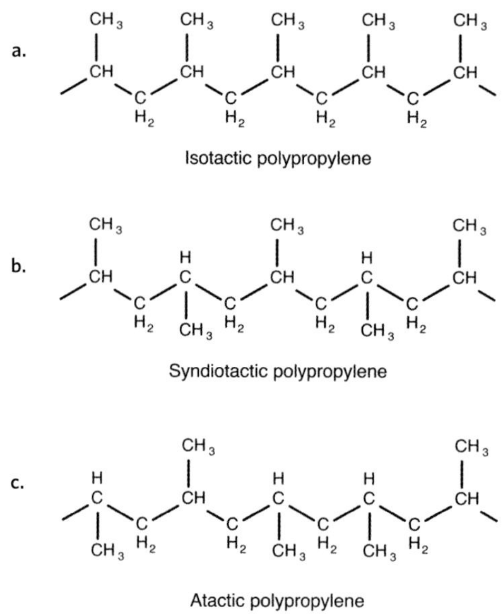

FIGURE 1.20
Configurations of polypropylene: isotactic (a), syndiotactic (b), and atactic (c).

Thermosets

Thermosetting plastics, or thermosets, are network polymers that consist of long molecules with complex cross-linkage. They generally require both heat and pressure to mold them into shape. The three-dimensional network polymer chains require bi- and trifunctional groups in the monomers.

Thermosets set irreversibly when heated; while strong and durable, they are also hard and brittle, and cannot be easily reprocessed. They might be compared with baked cookie dough, which cannot be returned to its previous state. Examples of thermosets are epoxies, unsaturated polyesters, urea-formaldehyde, and melamine-formaldehyde.

Unsaturated polyester, formed by condensation reaction of a polyol and an unsaturated dibasic acid or dibasic anhydride, is an example of a thermosetting polymer. Styrene is added as a cross-linker. The reaction is initiated by an organic peroxide such as benzoyl peroxide or methyl ethyl ketone peroxide. An unsaturated polyester cannot be remelted and reformed (Brydson 1999; Elias 2003) (fig. 1.21).

FIGURE 1.21
Unsaturated polyester, a thermosetting polymer.

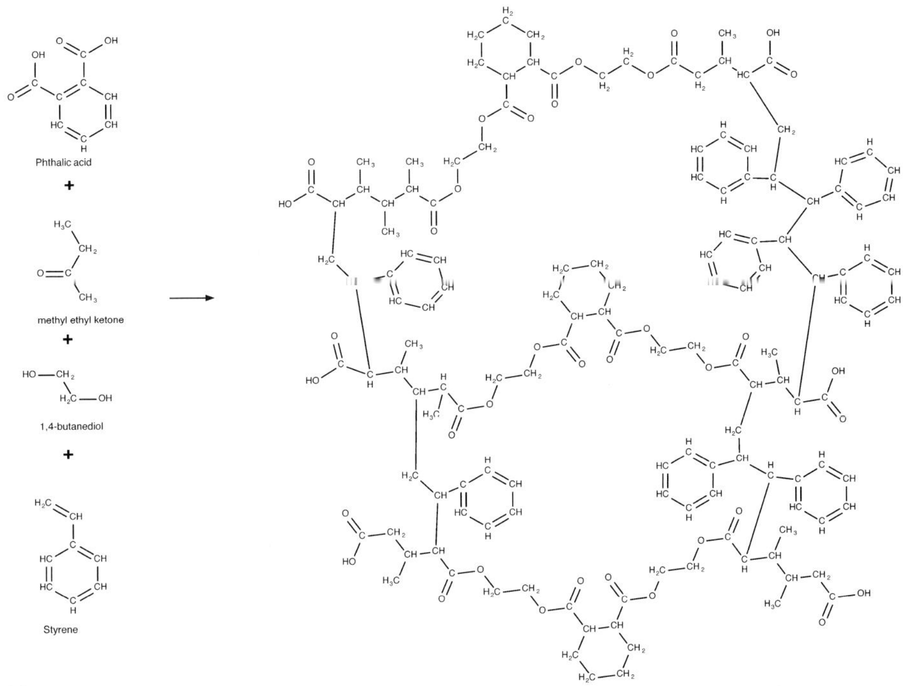

Elastomers

An elastomer is a rubbery polymer; *elastomer* is a shortening of *elastic polymer.* The long molecules that make up an elastomeric material are normally coiled irregularly but straighten and stretch out when force is applied. Then, when the force is released, the molecules return to their usual compact, random arrangement. Because of their generally weak intermolecular forces, elastomeric materials can stretch and recover with limited permanent set or distortion. They can be compared to gummy bears or licorice whips.

One of the best-known elastomers is rubber. A chemical process called vulcanization is used to harden rubber. Sulfur is added to polyisoprene,[3] causing the molecules to link with adjacent molecules at intervals along their length, forming a cross-linked elastic material that is stable at a wide range of temperatures. In this process, sulfur chain fragments attack the double bonds in the polymer chains, causing cross-linking (fig. 1.22).

Elastomers are generally cross-linked but have a low cross-link density. When elastomers have few cross-links, they resemble thermoplastics; those with more cross-linking resemble thermosets.

FIGURE 1.22
Cross-linking of polyisoprene by sulfur in a process called vulcanization.

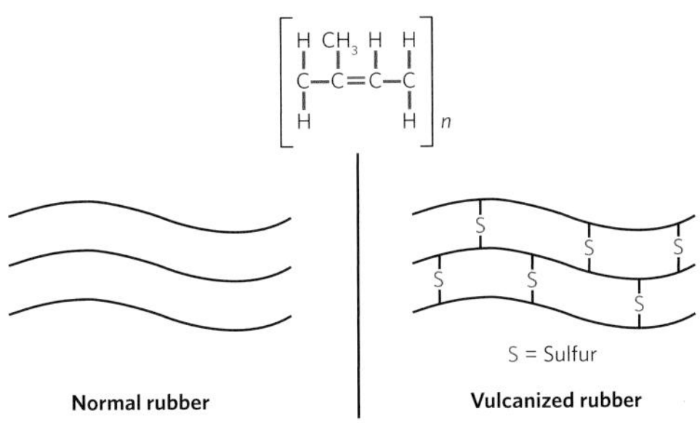

3. Polyisoprene, a polymer of isoprene (C_5H_8), is the primary component of natural rubber, balata, and gutta-percha.

Summary

- Polymerization takes place by:
  - Addition polymerization, in which a polymer is formed by an addition reaction, where many monomers bond together via rearrangement of bonds without the formation of by-products.
  - Condensation polymerization, in which a polymer is formed by a condensation reaction with the formation of by-products.
  - Ring-opening, or rearrangement, polymerization, which exhibits the general characteristics of addition polymerization but with an additional (ring-opening) step, without releasing by-products.
- Polymers:
  - Polymers can be homopolymers or heteropolymers (copolymers or terpolymers) or polyblends.
  - Copolymers, terpolymers, and polyblends are developed to improve the properties of a plastic, for instance, to enhance strength, flexibility, and/or service life.
- Thermoplastics, thermosets, and elastomers:
  - Thermoplastics are polymerized from bifunctional monomer groups, and linear or slightly branched polymer chains are formed.
  - Thermosets are polymerized from trifunctional or even more complex monomer groups, and three-dimensional cross-linked polymer chains are formed.
  - Elastomers are long-chain polymeric materials with specific elastic properties.
- Semicrystalline versus amorphous:
  - Whether a polymer has side chains or is branching determines whether it is crystalline or amorphous.
  - Small side chains yield crystalline structures, whereas larger side chains yield amorphous structures.
  - Amorphous polymers are transparent, and semicrystalline polymers are translucent or opaque.

Suggested Reading

Charles E. Carraher Jr., *Carraher's Polymer Chemistry: An Introduction*. 10th ed. (Boca Raton, FL: CRC Press, 2018).

Kishan Khemani and Carmen Scholz, "Introduction and Overview of Degradable and Renewable Polymers and Materials," in *Degradable Polymers and Materials: Principles and Practice*, 2nd ed., ed. Kishan Khemani and Carmen Scholz, 3–10 (Washington, DC: American Chemical Society, 2012).

Chapter 2 | History and Types of Plastics

This chapter begins with a short history of plastics.[1] To gain insight into the significance and versatility of plastics, we'll then look at two major categories of plastics—thermoplastics and thermosets—which encompass various subcategories: natural plastics; modified natural, or semisynthetic, plastics; and fully synthetic plastics, as well as biobased and biodegradable plastics.

2.1 HISTORY OF PLASTICS

The story of plastics is the story of industry and society from the nineteenth century to the present day. Many advances in domestic, industrial, commercial, and technological design could not have been achieved without the development of plastics (Katz 1984, 1994; Mossman 1997).

Nineteenth-Century Plastics

Historically, humans have relied on materials from nature to make objects. Long before the development of commercial plastics, various natural materials displayed unique properties that enabled them to be plasticized and molded and carved for use in a variety of products, such as tortoiseshell combs, gramophone records made from shellac, gutta-percha picture frames, and horn knife handles. When precious, expensive materials such as tortoiseshell and ivory became scarce at the end of the 1800s, there was a need to replace those natural materials with less expensive alternatives.

Experimentation into plasticizing rubber and the use of cellulose nitrate was conducted in England and the United States during the latter half of the nineteenth century. The birth of the plastics industry occurred in 1862, when the Englishman Alexander Parkes invented a new material he called Parkesine (fig. 2.1). Made from cellulose and nitric acid, it was the first commercial plastic. However, it was not Parkes but the American John Wesley Hyatt who obtained the patent for Celluloid, the trade name for cellulose nitrate (Mossman 1997). Celluloid retained its status as the sole manufactured plastic for the rest of the nineteenth century—"By 1900 the only plastics materials available were shellac, gutta percha, ebonite and celluloid" (Brydson 1999, 3–4)—but concurrent experiments led to materials such as casein-, phenolic-, and urea-based plastics, and cellulose acetate in the early twentieth century.

The number of synthetic materials continued to increase from that point forward, especially after Leo Baekeland's discovery of the fully synthetic phenol-formaldehyde in 1907. Baekeland called his plastic Bakelite, and it was described as "a material of a thousand uses" (Katz 1994; Mossman 1997). The limiting factor of Bakelite was its dark tones and narrow

1. More detailed information can be found in the timeline of plastics published by the British Plastics Federation (see www.bpf.co.uk) and in books about the conservation of plastics (Quye and Williamson 1999; Shashoua 2008; Waentig 2008). See the Suggested Reading list at the end of the chapter.

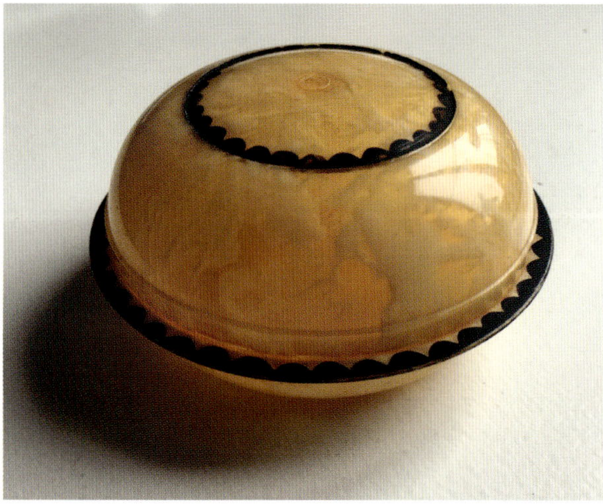

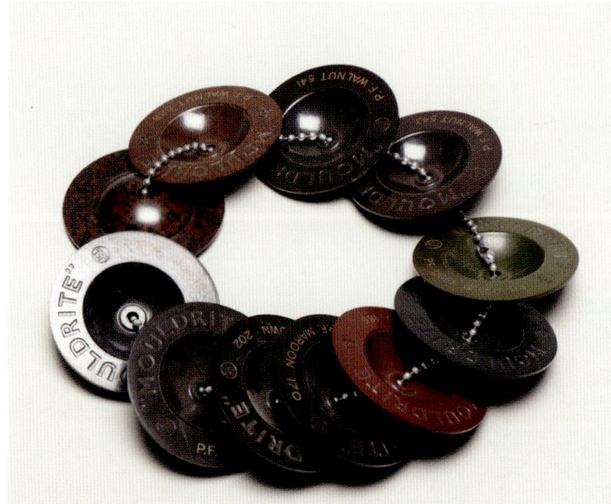

FIGURE 2.1
Imitation marble box made of cellulose nitrate.

FIGURE 2.2
Dark green, brown, and red phenol-formaldehyde (Bakelite) sales samples.

FIGURE 2.3
Naum Gabo (Russian, 1890–1977), *Construction in Space (Crystal)*, 1937–39. Cellulose acetate, 22 × 27 × 18 cm. London, Tate, T06978.

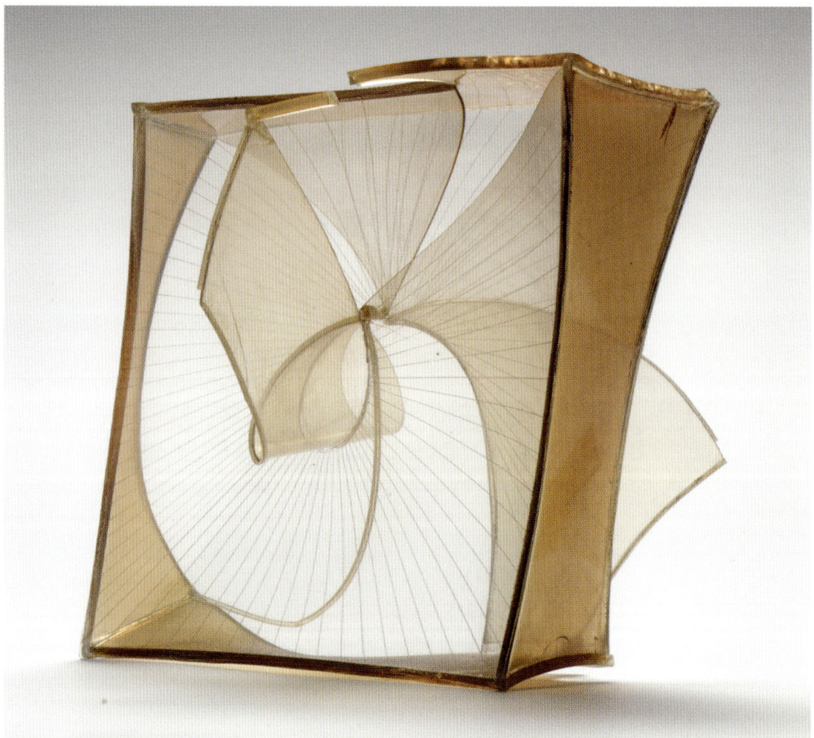

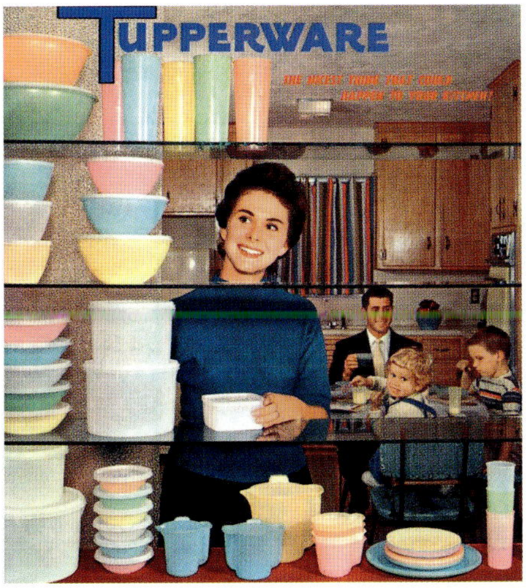

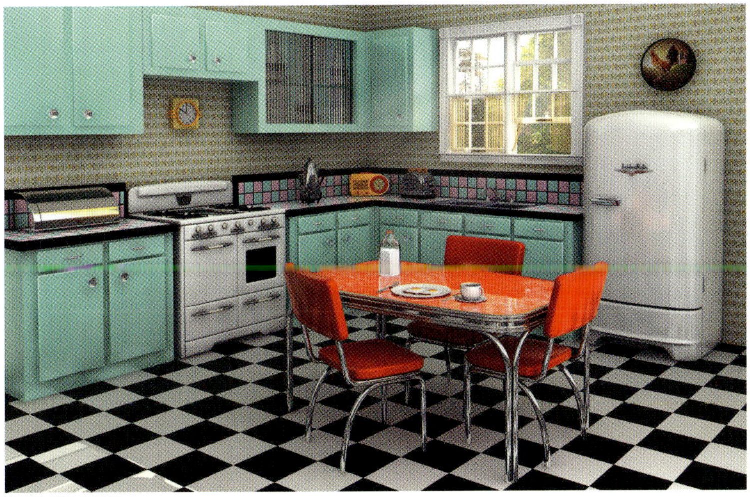

FIGURE 2.4
Tupperware polyethylene kitchenware.

FIGURE 2.5
Kitchen featuring table with Formica (melamine-formaldehyde) surface.

range of colors (fig. 2.2). Overcoming this problem led researchers to the development first of light-colored urea-formaldehyde, in the early 1920s, and then melamine-formaldehyde, in the early 1930s. At the same time that urea-based plastics were being developed, cellulose acetate was making its debut (fig. 2.3). The discovery of suitable plasticizers in 1927 led to the introduction of this material as a nonflammable alternative to celluloid, and the introduction of injection mold manufacturing made it possible to mass produce plastic objects (Katz 1984).

Plastics from 1900 to 1945 and Beyond

During World War I, the demand for rubber led to the development of the first synthetic rubber and many other synthetic plastics, especially after the discovery by the organic chemist Hermann Staudinger that plastics are composed of long-chain molecules (Staudinger 1920). In the early days of the plastics industry, the main source of chemicals used to produce plastics was coal tar. This changed in the 1930s, when petroleum became the main source instead. This not only brought down the price of plastic products but also led directly to the development of new polymers (Katz 1984; Kaufman 1968).

Between 1930 and 1940, industrial applications brought about four of today's major types of thermoplastic: polyolefins, polyvinyl chloride, polystyrene, and polymethyl methacrylate. In that same period, polyethylene began to be produced by Imperial Chemical Industries in 1933. Uses for the semiflexible plastic were soon developed, and it is well known today in applications such as bottles, dispensers, Tupperware containers, and packaging films (fig. 2.4).

World War II heightened demand for plastics, as natural materials were in short supply, leading to major research into polymer chemistry and the large-scale production of materials such as polyamide (nylon) and new types of laminates. Most of the plastics first used in the war industry later became available for use in daily life.

The use of catalysts and the availability of inexpensive petroleum feedstock led to the rapid growth of synthetic plastics, and after 1945 the "poly era" really began (Sparke 1990). In addition to those plastics already mentioned, decorative laminates such as Formica became popular—made of melamine-formaldehyde, it was used for kitchen tables and chairs (fig. 2.5), in hotel bars, and in cafeterias and movie theaters (Katz 1984).

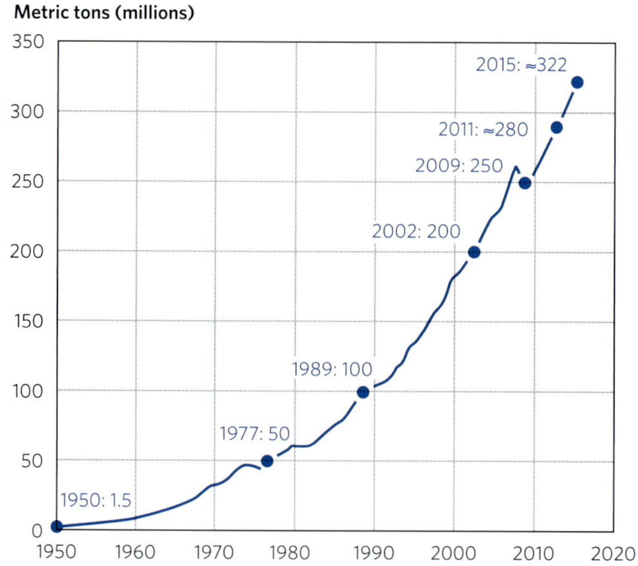

FIGURE 2.6
World production of plastics, 1950–2015, in millions of metric tons.

From 1950 onward, demand for plastics rose at an ever greater pace, albeit with some hiccups, typically coinciding with economic recession (fig. 2.6). The period of 1978–82 marked one such slowdown. The energy crisis that had begun in 1973 resulted in a shortage of materials, and by 1979 oil prices had quadrupled. The economic recession that followed temporarily affected the demand for plastics. Growth resumed from 1983 to 2008, when another economic downturn again led to a decline in plastics production. Nonetheless, since the 1980s plastics have been the most used type of material worldwide (Glenz 1984, 6–8; Ward 1997). And since 2009, the growth in production and consumption of plastics has continued to increase, in part due to innovative efforts by industry.

Plastics in Art and Design

Historically, plastics were not just found in industrial and household items; artists and designers also discovered the advantages of these new materials. They were inexpensive and ready to use, and equally important, they could express the materiality that artists were looking for.

Artists such as Naum Gabo, László Moholy Nagy, and Antoine Pevsner started using cellulose nitrate, cellulose acetate, and polymethyl methacrylate in the early 1930s. Thirty years later, in the 1960s, there was a boom in the use of plastics, as Italian artists such as Piero Gilardi, Gaetano Pesce, Piero Manzoni, and others recognized the possibilities (Chiantore and Rava 2005). One material that found favor with artists and designers was polyurethane foam. At once "flexible and rigid, it freed furniture from the restrictions of traditional construction" (Katz 1984, 13–14). The development of soft and hard foams with a molded skin resulted in furniture taking unconventional forms—including rocks, cacti, blades of grass, and other shapes—produced by the company Gufram since about 1970 (Katz 1984, 13–14) (fig. 2.7).

Plastics have found acceptance all around in the world and especially among Italian and German industrial designers, who are known for their elegant innovations. Plastics have enabled them to undertake projects that could not have been realized with any other material. Designers such as Anna Castelli Ferrieri, who worked for furniture maker Kartell,

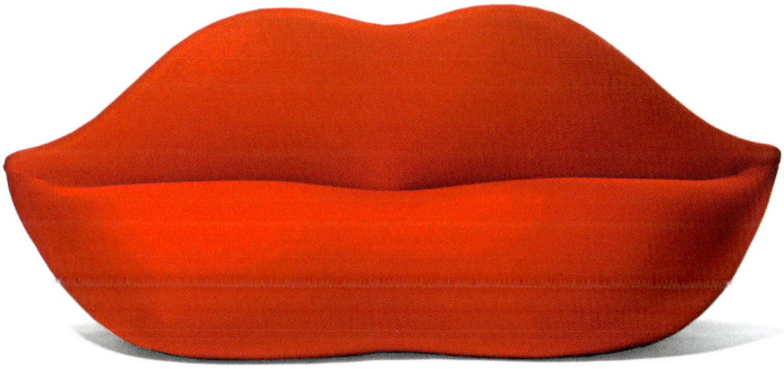

FIGURE 2.7
Studio 65, Bocca sofa (a.k.a. Red Lip sofa, Marilyn sofa), manufactured by Gufram since 1970. Polyurethane foam, 80 × 212 × 85 cm. Gufram, Italy.

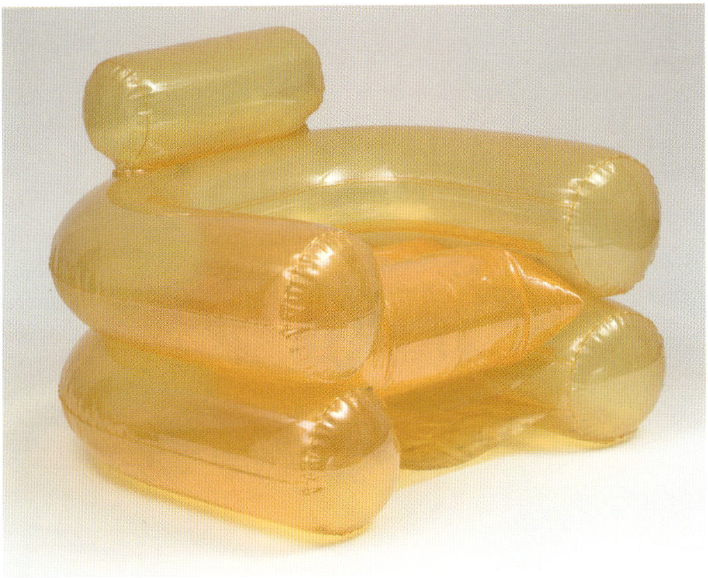

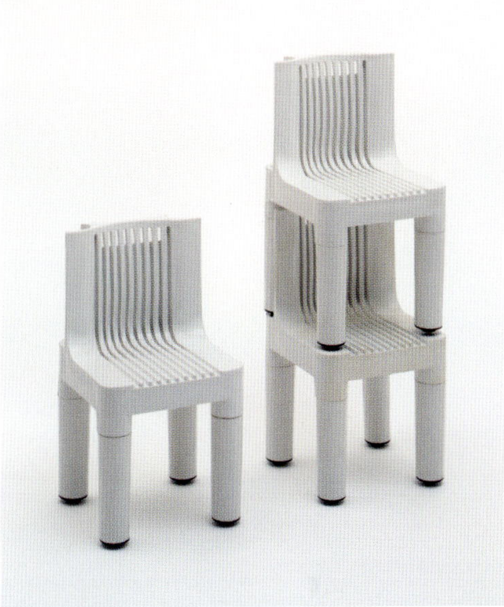

FIGURE 2.8
Paolo Lomazzi (Italian, b. 1936), Donato D'Urbino (Italian, b. 1935), and Jonathan De Pas (Italian, 1932–1991), Blow inflatable chair, manufactured by Zanotta S.p.A. Italy. Polyvinyl chloride, 83.8 × 119.7 × 102.9 cm (inflated). New York, The Museum of Modern Art, SC3.1972.2.

FIGURE 2.9
Marco Zanuso (Italian, 1916–2001), Child's stacking chairs, model 4999, 1964, manufactured by Kartell S.p.A., Italy. High-density polyethylene, 19¾ × 10¾ × 10¾ (each). New York, The Museum of Modern Art, 367.1993.1-3.vw2.

and Dieter Rams, once head of product design at Braun, represented the "form follows function" credo in their search for perfect performance. The Blow chair, manufactured by Zanotta in 1967, was made from inflatable polyvinyl chloride (fig. 2.8) and had characteristic Italian design flair, as did the stacking chairs (fig. 2.9) made from high-density polyethylene manufactured by Kartell in 1967 (Sparke 1990).

Over the last century, plastics have enabled dramatic changes in our lives, including enabling much of current telecommunications and information technology—the "global network of invisible forms that has come so quickly to dominate our lives" (Sparke 1990, 144–52). In the absence of plastics, it would not be possible to "enjoy a Formula 1 race, decorate our homes with reasonably priced furniture, listen to music, enjoy our iPads, iPhones, perform surgery," as well as myriad other activities we consider essential to daily life (Campos 2007, 8–9). That is why plastics can be found today in nearly all collections: ethnographic, textile, modern (three-dimensional, nontraditional) and applied art, science, and aerospace, to name but a few. Knowing and understanding this wonderful "new" material is the first step in the caretaking of our cultural heritage collections that contain plastics.

2.2 TYPES OF PLASTICS

FIGURE 2.10
Examples of logos used to identify compostable plastics: European Union (a), Biodegradable Product Institute (US) (b).

Plastics may be of natural or synthetic origin, or they can be a hybrid of the two created by modifying natural materials, which are called semisynthetics. Raw materials used for processing synthetic polymers are derived from fossil fuels (petroleum, natural gas, and coal). In semisynthetics, wood pulp and milk form the building blocks of the polymers.

Natural polymers comprise polysaccharides (e.g., cellulose, amylose), proteins (components from milk), and lignins (components of wood and plants). Some natural polymers are used commercially, either directly, such as cellulose from cotton fibers, or after chemical modification, for instance, cellulose acetate, cellophane (regenerated cellulose), and cellulose nitrate.

Synthetic plastics are usually divided into two groups according to their physical or chemical hardening processes, thermoplastics and thermosets, as mentioned in chapter 1. These categories can be further subdivided into semisynthetics and synthetics. Today, we might speak of semisynthetic plastics using the terms *biobased* or *renewable* plastics, because one of the monomers is made from renewable cellulose feedstock. Biobased thermosetting plastics are being developed from vegetable oil–based feedstock, but these are still in the early stages of research and development.

The labels *biopolymer* and *bioplastic* are widely used, often together, to describe two different concepts. To avoid ambiguity, both terms should be avoided in favor of the more accurate *biobased plastics* (British Plastics Federation n.d.). As we will see later in the chapter, biobased plastics can be differentiated between plastics made from renewable resources and those that are biodegradable or compostable. In Europe, only biocompostable plastics are allowed to carry the compostable symbol (fig. 2.10a); other parts of the world may use other symbols to distinguish between types (fig. 2.10b).

Thermoplastics

Thermoplastics are a family of plastics that can be semicrystalline or amorphous. They have long, mostly linear chains and slightly branched polymers and can melt when heated and harden when cooled. Table 2.1 gives an overview of the thermoplastics most likely to be found in museum collections, including the year they were brought on the market, registered, or patented. (More detailed information about individual thermoplastics is provided in the fact sheets in part III, chapter 13.)

Thermoplastics can be arranged according to increasing performance and crystallinity. In that system, they are often divided into commodity, engineering, and high-performance thermoplastics.

Commodity plastics are thermoplastics produced inexpensively and in high volumes for common disposable products and durable goods. They are mostly used for household items and other everyday purposes, which explains why they often appear in museum collections. These plastics include polyethylene, polypropylene, polystyrene, polyvinyl chloride, polyethylene terephthalate, styrene-acrylonitrile, acrylonitrile-butadiene-styrene, cellulose acetate, cellulose acetate-butyrate, and cellulose acetate-propionate, to name a few. Some of these thermoplastics, such as polypropylene, polyethylene, and polyethylene terephthalate, are suitable for recycling, which involves the reheating and reshaping of the material.

Table 2.1. Thermoplastics in museum collections

Polymer	Trade name or common name	Date introduced
Semisynthetic/biobased		
Cellulose nitrate (CN)	Celluloid	1862
Ethyl cellulose (EC)	Ethocel	1912
Cellulose acetate-propionate (CAP)	Cellidor	1924
Cellulose acetate (CA)	Lumarith	1927
Cellulose acetate-butyrate (CAB)	Tenite II	1929
Synthetic		
Polyvinyl chloride, rigid, unplasticized (PVC-U)	Vestolit	1912
Polyvinyl chloride, plasticized (PVC-P)	Tygon	1926
Polystyrene (PS)	Bextrene	1930
Polymethyl methacrylate (PMMA)	Perspex, Plexiglas, Lucite	1933
Polyvinylidene chloride (PVDC)	Saran	1933
Polyethylene (PE-LD)	Marlex	1933
Polyamide-6,6 (PA-6,6)	Nylon-6,6, Zytel	1935
Polyamide-6 (PA-6)	Nylon-6, Celanese	1938
Styrene-acrylonitrile (SAN)	Luran	1938
Acrylonitrile-styrene-acrylate (ASA)	Luran-S	1940s
Polyethylene terephthalate (PET)	Melinex, Mylar	1941
Polystyrene expanded (EPS)	Styrofoam, Styropor	1941
Acrylonitrile-butadiene-styrene (ABS)	Lustran, Novodur	1948
Polyvinyl chloride-polyvinyl acetate (PVC/PVAC)	—	1948
High-impact polystyrene (HIPS)	Bextrene	1949
Polypropylene (PP)	Hostalen PP, Moplen	1951
High-density polyethylene (PE-HD)	Paxon	1953
Polycarbonate (PC)	Makrolon, Lexan	1956

continues on next page

Table 2.1. Thermoplastics in museum collections *(continued)*

Polymer	Trade name or common name	Date introduced
Polyoxymethylene (POM)	Polyacetal, Acetal	1958
Polyamide-12 (PA-12)	Nylon-12, Vestamid	1960s
Ultra-high molecular weight polyethylene (PE-UHMW)	Dyneema	1965
Biodegradable		
Polylactic acid (PLA)	Plantic	1989
Polyhydroxybutyrate (PHB)	Biopol	1992
Polycaprolactone (PCL)	Shapelock	1965
Thermoplastic starch (TPS)	Bioflex	1993

Engineering, or technical, plastics are, in general, thermoplastics that possess improved mechanical properties compared to commodity plastics. Like commodity plastics, engineering plastics often appear in museum collections. They include polyamides, polycarbonate, polymethyl methacrylate, polyoxymethylene, and ultra-high molecular weight polyethylene.

High-performance plastics are plastics polymerized from more unusual monomers; they are more durable and costly than engineering or commodity plastics and are produced in smaller quantities. Their higher melting points and greater scratch resistance make them suitable for uses in aerospace industries and medicine.[2]

The term *functional plastics* is also used in categorizing plastics. These can be thermoplastics, thermosets, or elastomers, and may fall into any of the three categories discussed: commodity, engineering, or high-performance. Their distinguishing feature is that they typically have one very specific use. For instance, polyethylene vinyl alcohol, a copolymer of ethylene and vinyl alcohol, serves as an excellent oxygen barrier due to the large number of vinyl alcohol groups in its polymer chains. As a result, it is used in packaging films.

Chemical families | Another way of classifying thermoplastics is to subdivide them into chemical families based on shared common monomeric units (Elias 2003): polyolefins, styrenics, vinyls, acrylics, thermoplastic polyesters, polycarbonates, polyacetals, and polyamides (fig. 2.11).

2.. Fluoroplastics are a subset of high-performance specialty plastics with unique surface properties. They are commonly known for their use as a nonstick coating for cookware, though they have many other applications. Examples of fluoroplastics are polytetrafluoroethylene (found in Teflon and Gore-Tex) and polyvinylidene fluoride. Fluoroplastics are not commonly found in cultural heritage objects, except in textile collections, and thus are beyond the scope of this book.

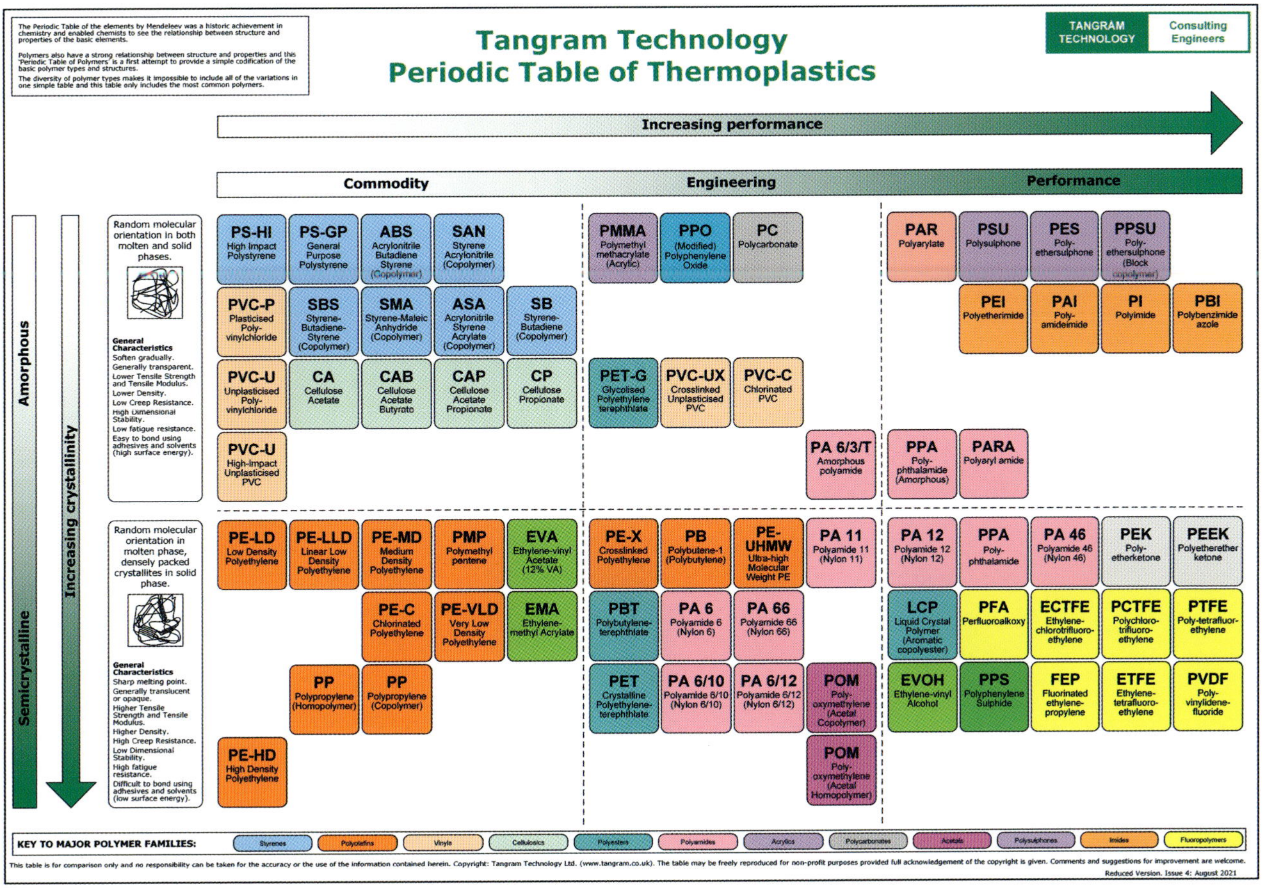

FIGURE 2.11
Periodic table of thermoplastics.

Polyolefins | Polyolefins have the simplest basic structure of any polymer, containing only carbon and hydrogen in their polymeric structures, and are polymerized from the monomers ethylene, propylene, and butylene, among others. Depending on the length of polymer chains and the amount of branching, a variety of polyolefins with different properties exist, such as low-density polyethylene (PE-LD), high-density polyethylene (PE-HD), ultra-high molecular weight polyethylene (PE-UHMW), and polypropylene (PP).

Polyolefins are primarily determined by their molecular weight and degree of crystallinity, ranging from 20% to 60% and higher. Fuel tanks, bottle caps, and plastic bottles are typical products made of PE-HD, while containers for liquids, tubing, and plastic wrap are typical for PE-LD. PP products include hinges, carpets, soap containers, and auto parts.

Both artists and designers made use of PE and PP due to their flexibility and low cost. Examples of these are polyethylene Tupperware containers, recognized as classics in plastics (see fig. 2.4), and the inexpensive and ubiquitous shopping bags incorporated in works of art by Andreas Slominski (fig. 2.12). Examples of polypropylene design are the Panton chair by Herman Miller/Vitra (fig. 2.13), in which polypropylene replaced earlier, less durable plastics from 1999 onward (Brydson 1999, 425–65; Kessler, van Oosten, and van Keulen 2004), and the use of polypropylene ropes in works by Wil Fruytier (see chap. 7).

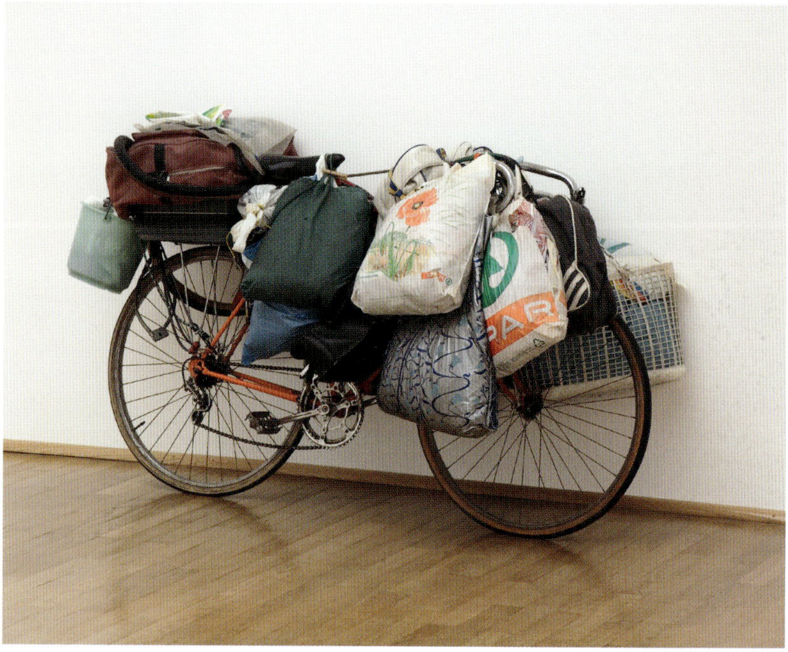

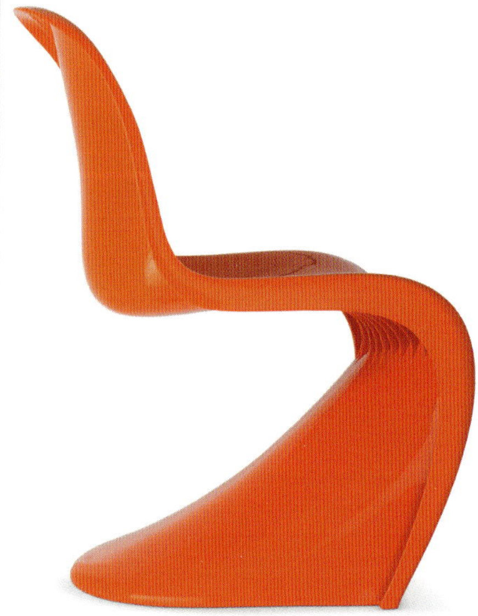

FIGURE 2.12
Andreas Slominksi (German, b. 1959), *Ohne Titel* (*Untitled*), 1991. Bicycle, polyethylene bags, mixed materials, ca. 115 × 175 × 85 cm. Frankfurt, Museum für Moderne Kunst, 2005/92.

FIGURE 2.13
Verner Panton (Danish, 1926–1998), Panton chair, ca. 1956–67, manufactured by Vitra. Polypropylene, 83 × 50 × 59.5 cm. Weil am Rhein, Vitra Design Museum, MPA-1036-02.

Styrenics | Polystyrene is a hard, rigid, transparent thermoplastic that emits a characteristic sound when dropped (Brydson 1999). Desirable properties of polystyrene—low cost, transparency, rigidity, and low water absorption—make it suitable for use in injection molding and vacuum forming (see chap. 5). However, polystyrene is also very brittle, and this has led to the development of rubber-modified high-impact polystyrene (HIPS) and the copolymers acrylonitrile-butadiene-styrene (ABS), acrylonitrile-styrene-acrylate (ASA), and styrene-acrylonitrile (SAN). Polystyrenes are used in packaging and containers, whereas impact-modified styrenics such as ABS and ASA are used for equipment housings, such as those for radios, telephones, and computers. Expanded polystyrene foam, a cellular polystyrene, is used as insulating material. It is also used by artists as the interior structure of works of art (fig. 2.14).

Vinyls | Plastics containing monomeric units derived from a vinyl chloride monomer and consisting of slightly branched polymer chains are called vinyls. Polyvinyl chloride (PVC) is an amorphous plastic that is used in several different forms, including the rigid, unplasticized PVC-U and the flexible, soft, external plasticized PVC-P. Plasticizers are added in PVC-P to enhance the flexibility of PVC-U. A disadvantage of this addition is that the plasticizers leach out during the service life of PVC-P products.

Applications for PVCs include pipes and fittings from PVC-U, and flexible film and sheet, tubes, flooring, and bottles from PVC-P. Other examples of plasticized PVC uses include shower curtains, children's toys, seat coverings, inflatable swimming gear, and designer chairs. One icon of plasticized PVC is *The Aeromodeller* by the Belgian artist Panamarenko (fig. 2.15).

Internal plasticized PVC, which avoids the problem of the leaching of plasticizer, is achieved by copolymerization of vinyl chloride with about 3%–20% vinyl acetate, creating PVC/PVAC, which is used to make bank and transit cards, among other things.

FIGURE 2.14
Jean Dubuffet (French, 1901–1985), *Table amoncellante I*, 1968. Polyester, polystyrene foam (Styrofoam), and paint, 43 × 21 × 11 cm. Amsterdam, Stedelijk Museum, BA 2621 (1-14).

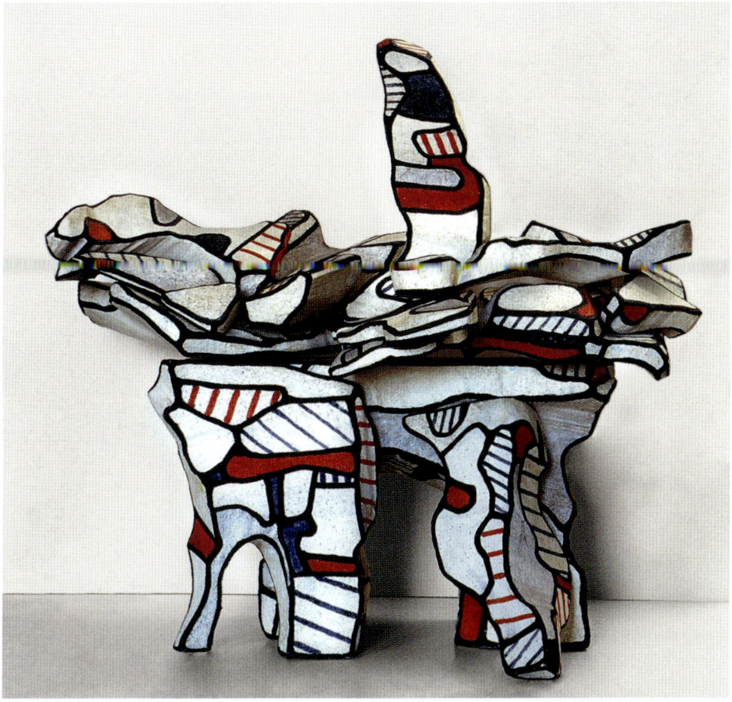

FIGURE 2.15
Panamarenko (Henri van Herwegen, Belgian, 1940–2019), *The Aeromodeller*, 1969–71. Polyvinyl chloride film and mixed media, 2700 × 600 cm (balloon), 200 × 600 × 300 cm (gondola). Ghent, Stedelijk Museum voor Actuele Kunst, WV46.

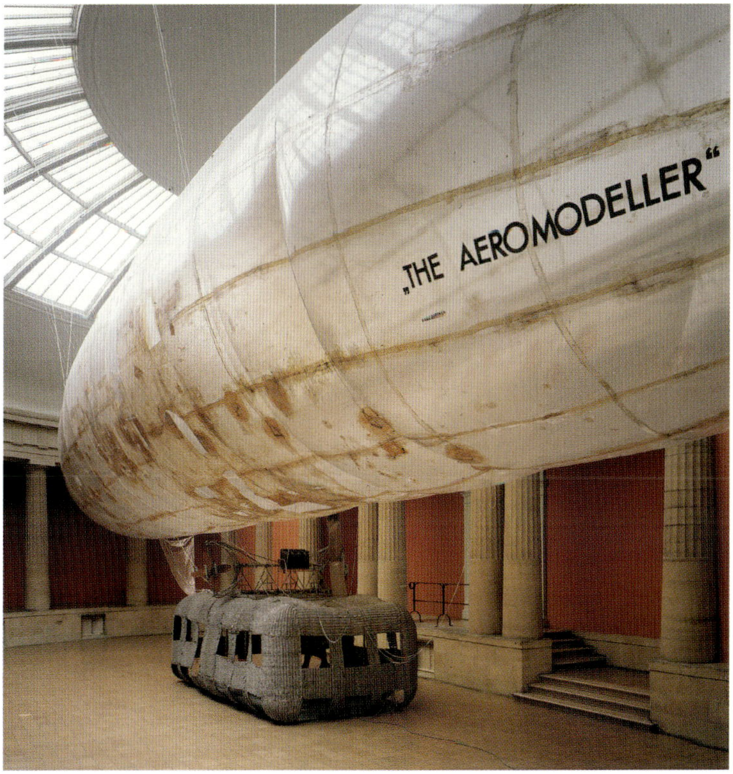

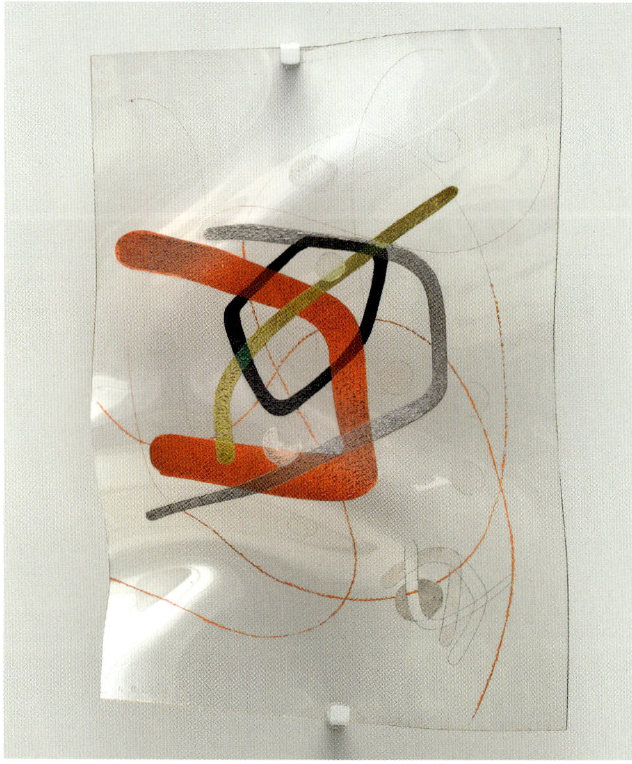

FIGURE 2.16
Lászlo Moholy Nagy (Austro-Hungarian, 1895–1946), *B-10 Space Modulator*, 1942. Oil and incised lines on Plexiglas, 42.9 × 29.2 × 6 cm. New York, Solomon R. Guggenheim Museum, 47.1063.

Polyvinylidene chloride (PVDC), brand name Saran, is polymerized from the monomer vinylidene chloride. It has a low permeability for gases and is used as barrier film (though it is no longer used in the eponymous food wrap).

Acrylics | Many acrylates exist, including polyacrylonitrile (used for fibers and for SAN—see "Styrenics," above), polyacrylates (an adhesive for polymethyl methacrylate [PMMA] and cyanoacrylates), and methyl methacrylate and n-butyl methacrylate (softer than PMMA and used in conservation as an adhesive). Commercially, the most important acrylate is the polymer PMMA, which is made by free radical polymerization from the monomer methyl methacrylate. It is an atactic, amorphous, transparent thermoplastic. Because of its transparency, it is used for many optical applications, such as lenses for optical instruments and contact lenses, as a substitute for glass in windows, and in costume jewelry. Under the brand names Perspex and Plexiglas, it is formed into rods and sheets, which are used to make protective shields and incorporated into functional and decorative objects by hobbyists. Artists including László Moholy Nagy (fig. 2.16), Naum Gabo, Gino Marotta, and many others have used Perspex for their artworks.

Thermoplastic polyesters | Thermoplastic polyesters are synthesized from ethylene glycol or butylene glycol and dicarboxylic acids (phthalic acids). Polyethylene terephthalate (PET) and polybutylene terephthalate (PBT) belong to the family of saturated polyesters and are used in garments, blankets, sleeping bags, and carpets (under brand names such as Dacron, Terlenka, and Trevira). PET is a semicrystalline polymer that is lightweight and known for its low gas permeability; these properties make it suitable for soft-drink bottles.

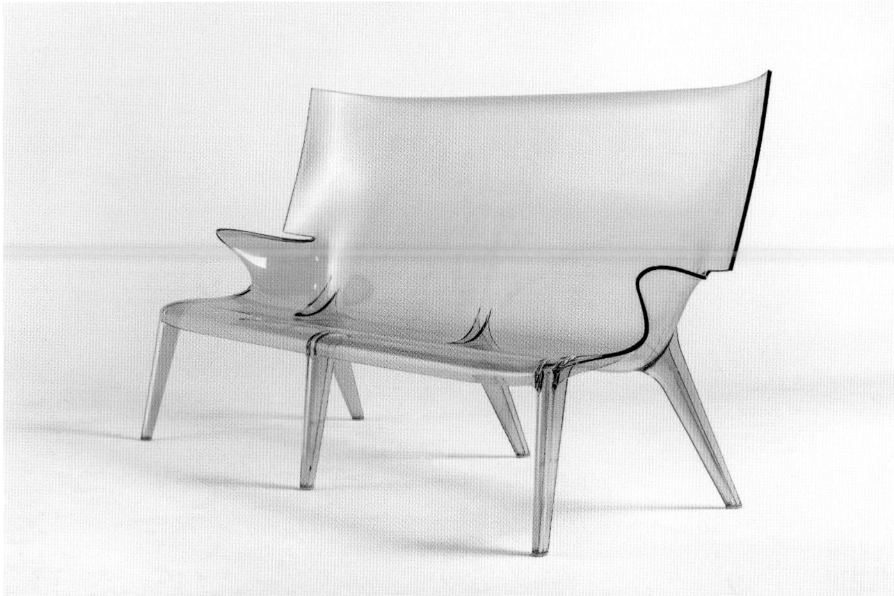

FIGURE 2.17
Philippe Starck (French, b. 1949), Uncle Jack sofa, 2014, manufactured by Kartell. Polycarbonate, 72 × 103 × 68 cm.

PET bottles are one of the most commonly recycled plastic objects; they can be converted into fabric. PET is used as packaging sheets, sleeves, and envelopes; by conservators it is known under the brand names Melinex and Mylar.

Polycarbonates | In general, polyhydroxy compounds with polybasic acids give rise to condensation polymers containing ester groups. Polycarbonate is prepared by reacting bisphenol A with phosgene under alkaline conditions. Polycarbonate is a linear, amorphous polyester that possesses high-impact strength, transparency, and good dimensional stability. It is known by conservators under the trade names Lexan and Makrolon. It is used for tableware such as plates and cups, containers for storing food, compact discs, safety glass windows, feeding bottles, and furniture (fig. 2.17).

Polyacetals | Polyoxymethylene, also called acetal or polyacetal, is a polymer made from formaldehyde. It has a linear, highly crystalline structure and good dimensional stability, and is resistant to chemicals, moisture, and temperature. Some dimensionally stable applications include screws and nuts, hinges, and zippers.

Polyamides | Polyamides are polymers created from condensation reactions of acids with diamines. When both the acid and the diamine contain six carbons, the plastic is called polyamide-6,6, or nylon-6,6. When a polyamide is formed after the ring-opening polymerization of ε-polycaprolactam (with six carbons), the polymer is called nylon-6. Nylon-12 is formed after the ring-opening polymerization of 12-dodecanoic (or lauric) acid.

Polyamides are semicrystalline, have a milky opaque appearance, and are tough and abrasion resistant. If at processing polyamides are cooled very rapidly, transparent films result. Due to their high durability and strength, polyamides are commonly used in sportswear and other textiles such as hosiery in works of art (fig. 2.18), and as aromatic polyamides in bulletproof vests (as Kevlar), carpets, sunglass frames, fishing gear, toothbrush bristles, and tennis rackets.

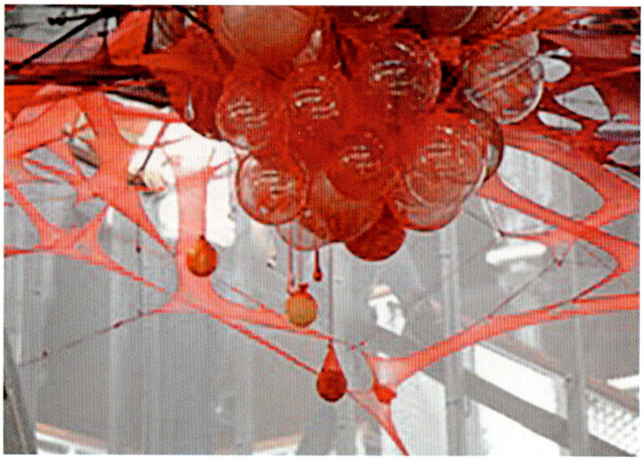

FIGURE 2.18
Madeleine Berkhemer (Dutch, 1973–2019), *Garden of Delight II*, 2010. Installation made of nylon pantyhose and glass balls. Rotterdam, Het Nieuwe Instituut.

Biobased Thermoplastics

Current synthetic plastic—upon which we have become so dependent—is largely derived from petroleum, through chemical extraction and synthesis. Given that petroleum is a nonrenewable resource, many plastics manufacturers are looking at biobased plastics as sustainable alternatives (Freinkel 2011). A key area of research in plastics technology over the past decade has been the development of biobased plastics, in which some or all monomers are derived from renewable ("green") feedstock, to replace the petroleum.

Biobased plastics are made from a variety of natural materials, including corn, sugar, canola oil, castor seeds, potatoes, rice, tapioca, palm fiber, wheat straw, potato starch, switch grass, and wood cellulose. The percentage of biobased ingredients in these plastics can vary widely and is measured by the ratio of biobased carbon to fossil-based carbon.

Cellulose-based plastic | Cellulosic plastics all are derived from the natural polymer cellulose. After chemical modification, cellulose is converted to the cellulose esters: cellulose acetate, cellulose nitrate, cellulose acetate-butyrate (CAB), cellulose acetate-propionate (CAP), ethyl cellulose, and regenerated cellulose (cellophane). Uses of cellulose nitrate include mass-produced dolls and other toys, the notoriously flammable "celluloid" movie film, and handbags, and it has been applied as a coating to imitate leather.

Cellulose esters are effective polymers for plastics, and until 1950 they constituted an important group of thermoplastics. Cellulose acetate found wide application in the manufacture of nonflammable "safety" films, molding and extrusion compounds, fibers, and lacquers (Gilbert 2017, 617–30). Uses of cellulose acetate include children's toys, including dolls' heads; handbags; geometry sets; and radio housings.

Over time, most cellulosic plastics have become less important to the plastics industry than the polyolefins, PVCs, and polystyrene. However, CAB and CAP appear to have retained their market because of their advantages in appearance and strength, in spite of their higher cost (Brydson 1999). Uses of CAB and CAP include the handles of screwdrivers and other tools, and in eye- and sunglass frames (fig. 2.19).

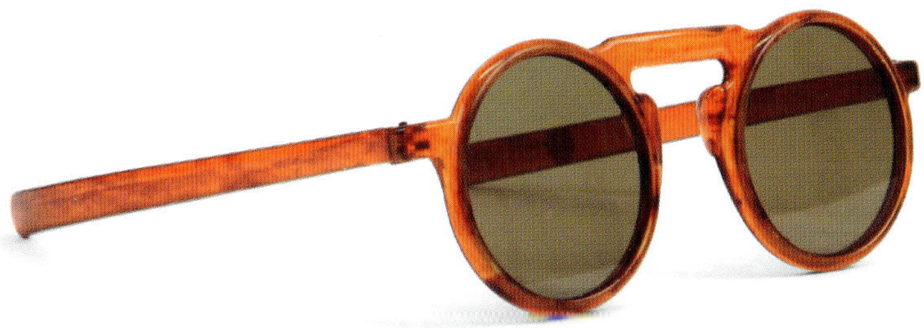

FIGURE 2.19
Orange-framed sunglasses, ca. 1950, manufactured by Twinco Ltd., United Kingdom. Injection-molded cellulose acetate and glass, W: 13 cm; D: 13 cm; Diam: 4.3 cm. Bournemouth, Arts University, Museum of Design in Plastics, PHSL: 249.

Biodegradable thermoplastics | Biobased plastics solve the problem of diminishing petroleum feedstock available for plastics manufacturing, but they leave another serious issue unaddressed. Neither petroleum- nor biobased plastics rapidly degrade in the environment. The impact of the last fifty years of plastics production, use, and disposal is now widely recognized. Plastics threaten species maintenance, the very life of the ocean, and the environment as a whole.

One approach to solving this problem has been the development of biodegradable plastics. Truly biodegradable, or biocompostable, plastics can decompose into carbon dioxide, methane, water, and inorganic compounds, or can be rendered into biomass through the enzymatic action of microorganisms (Gross and Kalra 2002). Several types of biodegradable plastics are currently made.

Thermoplastic starch | All biodegradable plastics are used for packaging purposes, including thermoplastic starch (TPS), which is obtained by combining potato starch with water and plasticizers (glycerol or sorbitol, among others). TPS is used for making biodegradable cutlery, plates, and cups.

Starch is an inexpensive, renewable material derived from corn and other crops. Starch-based plastics are produced by blending or mixing them with synthetic polymers, for instance, polyethylene. By adjusting the synthetic blend component and the proportion of starch, the properties of biobased plastic can be regulated (Gross and Kalra 2002).

Polyhydroxybutyrate | Polyhydroxybutyrate (PHB) is the best-known biodegradable plastic. Polyhydroxy alkanoates are made by enzymatic action from a renewable resource, often a sugar, and PHB is polymerized from glucose by the bacterium *Alcaligenes eutrophus* (Harmsen and Hackmann 2012). Like other biodegradable plastics, it is mostly used for packaging and disposable items such as cutlery and picnic plates.

Polylactic acid | Polylactic acid (PLA) is the most common biodegradable plastic in use today. Lactic acid, the monomer of PLA, is obtained by fermentation of sugars or cornstarch, and polymerization produces PLA. Once used mostly for packaging material like bags and films, today PLA is one of the most popular materials used in 3D printing (see "Additive Manufacturing" in chapter 5).

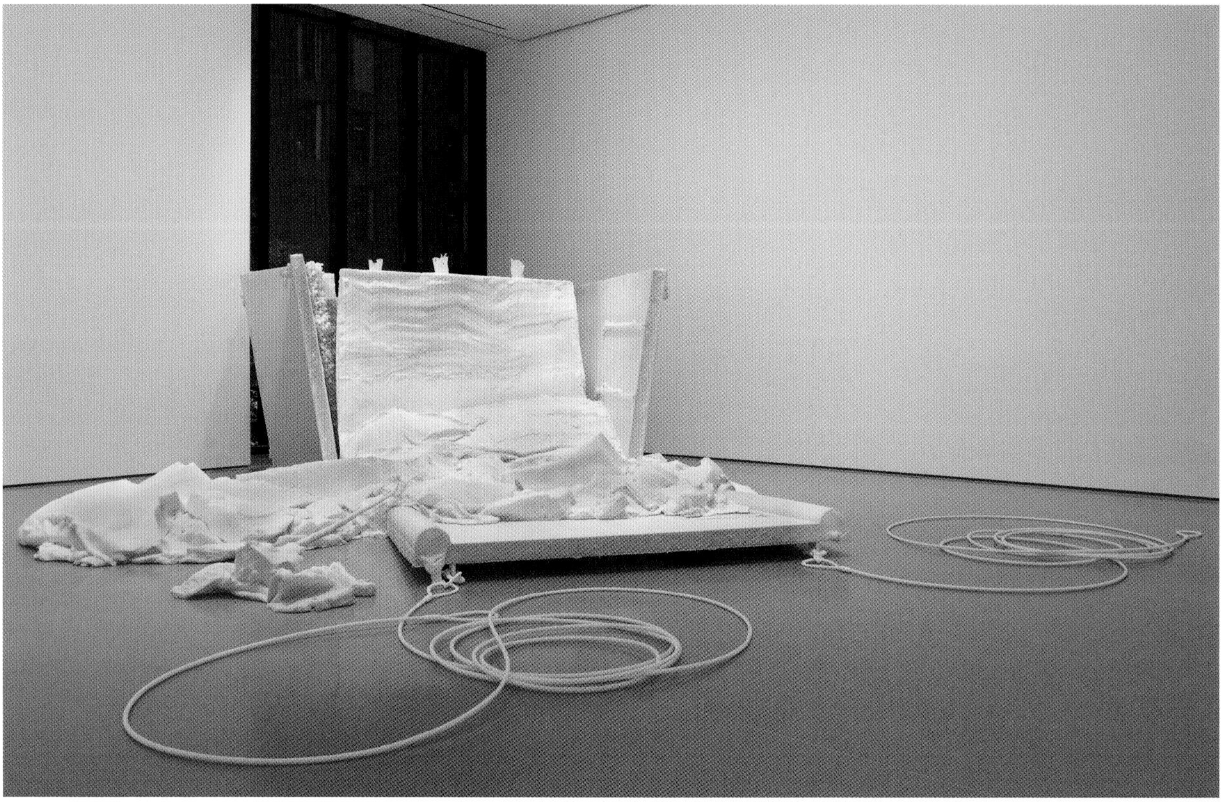

FIGURE 2.20
Matthew Barney (American, b. 1967), *The Deportment of the Host*, 2006. Cast polycaprolactone thermoplastic, self-lubricating plastic, ca. 264.2 × 617.2 × 76.2 cm. New York, The Museum of Modern Art, Gift of Maja Oeri and Hans Bodenmann, 357.2006.

Polycaprolactone | Polycaprolactone (PCL), polymerized by ring-opening polymerization of ε-caprolactone, is a biodegradable polyester-based plastic made from petroleum feedstock. It has a melting point of about 60°C and is used for splinting and as modeling material (Labet and Thielemans 2009; Woodruff and Hutmacher 2010). The American artist Matthew Barney uses PCL for his works of art (fig. 2.20).

Thermosets

Thermosets are a group of plastics that undergo a chemical cross-linking reaction when heated, creating a three-dimensional network that makes them stronger and more stable than thermoplastics. Once they are heated and formed, these plastics cannot be remelted, reformed, or recycled. Common thermosets are phenol-formaldehyde, amino formaldehydes, unsaturated polyesters, epoxies, and polyurethanes. Table 2.2 gives an overview of the thermosets most likely to be found in museum collections, including the date they were brought on the market, registered, or patented. (More detailed information about individual thermosets is provided in the fact sheets in chapter 14.)

Phenolics | Phenolics are produced by condensation of a mixture of phenols with an aldehyde. Phenolics were the first fully synthetic plastics of the twentieth century, and continue to be used today for a wide variety of applications such as molding powders, lamination, adhesives, binders, surface coatings, and impregnating material (Brydson 1999, 652, 658–62). Examples of objects made of phenolics are radio cabinets, gramophone housings, lamp holders (light bulb sockets), and household appliance housings.

The trade name Bakelite is still known for condensation products of phenols with formaldehydes. Acid catalysis with an excess of formaldehyde led to cross-linked Novolak, the first substitute for shellac in the early twentieth century, and in the 1930s cast phenolic (Catalin) was introduced by the American Catalin Corporation in a rainbow of colors. Costume jewelry, egg cups, fountain pens, and, famously, Fada and Motorola radio housings (fig. 2.21) were made of it.

Table 2.2. Thermosets in museum collections

Polymer	Trade name or common name	Date introduced
Semisynthetic/biobased	—	—
Casein-formaldehyde (CS)	Galalith, Ivoride	1897
Synthetic	—	—
Phenol-formaldehyde (PF)	Bakelite	1907
Urea-formaldehyde (UF)	Bandalasta, Beetle	Early 1920s
Melamine-formaldehyde (MF)	Melamine, Formica	1933
Unsaturated polyester (UP)	GRP, FRP	1933
Epoxy (EP)	Epikote	1934
Polyether urethane (PUR ether)	—	1937
Polyester urethane (PUR ester)	—	1957

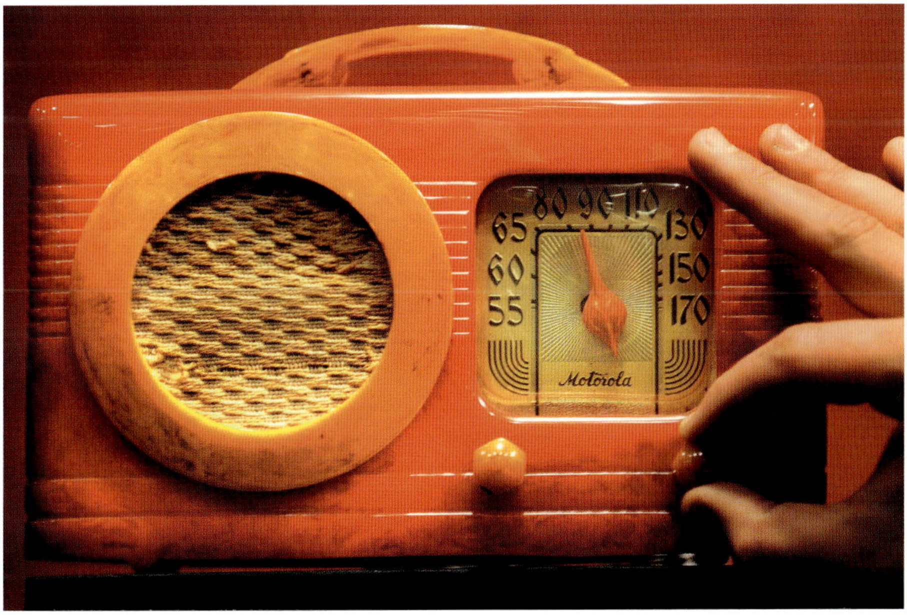

FIGURE 2.21
Motorola, a 1950s radio and American Art Deco icon.

Amino plastics | Amino plastics are condensation products of formaldehyde and monomers containing nitrogen and hydrogen, such as amines or amides. At the end of the nineteenth century and beginning of the twentieth, this process resulted in casein-formaldehyde and urea-formaldehyde; today it is used to make melamine-formaldehyde.

Casein-formaldehyde | Casein, a semisynthetic plastic, is made from milk curds that are dried and mixed with water to produce a dough that is extruded and then treated with formaldehyde, producing a material that resembles horn or bone. Casein can be molded into a variety of shapes and can be surface dyed or colored by the addition of pigments. In the late nineteenth century and beginning of the twentieth century, casein-formaldehyde was very popular, used to make combs, toothbrushes, and knitting needles. Today it is mostly used for buttons and buckles.

Urea-formaldehyde | Urea-formaldehyde is the condensation reaction product of urea and formalin. During the condensation polymerization process, the resulting excess water is evaporated. Urea-formaldehyde molding powders were used instead of phenol-formaldehyde at the beginning of the twentieth century because this plastic could be produced in various light colors. Urea-formaldehyde production decreased after the introduction of melamine-formaldehyde in 1933–35, but is still in use as an adhesive in wood-based panels. Examples of urea-formaldehyde include household items such as sockets and fittings for lamps, light switches, and housings for steam irons and electric shavers.

Melamine-formaldehyde | Melamine combined with formaldehyde using condensation polymerization results in the formation of the cross-linked product melamine-formaldehyde. Melamine-formaldehyde products are harder, have better scratch and stain resistance, and are more resistant to water and chemicals than those made of casein- or urea-formaldehyde. Although melamine products are more expensive than those made with urea, they are widely used for surface coatings, tableware, and decorative laminates such as Formica (see fig. 2.5). Typical objects made of melamine-formaldehyde plastics are cups and plates, such as those used for backyard barbecues and camping.

Unsaturated polyesters | Polyesters are formed by reacting polybasic acids and polyhydroxy alcohols. The term *polyester resin* refers to unsaturated resins based on unsaturated and saturated dibasic acids (such as phthalic anhydride, maleic anhydride, maleic acid, fumaric acid) and dihydroxy alcohols (such as 1,2-propylene glycol and diethylene glycol). These unsaturated polyesters are capable of cross-linking with unsaturated monomers such as styrene.

The main use for unsaturated polyester resin is in the production of reinforced plastics; glass fiber, carbon fiber, and aromatic polyamides (Kevlar) are the most used reinforcements. Fiberglass-reinforced polyester (commonly referred to as simply fiberglass or GRP) is a popular material used to make chairs, desks, boats, and mobile homes, but it is also used by many artists and designers such as Duane Hanson, Joep van Lieshout, Matti Suuronen, Marta Pan (fig. 2.22), and Ray and Charles Eames.

Epoxies | Epoxies are thermosetting plastics derived from bisphenol acetate (also called bisphenol A or just BPA) and epichlorohydrin; they are cured by adding catalysts or reactive hardeners. Recent developments include epoxies modified to contain thermoplastic processing properties. Epoxies are widely used for coatings and adhesives and for making jewelry and casting objects. In some cases, they are also used by artists. Despite epoxy's drawback of fast yellowing, Isa Genzken and Jean Shin (fig. 2.23) have used it in their works.

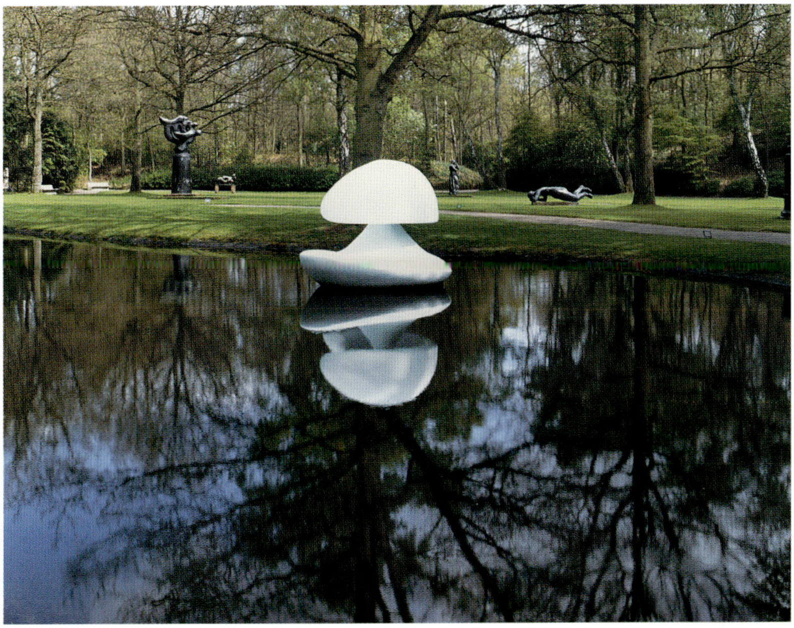

FIGURE 2.22
Marta Pan (Hungarian, 1923–2008), *Sculpture flottante*, 1960–61. Fiberglass-reinforced polyester resin and aluminum, 3200 × 1400 cm (pond), 216 × 226 × 185 cm (object). Otterlo, Kröller-Müller Museum, KM 115.762.

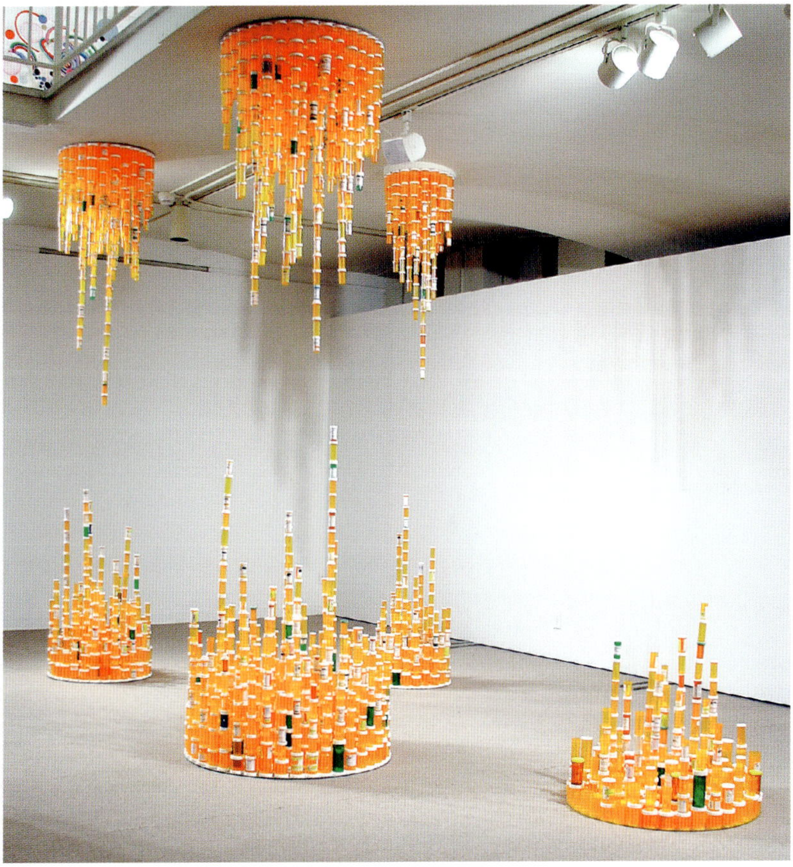

FIGURE 2.23
Jean Shin (South Korean, b. 1971), *Chemical Balance II*, 2005. Plastic pill containers, mirror, and epoxy, Diam: 45–96 cm. Albany, State University of New York, University Art Museum.

Thermosetting polyurethanes | Polyurethanes are a group of polymers formed by polymerization in two steps. First, low molecular weight prepolymers are formed from hydroxyl components and diisocyanates. The hydroxyls have two (di-) or more (poly-) reactive groups, and the diisocyanates have at least two reactive sites. Further, the hydroxyl components may be either poly*ester* or poly*ether* polyols (polyfunctional alcohols) (van Oosten 2011).

Second, the low molecular weight prepolymer is converted into a high molecular weight polyurethane polymer by further reaction with diol or diamine chain extenders. Catalysts and additives are also incorporated at this stage of the process.

The type of polyols (functionality), diisocyanates (aliphatic or aromatic and functionality), the chain extenders (short diols and diamines), catalysts, and additives used all determine whether the resulting polyurethanes' polymer chains are cross-linked and how much—that is, whether they possess thermoplastic properties (no cross-linking), elastomeric properties (some cross-linking), or thermosetting properties (fully cross-linked) (see fig. 1.17).

Examples of poly*ester* polyols are the reaction products of dibasic acids (such as adipic acid or sebacic acid) and diols (such as ethylene glycol or propylene glycol). Examples of poly*ether* polyols are the reaction products of either propylene oxide or ethylene oxide with a glycol such as diethylene glycol. Although polyurethanes themselves are relatively stable when immersed in water, polyurethane ester foams are more susceptible to hydrolysis than polyurethane ether foams, whereas polyurethane ether foams are more easily photooxidized.

Due to the variety of polyols, diisocyanates, and chain extenders available, many combinations are possible as starting components for polymerization. This results in polyurethanes that range from the thermosetting foams that are the subject of this section[3] to hard elastomers to thermoplastic fibers, adhesives, and coatings (which are beyond the scope of this book). Elastomeric polyurethanes are discussed in chapter 3.

Most thermosetting polyurethane (PUR) foams are polymerized by polyaddition of di- or polyfunctional isocyanates to three or more hydroxy groups containing polyols. The type of polyol and diisocyanate, chain extenders, and additives and water determines whether a PUR foam is flexible (resilient, open-cell structure), semirigid (partially open cells), or rigid (closed-cell structure). The foaming process in the manufacture of flexible foams is the liberation of carbon dioxide formed by the reaction of diisocyanate and water, and rigid foams are processed by adding blowing agents to the reaction mixture to form gas bubbles.

Flexible PUR foams are used for mattresses, for seating for upholstered furniture, and to line carrying cases for instruments and tools. Rigid PUR foams are used as structural foams and for thermal and acoustic insulating purposes. Many artists, including Piero Gilardi, César, and John Chamberlain, and designers like Joe Colombo and Gaetano Pesce (fig. 2.24) have used PUR foams for their objects.

3. More than 75% of the polyurethanes produced are foams—flexible, semirigid, or rigid—so polyurethanes are strongly associated with foams. As polyurethane foams have thermosetting properties, they are discussed in this chapter along with other thermosets.

FIGURE 2.24
Gaetano Pesce (Italian, b. 1939), *UP7/Il Piede*, 1969. Polyurethane foam, 80 × 65 × 165 cm. Weil am Rhein, Vitra Design Museum, MIT-1064.

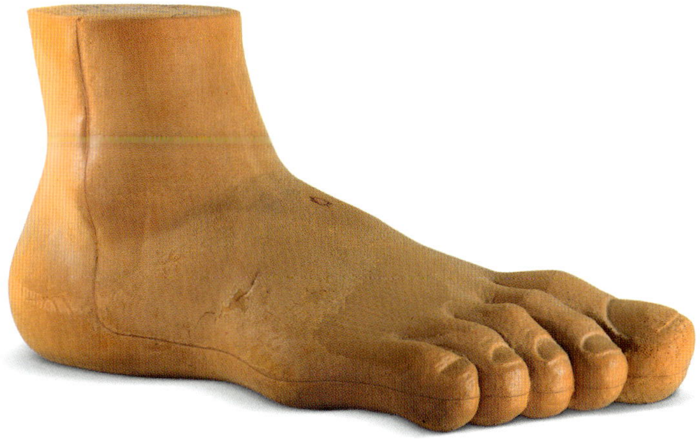

Summary

- Plastics are divided into two main groups: thermoplastics and thermosets.
- Thermoplastics can be grouped into commodity, engineering, and high-performance plastics.
- Thermoplastics have mostly linear, slightly branched polymer chains and can be semicrystalline or amorphous.
- Due to cross-linking, thermosets are stronger and more stable than thermoplastics.
- Thermoplastics and thermosets can be made from fossil-based feedstock (petroleum, coal, natural gas) or from renewable feedstock, in the case of biobased plastics.
- Biodegradable plastics are thermoplastics made from renewable feedstock, with the exception of polycaprolactone, which is made from petroleum feedstock.

Suggested Reading

Cecilia Cecchini and Marco Petroni, eds, *Plastic Days: Materials and Design* (Cinisello Balsamo, Milan: Silvana Editoriale, 2015).
Andrea DiNoto, *Art Plastic: Designed for Living* (New York: Abbeville Press, 1984).
D. P. Greenwood, *Modern Design in Plastics* (London: J. Murray, 1983).
Museum of Design in Plastics, https://www.modip.ac.uk/.
Thea B. van Oosten, *PUR Facts: Conservation of Polyurethane Foam in Art and Design* (Amsterdam: Amsterdam University Press, 2011).
Anita Quye and Colin Williamson, eds., *Plastics: Collecting and Conserving* (Edinburgh: NMS Publishing, 1999).
Yvonne Shashoua, *Conservation of Plastics: Materials Science, Degradation and Preservation* (Amsterdam: Elsevier/Butterworth-Heinemann, 2008).
Friederike Waentig, *Plastics in Art: A Study from the Conservation Point of View* (Petersberg, Germany: Imhof, 2008).

Chapter 3 | Elastomers: Natural and Synthetic Rubbers and Thermoplastic Elastomers

Up until the 1950s, when synthetic rubbers were introduced into everyday life, the term *rubber* meant natural rubber. *Elastomer* was then used to distinguish synthetic rubber from natural rubber. Today, *rubber* and *elastomer* tend to be used interchangeably; both refer to a class of polymeric materials having elasticity as an exceptional property. While natural rubber is made from latex, a milky tree sap, synthetic rubbers and thermoplastic elastomers are made from petroleum by-products.[1]

Both natural rubbers and synthetic rubbers undergo vulcanization, a process that improves their properties. Natural rubber production gave way from the 1950s on, with use of synthetic rubbers gradually increasing until today, when synthetic rubbers production exceeds that of natural rubber.

3.1 HISTORY OF NATURAL AND SYNTHETIC RUBBERS

Rubber objects can be found in nearly all cultural heritage collections: from the first rubber objects made by the Maya civilization to nineteenth-century objects made of gutta-percha, balata, and hard rubber to the latest modern synthetic rubber objects (table 3.1).

Natural Rubber

The oldest natural rubber objects served as offerings to Maya rain gods. Rubber was first introduced to Europe by Christopher Columbus. When he returned from his second journey to America (1493–96), he brought back a latex rubber ball from the West Indies. Some three hundred years later, in 1820, the Englishman Thomas Hancock started experimenting with rubber and invented a masticator known as the Pickle—a machine for shredding rubber to make it easier to shape (Hancock [1857] 1920).

The Scottish chemist Charles Macintosh manufactured the first waterproof raincoat in 1824. He fused two layers of fabric together using rubber treated with naphtha (a by-product of tar), which left the rubber soluble and thereby made it possible to "rubberize" the fabric (J. Loadman 2014, 53).

In those days, rubber had a fundamental weakness in that it softened and became tacky with heat and stiff in the cold (M. J. R. Loadman 1993). When Charles Goodyear discovered how to cure rubber, in 1839, the quality of rubber objects improved enormously. Curing—more commonly called vulcanization—is the chemical process that links rubber and sulfur (Huke 1961; M. J. R. Loadman 1995). Goodyear discovered that mixing rubber with white lead and about 4% sulfur, and then heating and cooling it, resulted in a solid that retained its elasticity and resiliency at high temperatures while remaining pliable and

1. The prefix *poly* in the names of some rubber types is used somewhat arbitrarily; *polybutadiene rubber* is synonymous with *butadiene rubber*, as is *polychloroprene rubber* with *chloroprene rubber*. This book omits the prefix in most cases, although polyisoprene is an exception.

Table 3.1. Brief history of rubbers and elastomers

Event/material	Trade name or common name	Date introduced
Aztec and Maya civilizations discover rubber	—	6th century
Rubber introduced in Europe	—	1736
Macintosh develops waterproof raincoats	Macintosh	1824
Goodyear (USA) discovers vulcanization with sulfur	Natural rubber	1839
Hard rubber	Vulcanite, Ebonite	1839
Gutta percha	—	1850s
Balata	—	1850s
Methyl rubber (2,3-dimethyl butadiene)	Methyl rubber	1914–18
Butadiene rubber (1,3-butadiene) (BR)	Buna	1926
Styrene-butadiene rubber (SBR)	Buna-S (Germany), SBR (UK), GR-S (USA)	1929
Chloroprene rubber (CR)	Neoprene	1930
Beginning of the polyurethane rubber/elastomer industry	Vulkollan, Adriprene	1930s
Nitrile-butadiene rubber (NBR)	Nitrile rubber, Buna-N	1931
Copolymer of isobutylene and isoprene (IIR)	Butyl rubber	1937
Chlorinated rubber	Pliofilm	1938
Beginning of the silicone rubber (Si) industry	Elastosil, Silastic, Siloprene	Early 1940s
Styrene-ethylene-butadiene-styrene rubber (SEBS)	Kraton	1950
Development of synthetic "natural" rubber	Polyisoprene	1954
Fluoride elastomers (FKM, FPM)	Viton	1958
Ethylene-propylene-diene-monomer rubber (EPDM)	Vistalon, Dutral	1960s
Ethylene-vinyl acetate elastomer (EVAC)	Elvax, Ateva	1970s

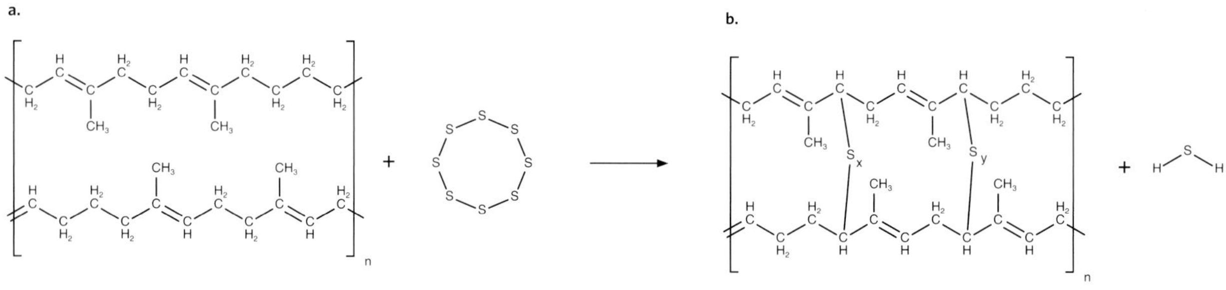

FIGURE 3.1
Vulcanized rubber: a three-dimensional (cross-linked) network. The process of vulcanization combines sulfur and cis-1,4-polyisoprene (latex) (a) to create cross-linked polyisoprene (b).

soft at low temperatures. We now know that sulfur reacts with unsaturated (double bond) hydrocarbon units in the polymer chain (Simmons 1921; van Rossem 1958, 65). One of the key features of vulcanization is that a few sulfur links are formed between the polymer molecules, making a three-dimensional, slightly cross-linked molecular network (fig. 3.1).

In 1876, the explorer Henry Wickham brought seeds of the *Hevea brasiliensis* tree from Brazil to England to ensure that the British rubber industry would not be totally dependent on Brazil for its natural rubber supply. The seedlings were raised at the Royal Botanical Gardens in Kew and shipped to British colonies in Southeast Asia, where large rubber plantations were established (Simmons 1921).

The advent of the modern, chain-driven, pneumatic-tire bicycle in the late 1880s coincided with the invention of the automobile. Together, they dramatically accelerated the demand for rubber. But during World War I, military demands and shipping disruptions limited the available supply of natural rubber, leading to the development of synthetic rubber. Nonetheless, natural rubber continues to be used today for hundreds of different products. It is still important in the worldwide automotive and aviation industries—in automobile, truck, and aircraft tires and in hoses and gaskets. It is also used to make consumer products such as toys, adhesives, gloves, balloons, waterproof clothing, hair ties, and athletic shoes.

Synthetic Rubber

Following the discovery of the vulcanization process and uses for vulcanized rubber in various applications, chemists experimented to discover the structure and composition of natural rubber. It was the British chemist William Augustus Tilden, in 1882, who, when heating turpentine, discovered the repeat unit of natural rubber: isoprene (2-methyl-1,3-butadiene), consisting of five carbon atoms and two carbon-carbon double bonds in the backbone chain of the monomer (fig. 3.2).

During World War I, the Allied powers imposed a rubber blockade against Germany. In response, Germany began to develop its own synthetic substitute, methyl rubber, which it then began producing for military purposes such as tires and hoses. Butadiene (1,3-butadiene), a compound similar to the isoprene repeat unit in natural rubber, proved to be the base unit for synthetic rubbers (fig. 3.3).

After the war, in 1926, Buna, a sodium-polymerized rubber, began to be synthesized from butadiene.[2] It was just the first of many other types of butadiene rubbers. At the

2. Buna takes its name from the first syllables of *butadiene* and *Natrium*, the German word for sodium.

FIGURE 3.2
Structural formula of isoprene (2-methyl-1,3-butadiene), the repeat unit of natural rubber, consisting of five carbon atoms and two carbon-carbon double bonds in the backbone chain of the monomer.

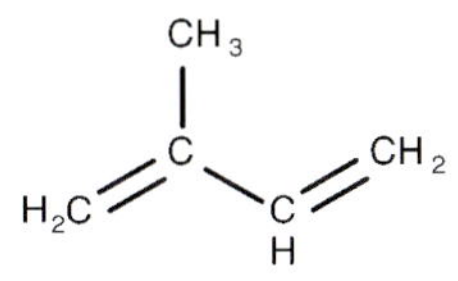

FIGURE 3.3
Structural formula of butadiene (1,3-butadiene), the repeat unit of synthetic rubber, consisting of four carbon atoms and two carbon-carbon double bonds in the backbone chain of the monomer.

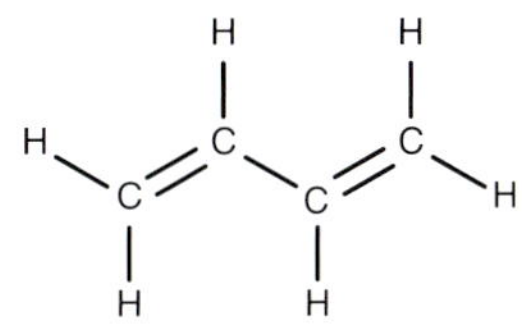

chemical company IG Farben, Buna-S, a synthetic rubbery copolymer of butadiene and styrene, was developed in 1929. This synthetic rubber was known as SBR (styrene-butadiene rubber) in the United Kingdom and as GR-S (government rubber-styrene) in the United States.

Plasticized polyvinyl chloride and polyurethanes, which have rubber-like, elastic properties, were developed in the 1930s and became the predecessors of the polyurethane elastomers we know today. Other synthetic rubbers developed between the First and Second World Wars include chloroprene, nitrile rubber, and butyl rubber.

In the 1950s, the growth of the electrical industry expanded the demand for rubber insulation. With the advent of Ziegler-Natta catalysts (see chap. 1), it became possible to produce commercially useful rubbers. It was just in 1954 that a synthetic rubber was synthesized that had the same chemical structure and nearly the same configuration (92%–96% cis) as natural polyisoprene rubber (99% cis). It was called synthetic polyisoprene, or Shell polyisoprene, in contrast to natural polyisoprene.

Since the 1960s, many developments have taken place, including thermoplastic olefin, styrenic copolymers, polyurethane elastomers, and thermoplastic polyester elastomers.

3.2 TYPES OF NATURAL RUBBER

Rubber can be obtained from several species of tropical plants, the primary one being *Hevea brasiliensis*. The milky juice (latex) is collected by slashing the tree diagonally (fig. 3.4). Latex contains 60% water and 4% various other substances, while the remainder is rubber. This rubber consists of an aqueous suspension of small particles about half a micron in diameter; the addition of formic acid causes the rubber to coagulate. The resulting raw rubber is processed into sheets that undergo a process of smoking and drying, which causes them to turn brown. If smoking is not part of the process and the rubber is only left to dry, then the sheets become a thinner, whiter material called crepe rubber. Applications of natural rubbers are listed in table 3.2.

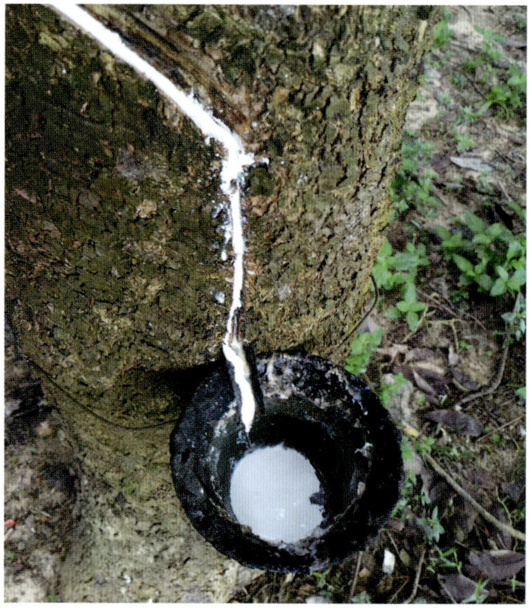

FIGURE 3.4
Tapping latex from a rubber tree, *Hevea brasiliensis*.

Table 3.2. Uses of natural rubber

Natural rubber	Gutta-percha	Balata	Hard rubber	Latex	Foam rubber
Tires Gas masks Safety goggles Insulation for electrical wiring Inflatable rubber boats Carpets Shoe soles Sealant strips Pressure hoses Hot water bottles Elastic bands Bathing caps	Picture frames Decorative inkstands Insulation on telegraph wires Filling for root-canal treatment	Cable insulation Coating on canvas to shed water Transmission belts	Pipe stems Dentures Knobs Battery housings Medical applications Victorian jewelery	Gloves Condoms Pacifiers (baby dummies) Toys Water-resistant fabrics (e.g., raincoats) Adhesives (e.g., rubber cement) Imitation leather Shoes	Mattresses Padding for upholstered furniture

Natural Rubber

Natural rubber is the polymer of the monomer isoprene (2-methyl-1,3-butadiene) (see fig. 3.2). When a monomer with two conjugated double bonds, such as isoprene, undergoes chain polymerization, one double bond remains in the chain. Double bonds in the polymer chain can show cis- or trans-isomerism, depending on where the methyl groups attach to the backbone.[3] Natural rubber is cis-1,4-polyisoprene (fig. 3.5), and in this con-

3. Cis- and trans-isomers are polymers that have different configurations due to the presence of a double bond (rigid structure) in the polymer chain. If the two groups attached to the double bond are on the same side of the polymer chain, the isomer is called cis. If the two groups attached to the double bond are on opposite sides of the polymer chain, the isomer is called trans.

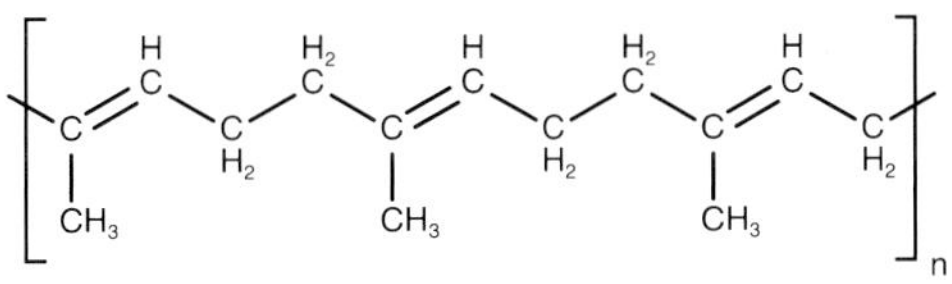

FIGURE 3.5
Natural rubber polymer (cis-1,4-polyisoprene).

figuration, nearly all methyl groups are on the same side of the polymer backbone (99% cis). This lack of symmetrical structure around the backbone does not allow for easy crystallization; therefore, natural rubber is an amorphous, rubbery material that is suitable for many uses.

One of the key features of rubber is its elasticity; however, as mentioned earlier, in its natural state it has the fundamental flaw of softening and becoming tacky with heat and stiffening when cold. These undesirable properties can be eliminated through vulcanization (M. J. R. Loadman 1995).

Gutta-Percha

Gutta-percha is the evaporated milky fluid from the family Sapotaceae; it is mainly obtained from the *Palaquium* tree of Southeast Asia. Like latex, it consists primarily of water and rubber particles. Under normal temperatures, gutta-percha is hard and not very elastic. It becomes moldable when heated to a temperature of 50°–60°C and toughens again when cooled. Gutta-percha is predominantly trans-1,4-polyisoprene (fig. 3.6), which has a regular structure that, as a result, allows crystallization (Roberts 1988).

Gutta-percha is obtained and processed in small quantities for decorative items (fig. 3.7). It is very resistant to water and has been widely used as insulation for underwater electrical equipment and cables. Other uses for the material are listed in table 3.2.

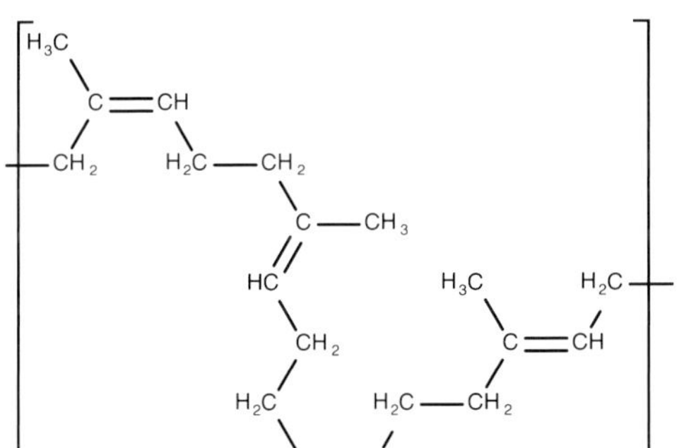

FIGURE 3.6
Gutta-percha (trans-1,4-polyisoprene).

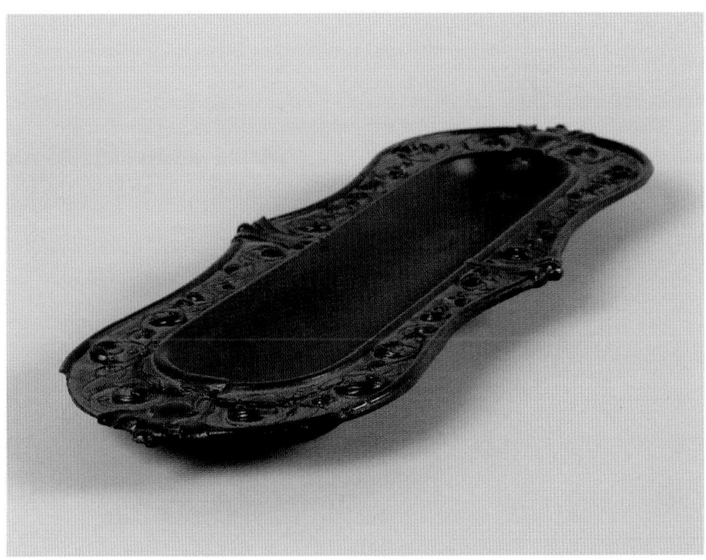

FIGURE 3.7
Pen tray, ca. 1970–79, manufactured by Telcon Plastics Ltd., United Kingdom, from a nineteenth-century mold designed by Richard Chambers. Compression-molded gutta-percha, 2 × 10.8 × 31.5 cm. Bournemouth, Arts University, Museum of Design in Plastics, AIBDC: 006890.

Balata

Balata is the evaporated milky juice from *Manilkara* and other trees in the Sapotaceae family of North and South America. It is composed of 5% water, 40% resin, and 40%–50% rubber. Raw balata sheets are brown, elastic, and leathery and consist of a mixture of cis- and trans-1,4-polyisoprene. The material becomes moldable when heated to a temperature of 50°C. Like gutta-percha, processed balata is rather inelastic, tough, and water resistant. Although it is obtained and processed only in small quantities, it is often used in belting applications (Kaminitz 1988), among other uses (see table 3.2).

Chicle

Chicle is a gum that consists of the evaporated milky juice (latex) of the sapodilla tree (*Manilkara zapota*), which grows on the Yucatán peninsula and in other regions of Central America (fig. 3.8).

The latex is tapped from the trees, and the coagulated material yields a thermoplastic material consisting of both trans- and cis-polyisoprene. First introduced as a substitute for rubber, chicle became the base for chewing gum, but it has been replaced almost entirely by synthetic polymers in that use as well (Brydson 1999, 866).

Hard Rubber

The term *hard rubber* is generally used when the amount of sulfur used in the vulcanization process is 5%–40%; that amount of sulfur causes the product to become hard and tough. This product is known by the trade names Vulcanite and Ebonite—the latter for its resemblance to ebony wood. From 1864 to 1881, the Goodyear Dental Vulcanite Company made dentures of Vulcanite (fig. 3.9). Additional uses are listed in table 3.2.

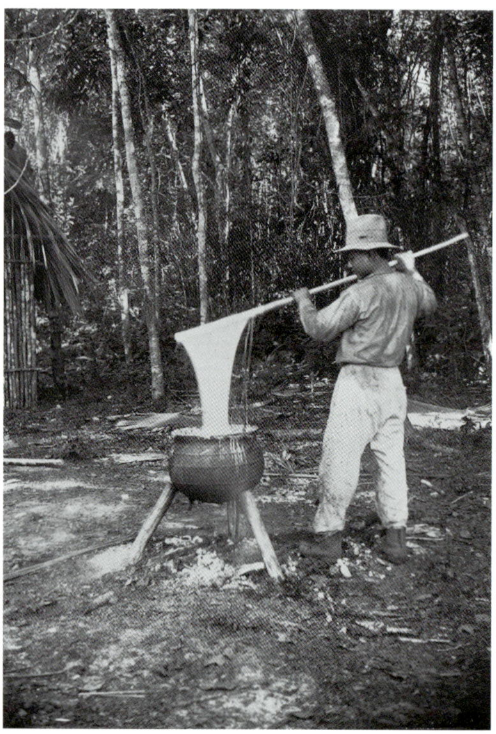

FIGURE 3.8
Chicle from *Manilkara zapota*.

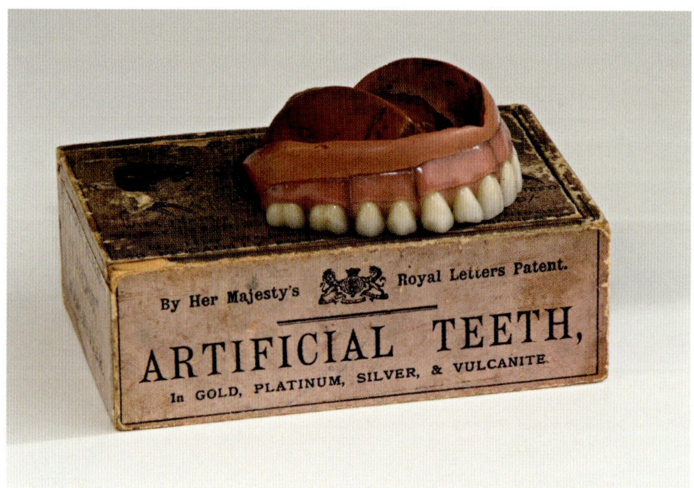

FIGURE 3.9
Vulcanite dental appliance.

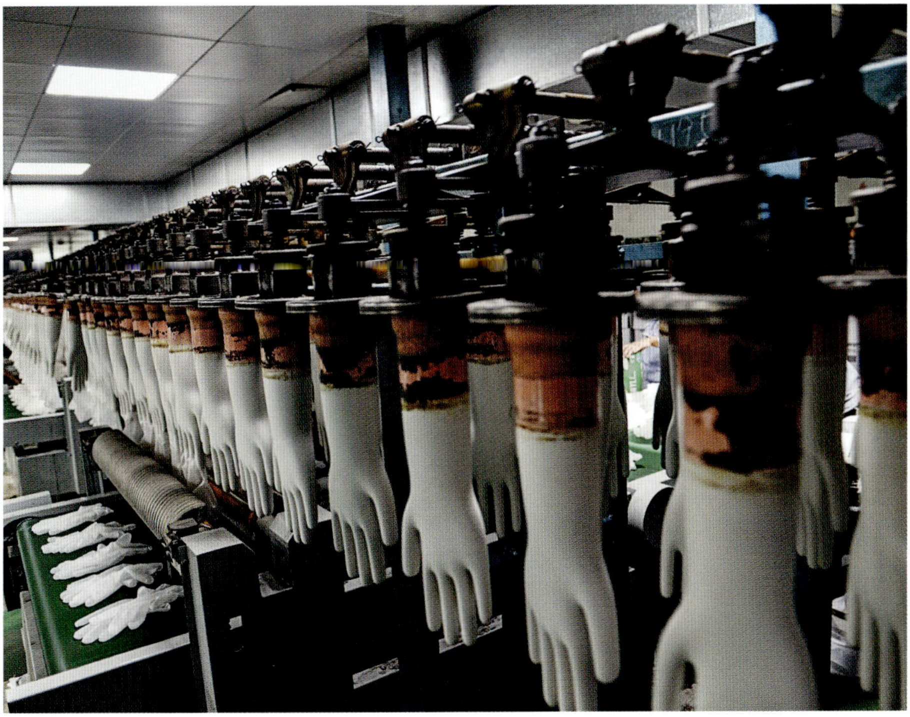

FIGURE 3.10
Dipping porcelain glove molds into latex.

Latex

Latex objects can be produced without the drying and other processing required for isoprene rubber, but they must still be vulcanized. The vulcanization process can occur at a lower temperature when an accelerator is added to the latex. Some objects are produced by dipping porcelain or aluminum molds of the desired shape into latex. A coating of latex adheres to the form and is stripped from it after vulcanization. The operation is repeated until the desired thickness is obtained (fig. 3.10).

According to van Rossem (1958, 31), about 10% of all natural rubber is shipped as latex. Once it is processed to a rubber content of approximately 60%, it is used for making dipped goods such as gloves and balloons, and medical devices such as grips, stent covers, and collection bags. (See table 3.2 for other common uses of latex.)

Foam Rubber

Latex is widely used for manufacturing foam rubber. The latex is poured into a frothing machine and mechanically whipped with air, which transforms it into foamed latex. The density of the latex foam depends on the amount of air that is added. After vulcanization, the article is removed from the mold and then washed and dried in circulating air ovens. Foam rubber was used in furniture upholstery until it was superseded by polyurethane foams in the 1960s (M. J. R. Loadman 1995).

Natural rubbers are still important industrial polymers; however, they compete today with a number of synthetic rubbers such as synthetic isoprene rubber, styrene-butadiene rubber, and isobutylene-isoprene rubber, as well as thermoplastic elastomers such as olefinic, styrenic, and polyurethane elastomers (Birley 1989; Brydson 1999).

3.3 TYPES OF SYNTHETIC RUBBER AND THERMOPLASTIC ELASTOMERS

Synthetic rubbers are polymers that are loosely cross-linked and amorphous, and remain elastic at temperatures well below room temperature. Synthetic rubbers have a low tensile modulus (little stiffness) and very high elongation at break (good ability to deform), compared to plastics (see "Flexural Strength" in chapter 4). Synthetic rubbers can be divided into three classes: diene rubbers, which have double bonds in the polymer chains; nondiene rubbers, with no double bonds in the polymer chains; and thermoplastic elastomers, which have hard and soft segments in the polymer chains, giving them elastic properties.

Diene Rubbers

Diene rubbers are polymerized from monomers having two conjugated double bonds. At polymerization, one double bond is used for linking and the other is left in the polymer chain and can be used for cross-linking with sulfur (vulcanization). Typical examples are described below.

Butadiene rubber | Butadiene rubber is polymerized from the monomer 1,3-butadiene using Ziegler-Natta catalysts. Elasticity, wear, and temperature resistance of butadiene rubber is better than that of natural rubber. Butadiene, when added to the polymerization process of acrylonitrile and styrene, improves the impact resistance of the resulting acrylonitrile-butadiene-styrene copolymer.

Chloroprene rubber | Chloroprene rubber, also known as neoprene, is produced by polymerization of the monomer chloroprene (2-chlorobuta-1,3-diene). As both its chemical and environmental resistance are good, industrial applications of neoprene include tanks for oil and chemical transport, and consumer goods such as safety gloves and wet suits.

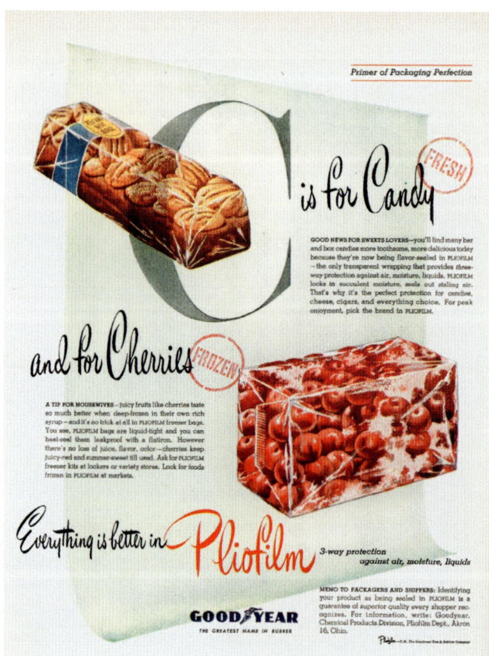

FIGURE 3.11
Advertisement for Pliofilm—rubber hydrochloride used for packaging.

Synthetic isoprene rubber | As mentioned above, isoprene rubber is a synthetic analog of natural rubber. Although attempts to manufacture rubber were made as early as the late nineteenth century, it was not until the 1950s, after the advent of Ziegler-Natta catalysts, that it became possible to produce commercially useful isoprene. Synthetic isoprene has more trans-1,4-polyisoprene than natural rubber; the amount of cis-1,4-polyisoprene ranges from 92% to 6%. As a consequence, synthetic polyisoprene is not as strong as natural rubber but offers the advantage of constant production not dependent on cultivation and weather/climate conditions.

In 1938, a product called Pliofilm (fig. 3.11), made from rubber and hydrochloric acid, was patented. Rubber hydrochloride was used for packaging fruit and other food, as well as cigarettes and cigars (van Rossem 1958, 49). In the 1960s, cellophane replaced Pliofilm due to its better stability.

Nitrile-butadiene rubber | Nitrile-butadiene rubber, also known as nitrile rubber and Buna-N, is a copolymer of acrylonitrile and butadiene. Properties of vulcanized nitrile rubber depend on the nitrile content: the greater the nitrile content, the better the aging, wear, and oil resistance. Uses include hoses, seals, waterproof textiles, and nitrile rubber gloves; the latter are commonly used in conservation laboratories and studios, among many other places.

Styrene-butadiene rubber | Styrene-butadiene rubber (SBR), one of the most important synthetic rubbers, has a greater durability than natural rubber but is less resistant to high temperatures, depending on the amount of styrene that is copolymerized. SBR is a general-purpose rubber that is widely used in the production of tires, cables, hoses, carpets, and for many other purposes.

Nondiene Rubbers

Nondiene rubbers consist of polymer chains in which no double bonds are present; this means they are unable to connect with sulfur to cross-link the polymer chains, so traditional vulcanization cannot be used to cure them. Instead, other methods are used for curing, including using peroxides or adding small amounts of monomers containing double bonds, such as isoprene. Examples of nondiene elastomers are discussed below.

Butyl rubber | Isobutylene is a nonrubbery polymer; however, when copolymerized with 2% isoprene it can be vulcanized using the traditional cross-linking with sulfur to give it rubbery properties. Isobutylene-isoprene rubber is commonly called butyl rubber. It has an excellent oxygen and ozone resistance and is therefore commonly used in the production of inner tubes due to its very low permeability to air.

Fluororubbers | Fluororubbers are fluorocarbon-based synthetic rubbers that form an important class of specialty elastomers. As with most nondiene rubbers, fluororubbers can be vulcanized using peroxides. They are more expensive than general-purpose rubbers and are used for O-rings and other engineering purposes in the aerospace industry. Two common fluororubbers (typically referred to by their abbreviations) are FKM (fluororubbers of the polymethylene type) and FPM (fluororubbers of the polypropylene type, better known as Viton). These are used mostly in automotive and industrial applications (Brydson 1999).

Silicone rubber | Silicone rubbers have linear polymer chains consisting of silicon and oxygen atoms; vulcanization takes place at room temperature with peroxide. Silicone rubbers maintain their intrinsic properties over a wide range of temperatures (−50°C to 250°C), and because of their antistick properties and water repellency they have a variety of applications in consumer industries as well as medicine and aviation (Brydson 1999).

Thermoplastic Elastomers

Today, many products are made from thermoplastic elastomers due to their combination of hard, rigid repeat segments (which allow the material to be processed like a thermoplastic) and soft, rubbery repeat segments (which give it the elasticity of rubbers) (fig. 3.12). Since the 1950s, polymer chemists have developed the thermoplastic rubbers discussed below, many of which have applications similar to general-purpose rubber.

Thermoplastic olefins | In olefinic polymers such as polyethylene, the number of double bonds is very low (sometimes zero), so little cross-linking can occur. Adding other hydrocarbon monomers to olefins makes it possible to produce rubbery products with overall better oxygen and ozone resistance than diene rubbers: an increased number of double bonds creates more cross-linking, which yields a more elastomeric product). Three thermoplastic olefins have been developed:

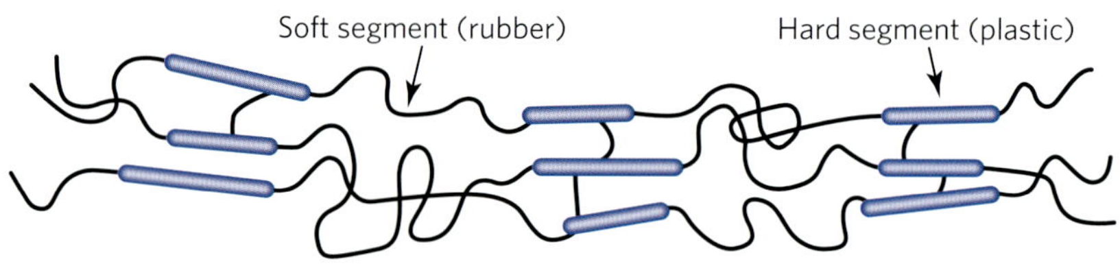

FIGURE 3.12
Hard and soft segments in thermoplastic elastomers' polymer chains.

- Ethylene-propylene monomer (EPM) was the first olefinic elastomeric; it has been produced since the 1950s but was not very successful. The lack of double bonds in the polymer chains means that aliphatic olefin elastomers cannot be vulcanized by sulfur, so cross-linking is achieved by peroxide curing. EPM rubbers are used for insulator applications and other industrial purposes.
- In ethylene-propylene-diene monomer (EPDM), a third monomer (a diene) is added to EPM to prevent premature peroxide vulcanization. The diene monomer provides double bonds for sulfur vulcanization. Such ethylene-propylene-diene monomer ternary copolymers (terpolymers) are designated EPDM rubber (Brydson 1999, 299–300). Due to the material's resistance to weathering, EPDM rubber uses include technical applications such as hoses, tubing, and seals for outdoor uses.
- Ethylene-vinyl acetate copolymer (EVAC) comes in three different forms that differ in vinyl acetate content and the way the materials are used. An EVAC with a vinyl acetate content up to 4% is considered a thermoplastic. EVACs with a vinyl acetate content up to 15% have a flexibility comparable to that of plasticized polyvinyl chloride; they are used as hot melt adhesives. Copolymers up to 45% vinyl acetate are rubbery and can be vulcanized using peroxides. Uses of rubbery EVAC include toys and automobile bumpers. Cellular EVAC (expanded foam) is used as padding in shoes and sports equipment.

Styrenic elastomers | Styrenic elastomers consist of polystyrene blocks (hard, crystalline blocks) and diene blocks (soft, elastomeric blocks). Styrenic block copolymers are used in shoe soles, adhesives, and toys, among other applications (Brydson, 1999). There are three types of styrenic elastomers:

- Styrene-butadiene-styrene triblocks (SBS) are made by polymerization of styrene and butadiene. First, a binary diblock copolymer is formed. Then, when all the butadiene monomer has been consumed, styrene monomer is added to produce the triblock copolymer styrene-butadiene-styrene (SBS). Properties of SBS include high tensile strength and good resistance to organic solvents and water.
- Polymerization of styrene and isoprene forms styrene-isoprene-styrene (SIS) triblocks. SIS properties are comparable to those of SBS; however, SIS is less costly.
- Hydrogenated SBS triblocks like styrene-ethylene-butadiene-styrene (SEBS) have a higher maximum service temperature and, because of the total absence of double bonds in the polymer chains, have better weathering properties than SBS and SIS. SEBS is used in toys as a substitute for plasticized PVC, in nonslip handles and the grips of toothbrushes, and for mouth guards.

Polyurethane elastomers | As discussed in chapter 2, polyurethanes are polymerized from di- or polyols reacting with di- or polyfunctional isocyanates and chain extenders (short diols or diamines), under the influence of catalysts.

Depending on the type of alcohols, isocyanates, and types of chain extenders, polyurethanes can be formed that have hard (cross-linked) segments (derived from chain extender and isocyanate, which lend thermosetting characteristics) and soft (less cross-linked) segments (derived from the polyols, which lend rubbery characteristics). The ratios, chain lengths, and chemical structures of the hard and soft segments determine the properties of the polymerized polyurethane elastomers: the more linear polymer chains (soft segments) in the polyurethane polymer, the more thermoplastic the material's properties; the more cross-linked polymer chains (hard segments), the more thermosetting its properties. Millable and malleable polyurethanes have both cross-linked hard segments and linear polymer chains (soft segments).

As with thermosetting polyurethanes, there are two chemical classes of polyurethane elastomers—poly*ester* and poly*ether* polyurethanes—and each has different properties. Polyethers are more hydrolysis resistant, while polyesters have good resistance to thermal and oxidative degradation.

Properties of polyurethane elastomers include high elongation, good tensile strength, and resistance to oils, solvents, and abrasion. Because of their strength, flexibility, and shock absorption, they are used as base materials for footwear, wheels, solid tires, rollers on transport and office equipment, and carpet underlay. Polyurethane elastomers are also used in works of art (see chaps. 9 and 11).

Thermoplastic polyesters | Thermoplastic polyester elastomers consist of hard (thermoplastic) segments and soft (rubbery) segments, and these segments are connected by intermolecular forces. They do not have to be vulcanized by sulfur. As they have no strong cross-linked bonds, they can be heated, melted, and recycled like thermoplastics. Polybutylene terephthalate and glycols form thermoplastic polyester elastomers, which are used in applications such as shoes, toys, and medical devices.

Though they are much more costly than other thermoplastic elastomers, thermoplastic polyesters are used as a replacement for oil-resisting rubbers, as they possess thermoplastics' property of good resistance to organic fluids and oils combined with the elastic property of rubber.

Summary

- The terms *rubber* and *elastomer* tend to be used interchangeably to mean a class of polymeric materials having elasticity as an exceptional property. The principal property of elastomers is elastic recovery after deformation via compression or tension. Even after an elastomer is stretched to many multiples of its original length, under ideal circumstances it will return to its original shape and size after removal of the tension.
- Elastomeric polymers are loosely cross-linked and amorphous, and remain elastic at temperatures well below room temperature.
- Many different types of rubber are available, but natural rubber, produced from the *Hevea brasiliensis* tree (latex), is still the most widely known.
- Synthetic rubbers have been developed that have better resistance to high temperatures, outdoor conditions, and fuels and oils than natural rubber.
- Thermoplastic elastomers have the performance and properties of rubber, but can be reprocessed and remolded like plastics.

Suggested Reading

Louise Bacon, "The Deterioration of Four Giorgi Flutes Made of Ebonite and a Possible Method for Their Conservation," in *Conservation Today: Papers Presented at the UKIC 30th Anniversary Conference, 1988*, ed. David Pearce (London: United Kingdom Institute for Conservation, 1988), 96–100.

Mary T. Baker, "Thermal Studies on Ancient and Modern Rubber: Environmental Information Contained in Crystallized Rubber," in *Resins: Ancient and Modern*, ed. Margot M. Wright and Joyce H. Townsend (Edinburgh: Scottish Society for Conservation and Restoration, 1995), 53–56.

Jack Mitchell Buist, *Ageing and Weathering of Rubber* (Cambridge: Heffer, for the Institution of the Rubber Industry, 1956).

Mark Gilberg and David Grattan, "Oxygen-Free Storage Using Ageless Oxygen Absorber," *Studies in Conservation* 39 (suppl. 2, 1994): 177–80, https://doi.org/10.1179/sic.1994.39.Supplement-2.177.

John Loadman, *Tears of the Tree: The Story of Rubber—A Modern Marvel* (Oxford: Oxford University Press, 2014).

A. van Rossem, *Rubber: Winning, Eigenschappen, Verwerking* (The Hague: Sevire, 1958).

Chapter 4 | Properties of Plastics and Elastomers

When considering plastics for conservation, their behavior under stress can be best understood if one is aware of their mechanical and physical properties.

How strong is the plastic? How much can it be bent without breaking? How stiff or brittle is it, and does it easily break when dropped on the floor? Is the plastic hard or soft? Can it be scratched easily? Why does a Bakelite beaker break when hit with a hammer and a polyethylene beaker does not?

Stiff, tough, strong, flexible, elastic, soft, ductile—all these terms come up when reading about properties of plastics and elastomers. But what do they mean? What properties do they relate to? One answer is that they all concern the behavior of these materials under some kind of force or stress.

The wide range of properties of the many plastics and elastomers available helps designers and manufacturers who are looking for the best polymers to serve their applications, whether those materials are being used in pharmaceuticals, aviation, automobiles, or day-to-day household products. Each application has a particular goal. For example, the aerospace industry typically requires high-performance engineering plastics and elastomers, whereas disposable household plastic bags have fewer and/or lower specifications, such as those for commodity plastics.

There are more than fifty different and unique families of plastics and just as many elastomers in commercial use today. In addition, each family has dozens of variations, which in aggregate translates to a vast number of available properties (Shashoua 2008, 1).

Of course, not all of these hundreds of polymers are found in cultural heritage objects, and consequently, many of them are beyond the scope of this book. In addition, plastic coatings, textile fibers, paints, and conservation materials are also not covered in this book, as many other resources exist for learning about the properties of those plastics and elastomers.[1]

In 2016, conservators of modern and contemporary art from various institutions were surveyed to determine which physical properties they considered useful in their conservation practice. They could choose from a list of fifty plastics properties selected from various manufacturers' data sheets and technical handbooks for plastics. Their feedback resulted in a list of twenty properties—divided into mechanical, physical, thermal, and optical properties—that was used as the basis for the fact sheets in part III of this book. The definitions of these properties, as well as the methods of measuring them, are explained in this chapter; some properties are mentioned only briefly, while others are described more extensively.

1. The properties of plastics and elastomers are given in texts such as the *Handbook of Polymers*, by George Wypych (2016), which lists more than a hundred properties of each of the two hundred most commonly used polymers. The MatWeb website (www.matweb.com) lists more than eighty thousand polymers, with data on about twenty to fifty properties of each polymer, all derived from manufacturer websites. (The graphs in this chapter are derived from the MatWeb database.) Many other polymer handbooks, encyclopedias, and websites with data on properties are readily available. See the Suggested Reading list at the end of the chapter for a few suggestions.

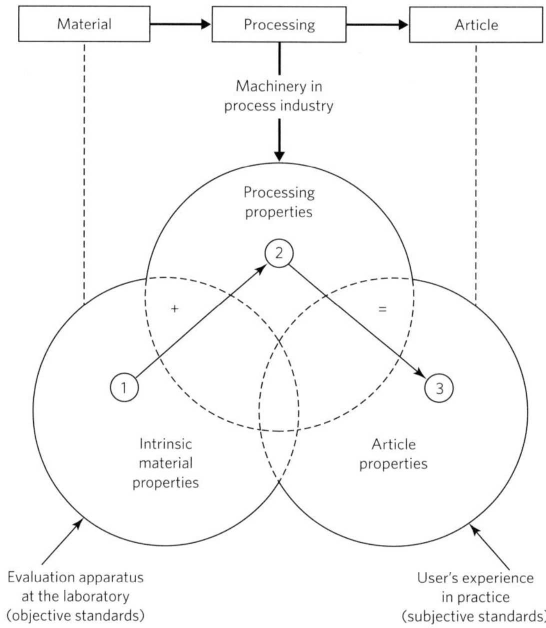

FIGURE 4.1
The relationship of intrinsic material properties, processing properties, and article properties. Adapted from van Krevelen 1990, 49.

The properties of plastics and elastomers can be divided into three interrelated categories: intrinsic properties, processing properties, and product or article properties (fig. 4.1). Van Krevelen (1990, 49) describes the typology of plastics properties:

- Intrinsic properties always refer to polymer composition, chemical composition, or the subunits and arrangement of these subunits.
- Product properties refer to an individual article and depend on the composition, size, and shape of the object.
- Processing properties are a combination of intrinsic properties and processing properties and are influenced by the method of manufacture.

In the case of a work of art, processing properties may be determined by the manufacturer, the artist, or both. Which of the two "creates" the object can affect the properties of the article substantially. Items produced by a factory nearly always have the same properties because they are made from a precise mix of ingredients (measured by instruments) in a controlled environment. In contrast, similar objects processed by an artist can differ from one day to the next due to the often imprecise addition of components and changing environmental circumstances (fig. 4.2). Moreover, both the chemical composition and physical structure of polymers can change over time, owing to aging or relaxation processes (such as creep, or cold flow). For example, the solvent resistance of new cellulose acetate is quite different than that of an aged cellulose acetate object (Lavédrine, Fournier, and Martin, 2012, 260).

Because polymers are long, complex chains, molecular weight, molecular weight distribution, and molecular size and structure all affect their properties. The chemical composition of a plastic (its compounding) also influences the forces of attraction between polymer chains and their effect on the physical properties of a plastic (van Krevelen 1990, 49), so it is understandable that there is enormous variation in properties.

In this chapter, graphs are used to show the data on a specific property for all selected plastics and elastomers. These graphs make it possible to compare the behavior of plastics

FIGURE 4.2
Piero Gilardi at work in his studio pouring latex for the underlayment of one of his Nature Carpets.

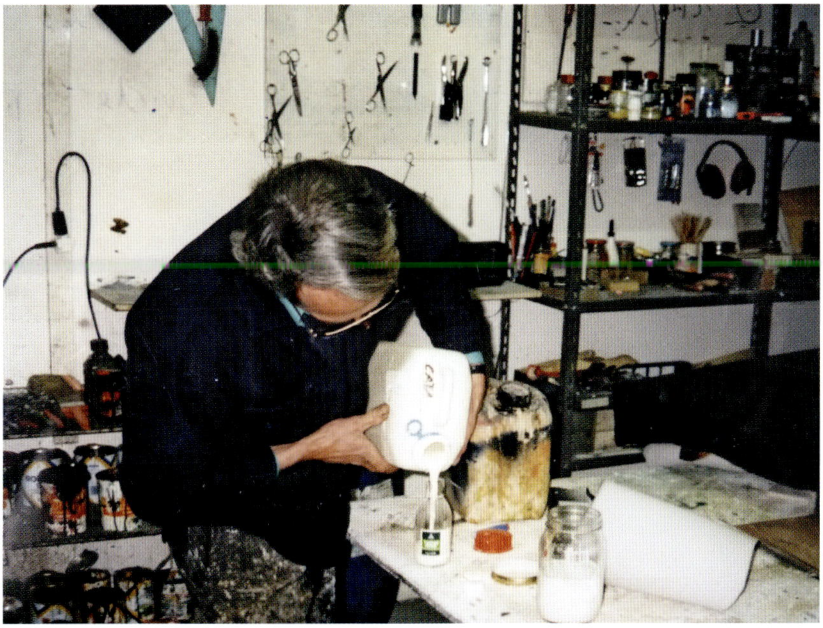

and elastomers; in them, it is obvious which plastic is stronger and which absorbs more moisture. Most properties of plastics and elastomers are expressed as a range of values rather than a single value. For comparison purposes, the *average* of a property is used, and this is the case with the data in the graphs. Also note that the values should be used for comparison only and not be regarded as absolute.

To test plastics and acquire data on their properties, numerous standard methods have been developed by organizations such as the International Organization for Standardization (ISO), ASTM International (formerly the American Society for Testing and Materials), and the German Institute for Standardization (DIN). Data from ASTM test methods were chosen for the graphs in this book.

4.1 MECHANICAL PROPERTIES

Mechanical properties of plastics consist of the various types of strength that plastics exhibit. It is essential to define the terms as they relate to plastics properties, to avoid confusion. (As technical terms are sometimes best understood in one's own language, translations of words related to properties are given in appendix C.) Because there is more than one type of force that can break a plastic, there are many types of strength that can be measured for a plastic object. Table 4.1 presents the various types of strength for plastics and their SI units.[2]

Plastics that are subjected to a force are said to be *stressed*, and the deformation (change in length) is known as *strain*. Strain is measured as the ratio of change in length of a plastic to the original length; it is expressed in µm/mm. Strain can be also expressed as a percentage, in which case it is called *percent elongation*.

2. Data in this chapter, as well as in the fact sheets in part III, are given in SI units, which can easily be converted to another system using tools available online.

Table 4.1. Strength categories of plastics

Type of strength	SI unit	Abbreviation
Tensile (stretch) strength	pascal/megapascal	Pa/MPa
Impact strength	joule per meter	J/m
Compressive (squash/crush) strength	megapascal	MPa
Flexural strength	megapascal	MPa
Torsional (twist) strength	newton meter	Nm
Shear strength[a]	pascal/megapascal	Pa/MPa[b]
Tear strength	kilonewton per meter	kNm^{-1}

a. Mostly used when testing adhesives and for elastomers.
b. Strength typically is better expressed in MPa; 1 MPa = 10^6 Pa.

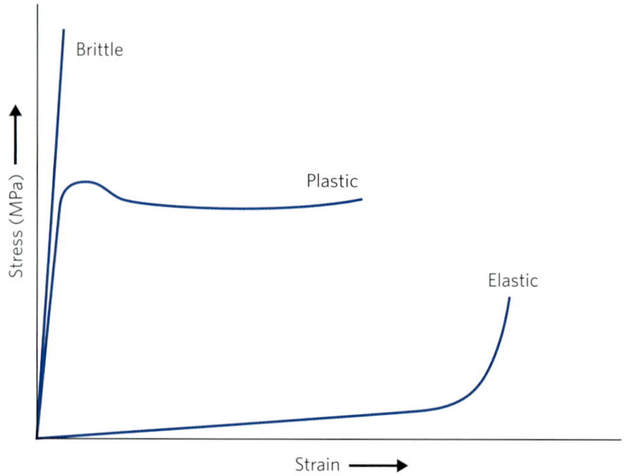

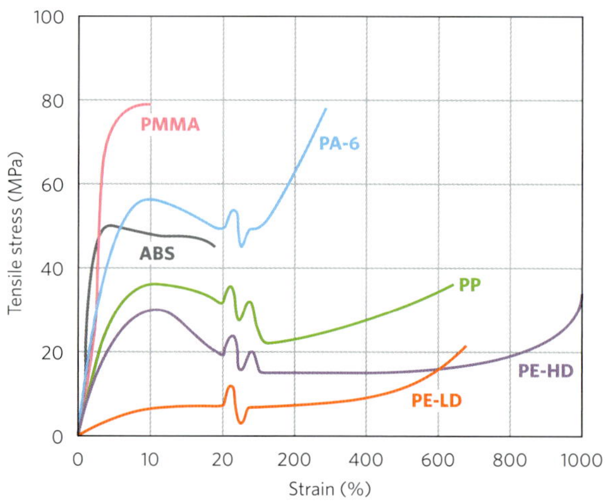

FIGURE 4.3
Basic stress-strain curves of polymers.

FIGURE 4.4
Stress-strain curves of various thermoplastics.

PMMA – polymethyl methacrylate
PA-6 – polyamide-6
ABS – acrylonitrile-butadiene-styrene
PP – polypropylene
PE-HD – high-density polyethylene
PE-LD – low-density polyethylene

In general, three main types of stress-strain curves can be distinguished, as shown in figure 4.3:

- **Brittle.** The first type is represented by thermoplastics such as polystyrene, which show a steep, almost straight line (Odian 1991). Plastics with this type of curve are inflexible materials, showing hardly any deformation.
- **Plastic.** The second type is represented by thermoplastics such as polyamides and polyethylenes, which show typical elongation and are strong. A decrease in stress and a deformation can be seen at a certain point in the graph. At constant stress, the deformation continues. At maximum elongation, the stress increases and the plastic breaks.
- **Elastic.** The third type, which is typical of elastomers, is elastic deformation. At relatively small stress, a comparatively large deformation will show, and in the end, with increasing stress, the elastomer will break.

All stress-strain curves are possible, and the shape of the curves serves as a method of classification according to their tensile properties: brittle, strong, or elastic (fig. 4.4). Typical materials with some of their mechanical properties are shown in table 4.2.

Table 4.2. Examples of materials and their mechanical properties

Material	Tensile strength (MPa)	Elongation at break (%)	Tensile modulus (GPa)
Stainless steel ball bearings	2000	Very little	200
Cellophane film	50-120	10-50	3
Nitrile rubber sheets	20-30	250-500	Very low
Fiberglass yarn	1400-2000	3-4	72
Nylon	50	150	2

Tensile Strength

Tensile properties are important indications of the strength of a plastic. Overall, tensile strength is subject to considerable variation, depending on manufacturer, molding technique, degree of polymerization, quantity and types of fillers and plasticizers used, temperature, and moisture in the plastic. Moreover, tensile strength experiments on plastics are also influenced by the form and size of the test sample and the speed at which the experiments are performed (Brown 1999).

To measure tensile strength of a plastic, a test sample of a specific size is clamped at each end; then the testing equipment pulls the test sample and records both force (stress) and elongation (strain), which are displayed on a stress-strain diagram. Tensile tests are performed using a universal tensile tester (fig. 4.5). While the test sample is stretched, the amount of force that the tester is exerting is measured—F, expressed in newtons (N). Once the force being exerted on the sample is known, that number is divided by the cross-sectional area of the test sample—A, expressed in cm^2 or m^2—to determine the tensile strength. Thus stress (tensile strength) is force divided by the cross-sectional area—Stress = F/A in Pa or MPa (1 Pa = 1 N/m^2).

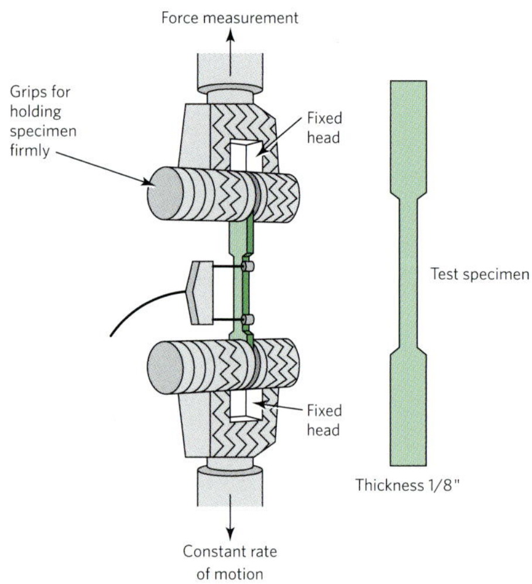

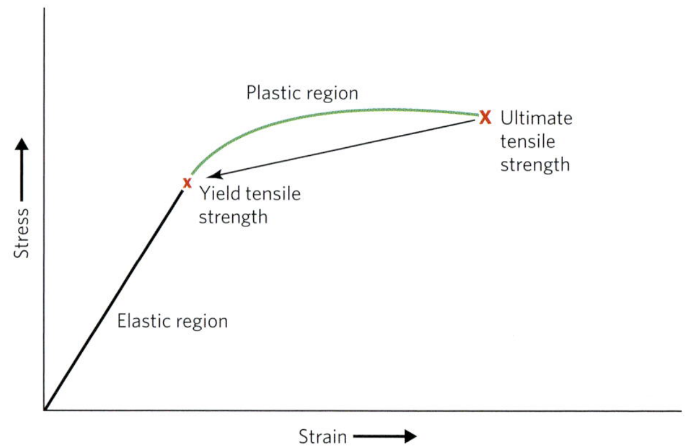

FIGURE 4.5
Universal testing machine for measuring tensile, compression, and shear strength.

FIGURE 4.6
Tensile strength curve with yield and ultimate tensile strength points.

Tensile strength at yield | The yield point is defined as the point at which a plastic changes from elastic deformation to plastic deformation (see the small red *x* in fig. 4.6). Once this point is exceeded, the plastic will no longer return to its original dimensions after the stress is removed. The yield strength can be determined by tensile testing using the ASTM D638 or ISO 527 test method.[3]

Tensile strength at break | The break point is what it sounds like: the point at which a plastic fails and breaks (see the large red *X* in fig. 4.6); this is called the ultimate tensile strength. The appropriate test for plastic films is ASTM D882; for other plastics, ASTM D638 or ISO 527 is used; and for elastomers, ASTM D412 is the standard test.

Figure 4.7 illustrates the tensile strength of various plastics and elastomers. Low-density polyethylene (PE-LD) and plasticized polyvinyl chloride (PVC-P) are the weakest, whereas nylon (PA-6), polycarbonate (PC), cellulose nitrate (CN), polymethyl methacrylate (PMMA), and polyoxymethylene (POM) are the strongest of the thermoplastics. Thermosetting materials all show high tensile strength, whereas the elastomeric materials show much less strength. Of all the materials shown, silicone rubber (Si) has the least tensile strength. Adding glass fibers or other reinforcing additives to the polymers increases tensile strength (fig. 4.8).

Tensile modulus | Various terms are used to describe the same property: tensile modulus, Young's modulus, modulus of elasticity, and E modulus, abbreviated as E. Tensile modulus is the rigidity of a plastic or elastomer and describes an object's resistance to being deformed when a force is applied. It is measured as the slope of a stress-strain curve within the linear range (fig. 4.9).

Tensile modulus values are given in gigapascals (1 GPa = 1 billion Pa); the higher the value, the stronger the plastic or elastomer. All thermosetting plastics are strong. Among the thermoplastics, the strongest are nylon, polycarbonate, polymethyl methacrylate, polyoxymethylene, and unplasticized polyvinyl chloride (PVC-U). Elastomers are the weakest materials in terms of rigidity (fig. 4.10).

---

3. A standard test method describes the procedure to determine a property of a material. The test method includes the test equipment, the test specimen, the test procedure, and calculations of the data derived from the test. ASTM D638 and ISO 527 are the standard methods used to test tensile strength.

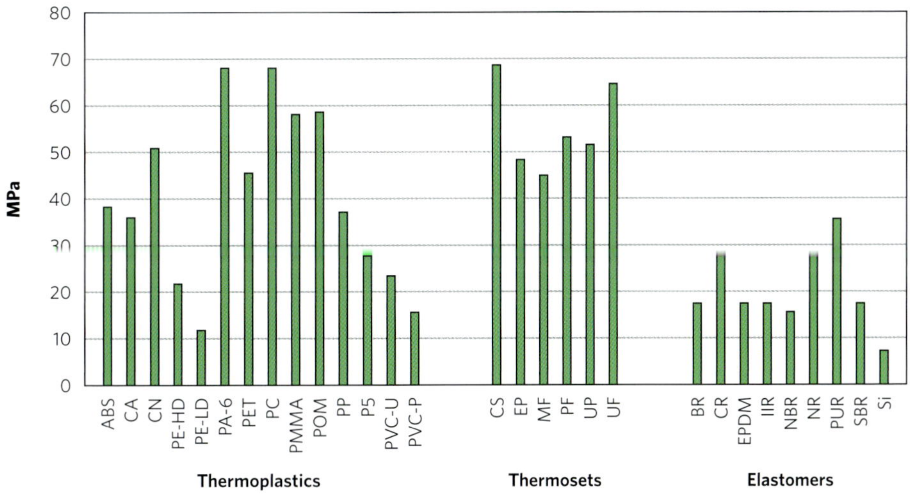

FIGURE 4.7
Average tensile strength of plastics and elastomers, determined using ASTM D638.

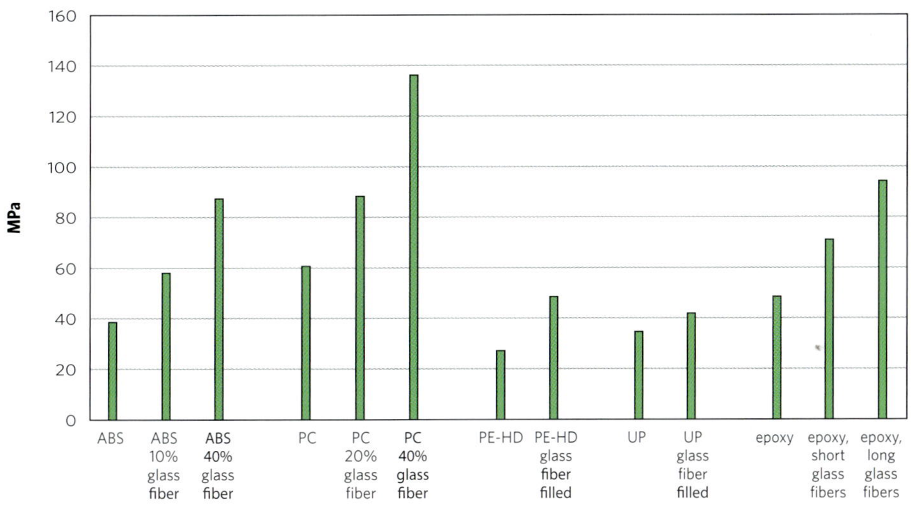

FIGURE 4.8
Average influence of glass fiber on the tensile strength of plastics, determined using ASTM D638.

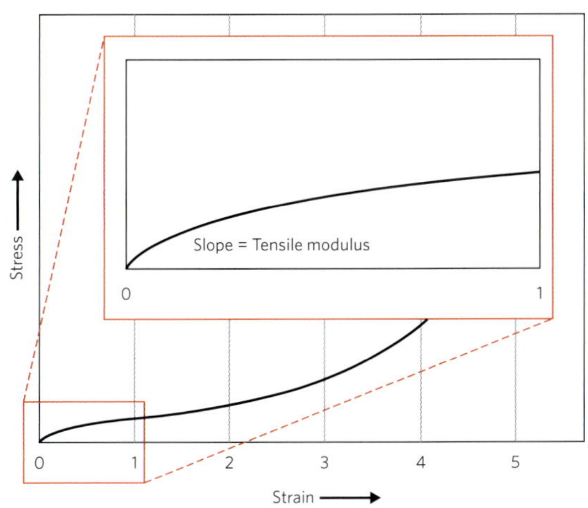

FIGURE 4.9
Tensile modulus curve.

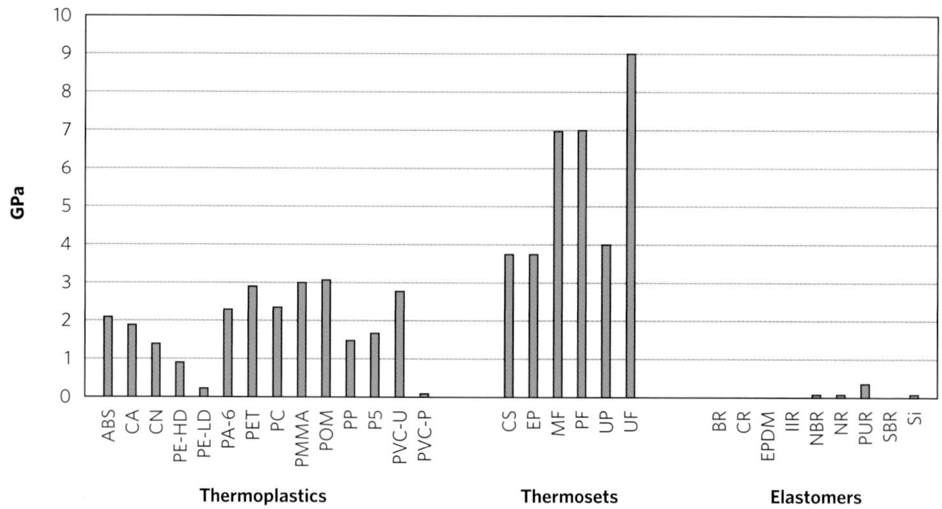

FIGURE 4.10
Average tensile modulus of thermoplastics, thermosets, and elastomers, determined using ASTM D638.

Tensile elongation | Tensile elongation, or elongation at break, is the measure of how much a plastic or elastomer can stretch without permanently being deformed. (*Ductility* is another term sometimes used to describe how much deformation a plastic or elastomer can withstand before breaking.) Tensile elongation is determined by testing in accordance with ISO 527 and ASTM D638.

The most common measure of tensile elongation is the percentage of change in length of a tensile sample after breaking; it is generally reported as percent elongation (% El). The most elastic materials, such as all elastomeric materials, have high values. Elastomers are able to stretch from 500% to 1000% and easily return to their original length. Rigid plastics, especially those that are fiber-reinforced, often exhibit values under 5% elongation. Polyolefins and plasticized polyvinyl chloride have higher values than other thermoplastics, while thermosets (except for epoxies) do not show elongation at break (fig. 4.11).

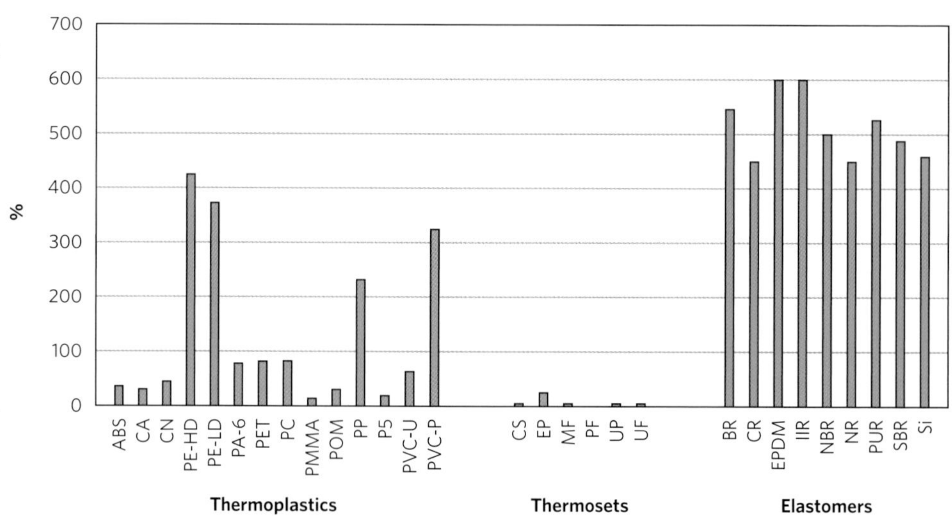

FIGURE 4.11
Average tensile elongation at break for thermoplastics, thermosets, and elastomers, determined using ASTM D638.

Impact Strength

The impact strength, or toughness, of a plastic is not truly a measure of strength but rather of the energy that can be absorbed by a plastic before it breaks. Energy, defined as the capacity to do work, is expressed as force × distance = newtons × meters = joules.

The thickness of the test sample also has to be taken into account, so impact strength is expressed in joules per meter (J/m) or joules per centimeter (J/cm).

There are two different methods of determining impact strength: Charpy (ISO 179) and Izod (ASTM D256). While both methods use a pendulum (swinging hammer) to strike the sample, in Charpy the test specimen is supported at both ends, whereas in Izod the test specimen is clamped on only one side.

Impact tests also specify whether the samples need to be unnotched or notched, and the number of notches in the latter case. In Izod the notch or notches are on the side of sample hit by the striker, whereas in Charpy any notches are on the opposite side (fig. 4.12). The Izod and Charpy impact values provide a useful indication of the material's toughness. Atmospheric temperature (warmth or cold) obviously plays a role in impact resistance, but other environmental factors do as well. For example, nylons (polyamides) generally exhibit greater impact strength in a moist state because of the plasticizing effect of absorbed moisture.

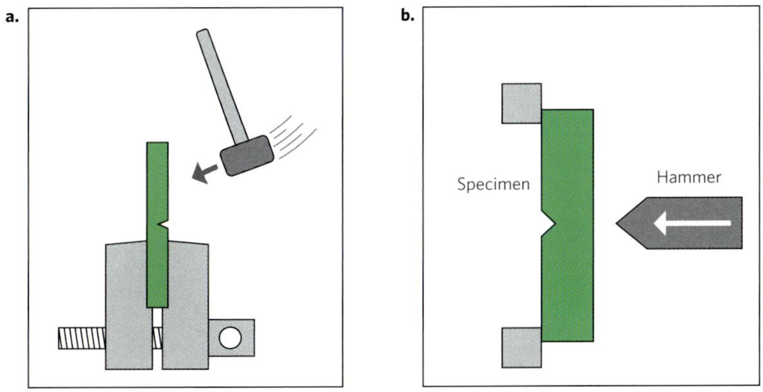

FIGURE 4.12
Impact tests: Izod (a) and Charpy (b).

The area under a stress-strain curve represents the energy required to break a plastic sample. As figure 4.13 shows, tough plastics (such as polyamides, polycarbonates, and unsaturated polyesters) can absorb a considerable amount of energy before fracture; brittle plastics (such as polystyrene) can absorb very little; and soft plastics (such as polyethylene and plasticized polyvinyl chloride) can absorb large amounts of energy before failure.

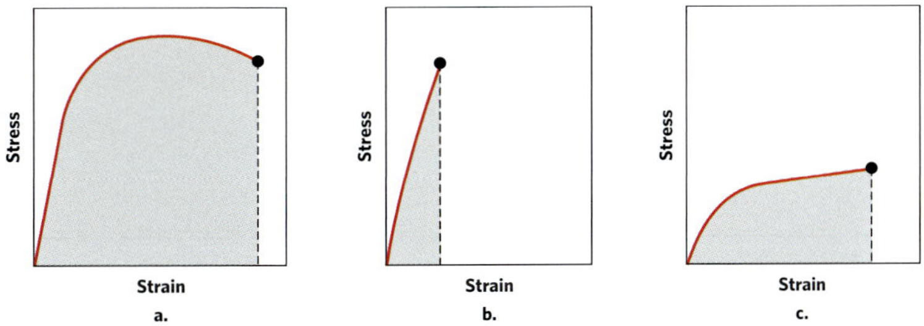

FIGURE 4.13
Impact strength of hard, tough plastics (a), brittle plastics (b), and soft, weak plastics (c). Adapted from van Krevelen 1990, 413.

FIGURE 4.14
Average impact strength of thermoplastics, thermosets and elastomers, determined using Izod notched test (ASTM D256).

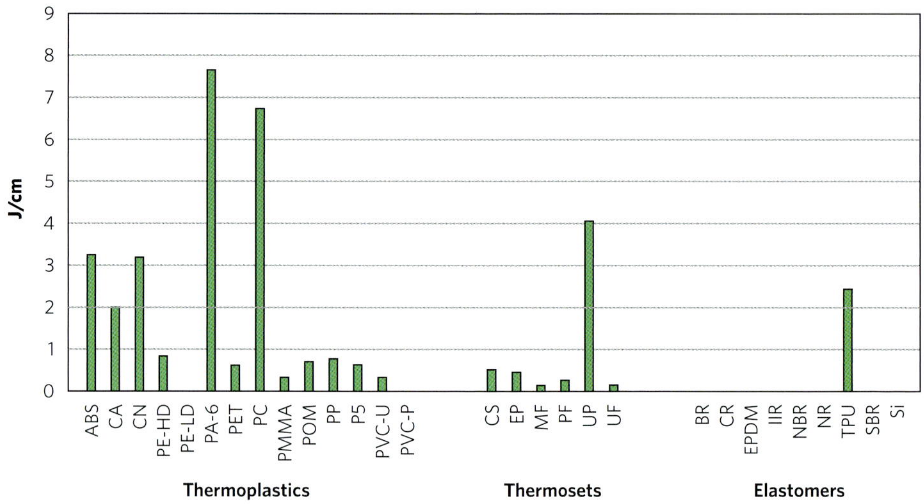

Thermoplastics such as nylon and polycarbonate have the highest impact strength, meaning that it is very difficult to break these plastics with a hit of a hammer (fig. 4.14). Unsaturated polyesters have the highest impact strength among the thermosets. Elastomers are not subjected to impact testing due to their inherent elasticity. As thermoplastic polyurethane elastomers (TPU) can range from extremely soft to very hard (see below), impact strength can be measured.

Compressive Strength

Compressive strength is the capacity to withstand loads, which typically reduce the material's height (fig. 4.15). Compare this to tensile strength, where the force is applied to elongate the plastic until it breaks. The standard tests for compressive strength testing are ASTM D695 and ISO 604.

Sometimes compressive *yield* strength is measured; this is defined as the stress measured at the point of permanent yield on the stress-strain curve. Other times *ultimate* compressive strength is used, which is understood as the stress required to rupture a test sample. In those cases where plastics do not rupture, the results may be reported as the

FIGURE 4.15
Compressive strength-testing schematic setup.

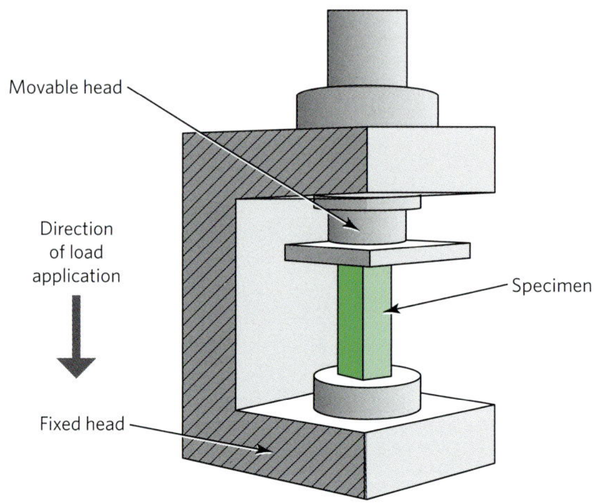

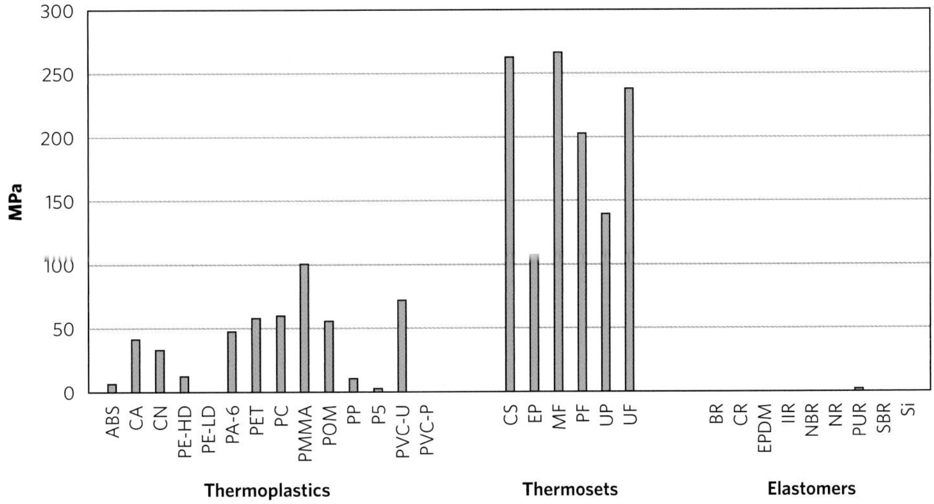

FIGURE 4.16
Average compressive strength of thermoplastics, thermosets, and elastomers determined using ASTM D695.

compressive strength at a specific deformation, such as 1%, 5%, or 10% of the test sample's original height or thickness.

Data on the compressive strength of selected plastics are shown in figure 4.16. Thermosetting plastics have high compressive strength. Thermoplastics such as polymethyl methacrylate, polyoxymethylene, polycarbonate, polyethylene terephthalate, polyamide, and rigid polyvinyl chloride have lower compressive strength than thermosets but still are difficult to compress. No breaking at all is seen when compressing polyolefins. (Elastomers are again omitted from the figure, for obvious reasons, except for hard polyurethane elastomers.)

Compression Set

Compression set tests are used to measure deformation for elastomers and for foamed plastics and foam rubber. While the compression set test for elastomers (ASTM D395) measures the loss in height of a specimen under a constant force, the test for foams (ASTM D3574) describes the loss in resilience—the ability of a foam to return to its original position after deformation—after applying a constant load (fig. 4.17). Both measurements are performed at a given temperature and measure the loss of thickness relative to the original thickness as a percentage.

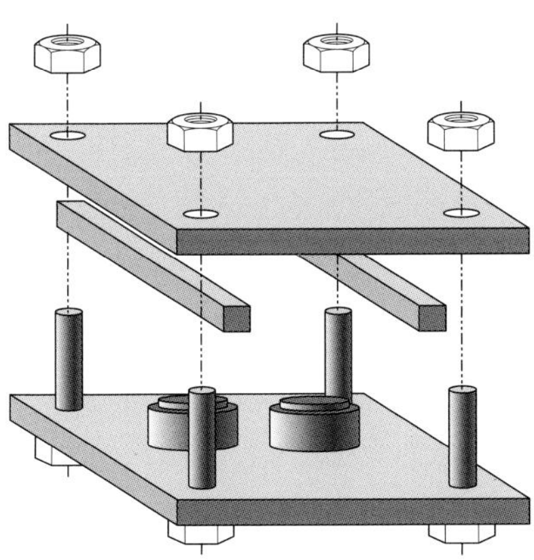

FIGURE 4.17
Compression set test.

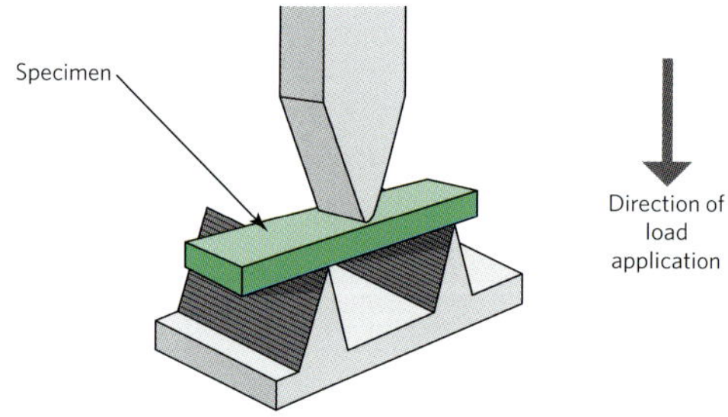

FIGURE 4.18
Flexural strength test.

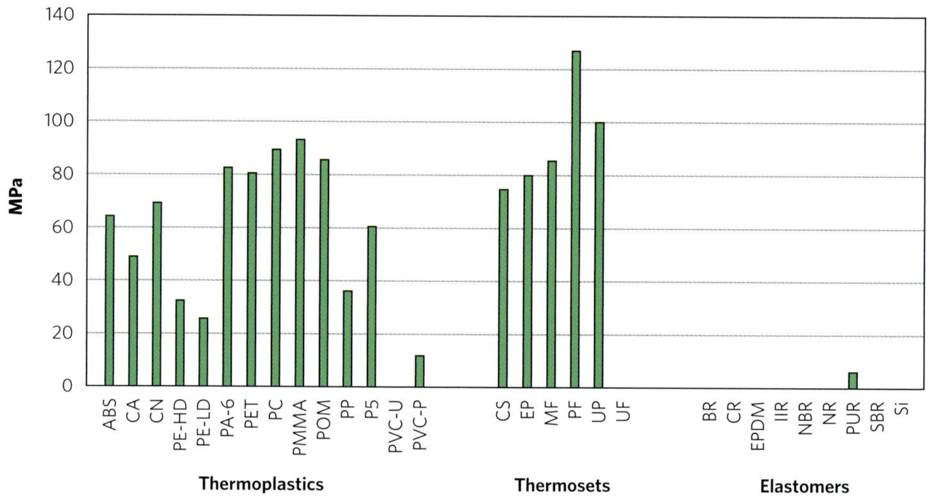

FIGURE 4.19
Average flexural strength of thermoplastics, thermosets, and elastomers determined using ASTM D790.

Flexural Strength

Stiffness and flexibility are opposing attributes often used to describe the behavior of plastics. Stiffness, or rigidity, is defined as the ability of an object to resist deformation in response to an applied force. Flexural strength—also known as modulus of rupture, bend strength, or fracture strength—is a mechanical variable defined as a plastic's ability to withstand deformation under load.

The flexural strength of a plastic represents the amount of stress the material is being subjected to at its moment of rupture (ASTM D790). Two stresses, tensile and compressive, are involved in bending a test sample, and are measured on a three-point bending setup (fig. 4.18).

Elastomers (with the exception of some types of hard polyurethane elastomers) do not have flexural strength because at bending they do not break. Thermosetting materials have high flexural strength before breaking, as do most thermoplastics. Only some polyolefins and plasticized polyvinyl chloride (which behave much like elastomers) do not have high values for flexural strength (fig. 4.19).

Shear Strength

Shear strength is the maximum load needed to produce a fracture. Complete separation can be measured using the test method described in ASTM D732 (fig. 4.20). The calculation of the shearing strength is the applied force (F, expressed in newtons) divided by the area of the sheared edge (m^2), and is expressed in Pa or MPa.

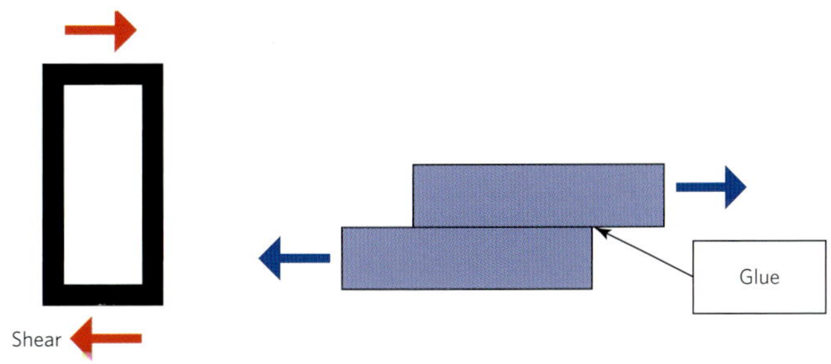

FIGURE 4.20
Shear strength testing.

Shear strength data are not available for many plastics because they are not designed to withstand shearing. Most shearing occurs at adhesion, where two surfaces are connected by an adhesive.

Tear Strength

Tear strength is the resistance of a plastic or elastomer to tearing when subjected to a force (fig. 4.21). Tear strength can be measured using a standard tensile testing machine, which measures the force per unit of thickness (kNm^{-1}) required to rupture or start a tear throughout the test sample (ASTM D624). Natural rubber and ethylene-propylene-diene-monomer rubber, and thermoplastic polyurethanes show the greatest resistance to tearing (fig. 4.22).

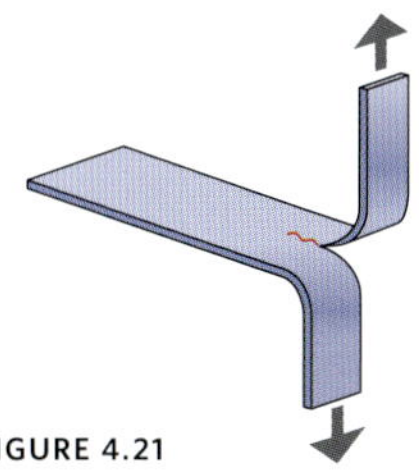

FIGURE 4.21
Tear strength testing.

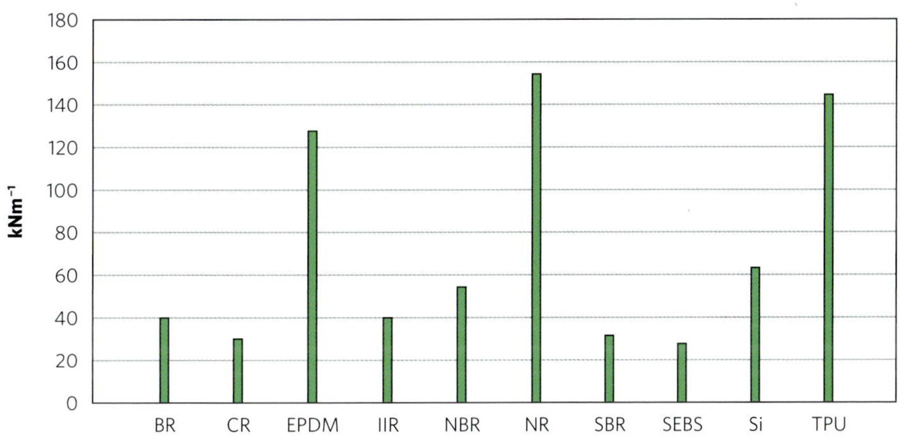

FIGURE 4.22
Tear strength of selected elastomers.

Torsional Strength

Torsional strength is the strength of a plastic when an object is subjected to a twisting load (fig. 4.23). An often-cited example of an object under torsional load is a polyamide screw as it is being screwed in. Data are available for specific engineering and high-performance plastics and are useful when using polyamide screws to connect broken plastics.

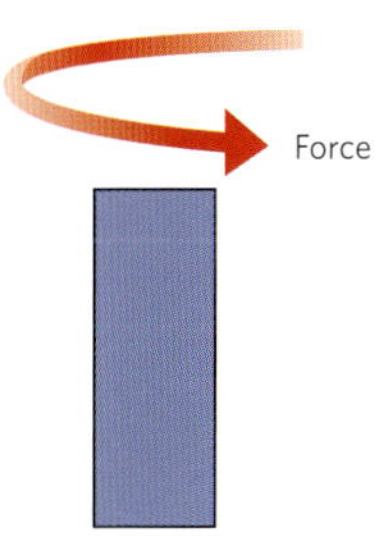

FIGURE 4.23
Torsional strength.

Hardness

Hardness is a measure of how resistant a plastic is to permanent indentation. It is measured by the depth of impression created on the plastic with a specified compressive force—by scratching, indentation, or rebound.

For plastics and elastomers, only the indentation measurements are relevant. Scratch is used primarily for coatings and minerals, while rebound is used for metals. Scratch resistance is measured by rubbing abrasives over a surface. Pencil hardness and "fingernail hardness" are assessed by observation of the damage at the end of the test (Horie 2010).

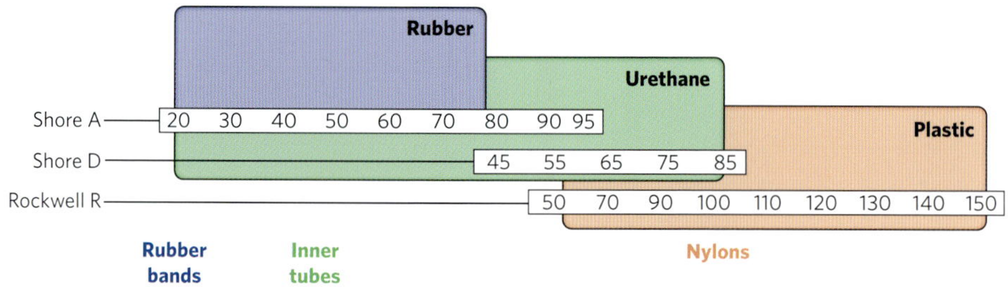

FIGURE 4.24
Hardness scales for rubber objects, polyurethanes, and selected plastics.

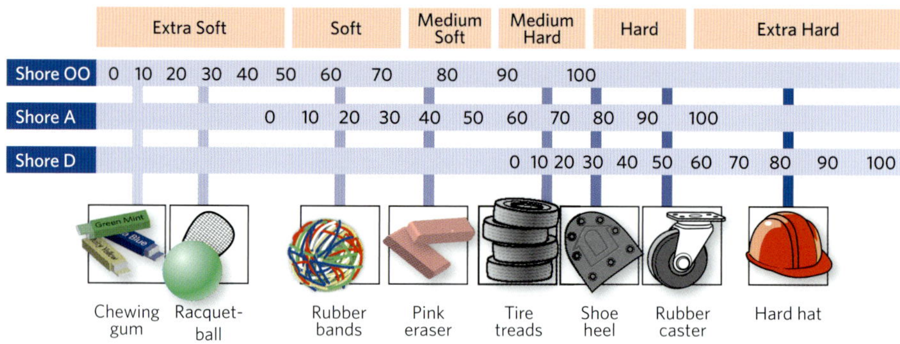

FIGURE 4.25
Shore hardness of elastomer objects.

Rubbing, pencil, and fingernail tests are used for coatings; however, they are not reliable and consequently are not found in technical databases of polymers.

Various hardness scales are used, but their results have no correlation to one another, although some overlap exists (fig. 4.24). For thermoplastics and thermosetting plastics, Rockwell R and M measurements are used (Crawford and Stephens 1985). For elastomers, the surface hardness is measured on the Shore A and Shore D scales—and for some elastomers the Shore OO scale (fig. 4.25). When giving results of hardness tests, the method of testing should always be given.

The Rockwell hardness testing method uses a cone-shaped diamond "indenter" to penetrate the plastic (ASTM E18 and ISO 2019), while the durometer method (ASTM D2240) uses a spherical or conical indenter, depending on the hardness to be measured (fig. 4.26). The hardness value is displayed on a gauge.

Different scales are typically used to measure the hardness of plastics and elastomers. Less hard materials are measured using Shore D (fig. 4.27), while the harder materials use Rockwell M (fig. 4.28).

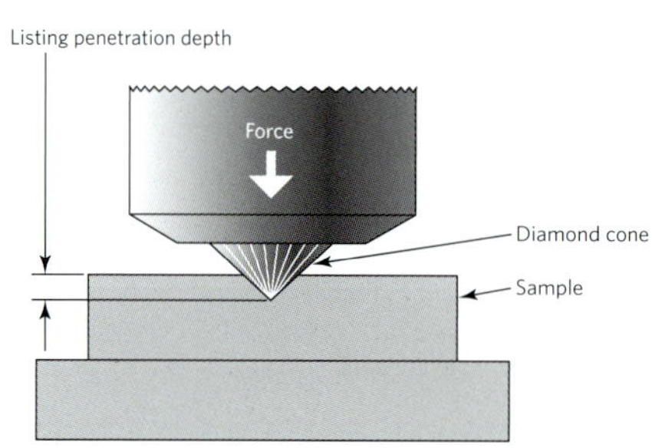

FIGURE 4.26
Rockwell hardness tester.

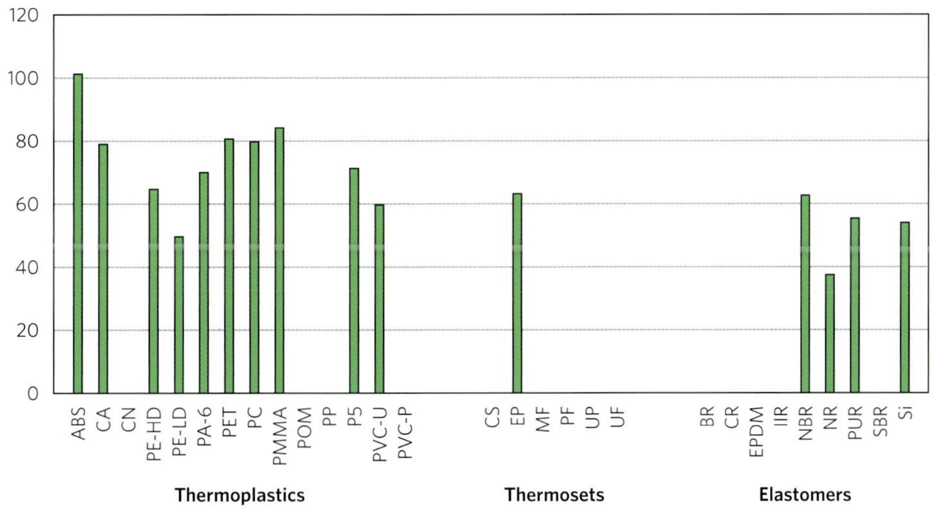

FIGURE 4.27
Average Shore D hardness of thermoplastics, thermosets, and elastomers.

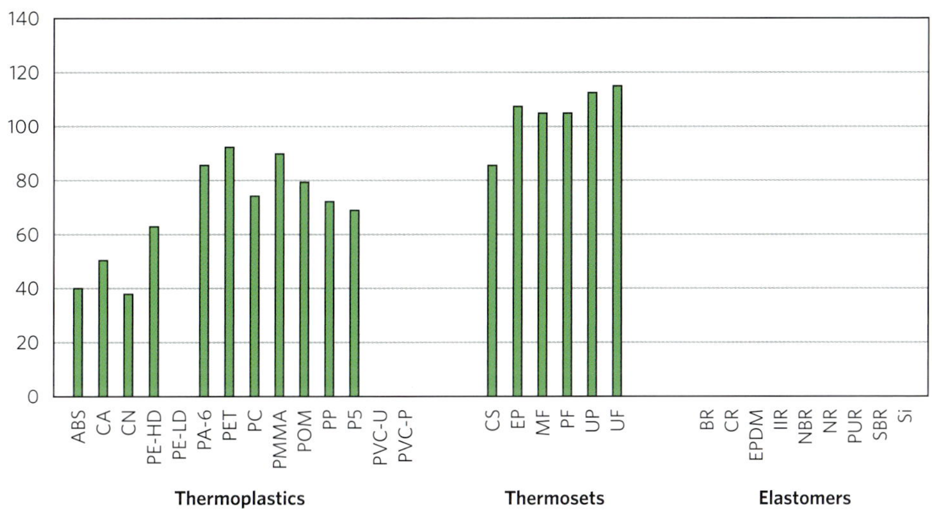

FIGURE 4.28
Average Rockwell M hardness of thermoplastics and thermosets, determined using ASTM D785.

4.2 PHYSICAL PROPERTIES

A physical property of a material is described as a static aspect of its physical state; physical properties can be measured. In contrast, chemical properties have to do with chemical composition, and they can change due to chemical reactions—which can have an influence on physical properties.

Physical properties include density, viscoelasticity, creep, shrinkage, molded-in stress, permeability, solubility, solvent stress cracking, surface tension/energy, and water absorption. Physical and chemical factors influence plastics and elastomers, leading to specific physical and/or chemical reactions such as environmental stress cracking.

Density or Specific Gravity

The terms *density* and *specific gravity* are frequently used interchangeably. Density is defined as the ratio of an object's mass to its volume (g/cm^3) at 23°C. It can be measured using either ASTM D792 or ISO 1183. Specific gravity is the ratio of the weight of a given volume of material at 23°C to that of an equal volume of water at the same temperature.

The density of plastics ranges from 0.8 to 2.0 g/cm^3. When comparing plastics with lead (11.3 g/cm^3) and aluminum (2.7 g/cm^3), it is obvious that plastics are lighter weight materials.

The density of a plastic can be modified by using fillers, additives, and other substances. The density of various plastics is given in the fact sheets in part III. It can be used to calculate or estimate the weight of an object, which is useful when objects could be exposed and/or handled without sufficient support and thereby torn by their own weight.

Viscoelasticity

Plastics under stress or load will exhibit viscoelasticity, a property in which both viscous and elastic characteristics are present. Viscoelasticity is unique to thermoplastics due to the structure of the polymer chains: they are not connected to the other, surrounding polymer chains.

Creep

The deformation of a plastic under a load or stress is called creep. Creep is the property that results from viscoelasticity. The rate of creep depends on the composition of the plastic, the duration of stress, ambient temperature, and the applied force—or its own weight (Riara, Merenga, and Migwi 2013).

Most plastics[4] are subject to extensive creep above their glass transition temperature (T_g): the temperature at which an amorphous plastic changes from a glassy, rigid state to a rubbery, soft state. Below that temperature, very little creep will take place, even after longer periods of time. Creep occurring at room temperature is called cold flow (see "Thermal Properties," later in this chapter). The deformation of a plastic object is permanent when it remains set upon removal of a fixed load, whether that load is applied or is simply the object's own weight.

Shrinkage at Molding

As semicrystalline thermoplastics cool after molding, the molecules in the crystalline areas arrange themselves more regularly and the plastic shrinks. Mold shrinkage is the ratio of the decrease in length to the original length (L_0) and is calculated as $L_0 - L_1 / L_0$ and can be expressed as a percentage. Typical shrinkage rates for an amorphous material are in the region of 0.4%–0.7%, while for semicrystalline polymers, rates can range from 1.0% to 3.0%.

Molded-in Stress

At processing, polymers are forced into molds at high pressure and shear rates. As the polymer enters the relatively cooler mold, it starts to cool. If this cooling happens too rapidly, the polymer chains may be "frozen" in a state of tension. Because polymers are

4. Exceptions are cross-linked, crystalline, and reinforced plastics.

poor heat conductors, uneven cooling (and therefore uneven shrinkage) may occur, resulting in molded-in stresses. Once out of the mold, the polymer chains will want to relax back to their original dimensions, and this may result in warped objects.

Permeability

Plastics can react to their surroundings by, for example, absorbing and desorbing moisture, oxygen, CO_2, and pollutants. Permeability is the property that allows molecules to diffuse through a plastic; a gas will move from an area of high concentration to one with a lower concentration across a plastic. Permeability is defined as the quantification of the permeate gas or vapor through a resisting material (Pauly 1999). Plastics that have high barrier properties to gases are used in food packaging to contain the quality (freshness) and flavor of edible products.

Numerous other applications exist for polymers with low, high, or tailored permeability, such as polymer coatings for controlled drug release and barrier films used to encapsulate cultural heritage objects to prevent oxygen permeation.

The permeation of a plastic can be measured through standardized methods (ASTM D1434). Figure 4.29 shows the gas permeability of some plastics to oxygen (O_2), carbon dioxide (CO_2), and moisture (H_2O).

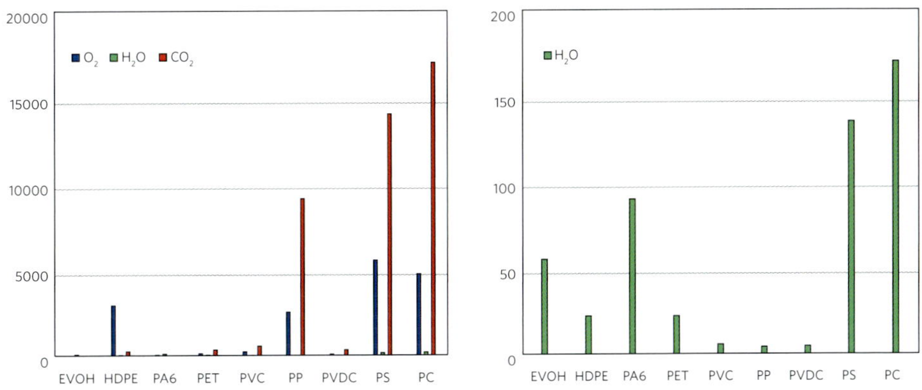

FIGURE 4.29
The gas permeability of oxygen (O_2), carbon dioxide (CO_2) and moisture (H_2O) for selected plastics.

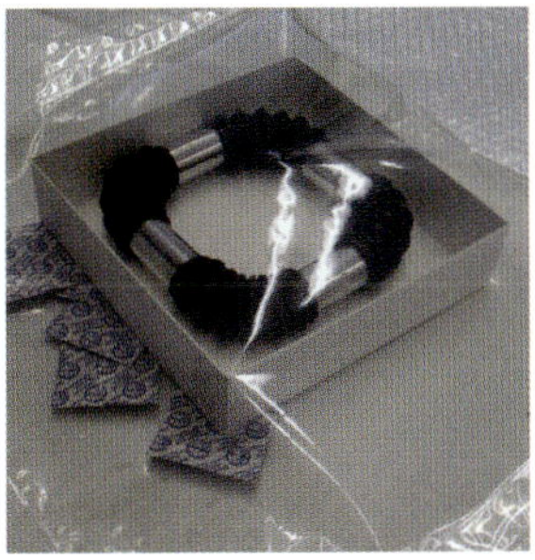

FIGURE 4.30
Encapsulated rubber bracelet.

Encapsulation of rubber objects in a barrier film with oxygen absorbers or oxygen scavengers (such as Ageless) to prevent oxidization of the rubber is achieved by using a barrier film with a low gas permeability for oxygen. Ethylene-vinyl alcohol (EVOH) is an excellent oxygen barrier, while high-density polyethylene (PE-HD) is not (see fig. 4.29). In contrast, polypropylene (PP) is a good moisture barrier and polystyrene (PS) is not. Barrier film used in packaging may contain several layers of different plastics to achieve the desired control of permeability. Figure 4.30 shows an encapsulated rubber bracelet with silver ornaments; the barrier film is a three-layer film that sandwiches a layer of EVOH between two layers of polyethylene.

Table 4.3. Solubility parameters of common polymers

Plastic	Hansen solubility parameter $cal^{1/2}\ cm^{-3/2}$	Hildebrand solubility parameter (δ) $(MPa)^{1/2}$
Polyamide 6,6	13.6	27.2
Cellulose acetate	11.4	22.8
Polyacetal	11.1	22.2
Epoxy	11.0	22.0
Polyethylene terephthalate	10.7	21.4
Cellulose nitrate	10.6–10.5	21–21.2
Polyvinylidene chloride	9.8	19.6
Polycarbonate	9.5	19.0
Polyvinyl chloride	9.4–9.5	18.8–19
Polyvinyl acetate	9.4	18.8
Polystyrene	8.5–9.7	17–19.4
Polyethylene	7.9–8.1	15.8–16.2
Polypropylene	7.9	15.8
Polytetrafluoroethylene	6.2	12.4

Note: Where the units of the solubility parameters are not given, for example in older books, it is usually safe to infer the non-SI unit $cal^{1/2}\ cm^{-3/2}$ is meant.

Solubility (Chemical Resistance)

Solubility parameters are generally used to describe a plastic's resistance to the effects of solvents, such as swelling, softening, or dissolving. The influence of solvent on a plastic depends on temperature, the plastic's chemical composition, the strength of its chemical bonds, and the solvent's ability to break those bonds. The Hansen solubility parameter ($cal^{1/2}\ cm^{-3/2}$) is calculated as the cohesive energy density, which expresses the energy of vaporization in calories per cubic centimeter. It conveys the degree of the intermolecular forces holding the molecules together and is most used for solvents.

The Hildebrand parameter (δ) is a more practical solubility parameter (and therefore the one most used for plastic) that also measures the degree of intermolecular forces, mostly between nonpolar polymers. It is expressed as $(MPa)^{1/2}$ and can be calculated as twice the value of the Hansen solubility parameter: $(MPa)^{1/2} = 2 \times cal^{1/2}\ cm^{-3/2}$.

The rule of thumb the parameter expresses is that a polymer (table 4.3) will dissolve in a solvent (table 4.4) with a *similar or lower* solubility parameter (Shashoua 2008). Some plastics are very stable, whereas others are quite unstable.

Table 4.4. Solubility parameters of common solvents

Solvent	Hansen solubility parameter $cal^{1/2} cm^{-3/2}$	Hildebrand solubility parameter δ $MPa^{1/2}$
Water	23.4	46.8
Methyl alcohol (methanol)	14.5	29.0
Ethyl alcohol (ethanol)	12.7	25.4
Isopropyl alcohol (IPA)	11.6	23.2
n-butyl alcohol	11.4	22.8
Acetone	10.0	20.0
Cyclohexanone	9.9	19.8
Ethylene dichloride	9.8	19.6
Methylene chloride	9.7	19.4
Trichloroethylene	9.3	18.6
Chloroform	9.3	18.6
Ethyl acetate	9.1	18.2
Toluene	9.0	18.0
Xylene	8.9	17.8
Carbon tetrachloride	8.6	17.2
Isopropyl acetate	8.4	16.8
Methyl isopropyl ketone	8.4	16.8
Cyclohexane	8.2	16.4
Turpentine	8.1	16.2
Methyl amyl acetate	8.0	16.0
Methylcyclohexane	7.8	15.6
Heptane	7.5	15.0

Note: Where the units of the solubility parameters are not given, for example in older books, it is usually safe to infer the non-SI unit $cal^{1/2} cm^{-3/2}$ is meant.

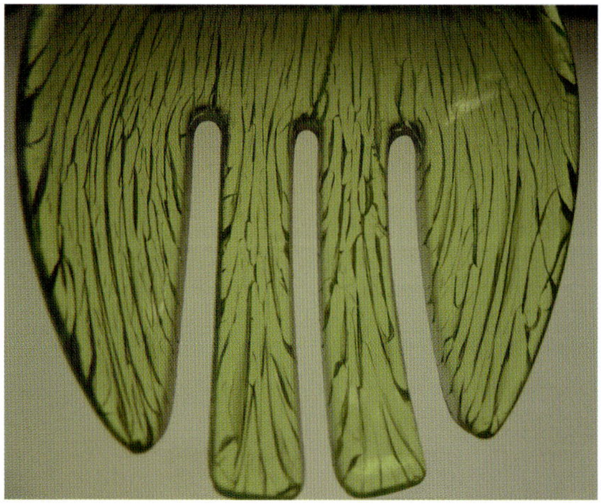

FIGURE 4.31
Solvent (water) stress crazing of polymethyl methacrylate salad fork from normal household use and dishwashing.

In practice, however, due to factors including the age of the plastic (crystallization), the hydrogen bonding of the molecular chains, and the addition of fillers, plasticizers, and the like, the values of the solubility parameter may be different than those for the pure plastic. For this reason, the values given in the fact sheets in part III are to be regarded only as guidelines and should be supplemented by tests.

Solvent Stress Cracking

Amorphous thermoplastics like polystyrene, polymethyl methacrylate, and polycarbonate are more sensitive to continuous load and chemicals than semicrystalline plastics, which can lead to crazing or stress-crack formation. Once formed, microcracks can grow under the influence of mechanical stress to become large crack networks (fig. 4.31). Solvent stress cracking (SSC) of an amorphous plastic is the result of exposure to a corrosive agent (e.g., solvent, adhesive, water) or tensile stresses above a threshold. If all of these factors are absent, SSC can be eliminated.

Stress due to processing can be relieved by stress-relief annealing, the process of slowly heating and cooling the plastic. The most direct way of controlling SSC is to remove or replace whatever is responsible for the problem. For example, if stress cracking of plastics is caused by solvents or water, removing the solvent or eliminating the water will solve the problem.

Surface Tension/Surface Energy

The polarity of the chemical structure of polymers is the factor that determines their surface energy. Surface tension or energy is an important factor when cleaning, retouching, or adhering objects. Surface tension and surface energy are equivalent. Surface tension is measured along a line in units of Nm^{-1}, while surface energy is measured in an area in units of Jm^{-2} (Kinloch 1987).

A normal adhering bonding requires about 50 mNm^{-1}, so for good adhesion of plastics it is often necessary to increase the surface tension of the plastic. This can be done using techniques such as corona treatment or flame treatment, thus improving wetting and adhesion.

The surface tension of a liquid with high surface tension, like water, can be lowered by adding wetting agents (or detergents, in the case of cleaning). Surface tension data for plastics and solvents are given in tables 4.5 and 4.6, respectively.

Table 4.5. Surface tension data for plastics

Material	Surface tension (mNm^{-1}) at 20°C
Polytetrafluoroethylene	18
Silicone	24
Styrene butadiene rubber	29.1
Polyethylene	31
Acrylics	32
Polystyrene	33
ABS rubber	36
Cyanoacrylates	37
Polymethyl methacrylate	39
Polyvinyl chloride	39
Polyethylene terephthalate	43
Cellulose (paper)	45
Polyamide	46
Polycarbonate	46
Epoxy	47
Phenol-formaldehyde	52
Urea-formaldehyde	61

Table 4.6. Surface tension data for solvents

Solvent	Surface tension (mNm^{-1}) at 20°C
Hexane	18.4
Dearomatized mineral spirit	20
Ethanol	23
Acetone	24
Ethyl acetate	24
Glycerol	58.3
Water	73

Marker pens that indicate surface energy can be used to learn about the state of the substrate (fig. 4.32); however, this method does not provide information about the proper adhesion of an adhesive on a substrate. Moreover, the inks may leave a mark on the object.

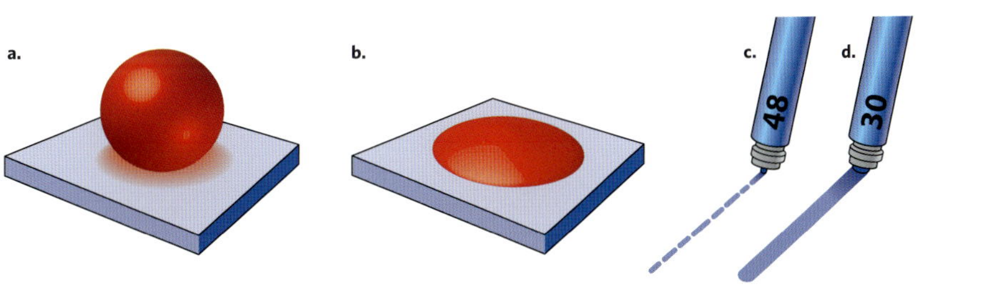

FIGURE 4.32
Low surface energy (a); high surface energy (b); marker pen indicating surface energy is less than 48 mNm^{-1} (c); marker pen indicating surface energy is 30 mNm^{-1} (d).

Water Absorption

Water is relatively harmless for most polymers; however, for some plastics and elastomers, absorption or desorption of moisture can be very harmful.

Many plastics absorb water over time, but that absorption is reversible through desorption. The rate of absorption/desorption varies with the type of plastic as well as with temperature and relative humidity. Similarly, the amount of water absorbed by different plastics varies, as do the effects of the absorbed water on the plastics themselves. The water moves into the free volume between the polymer chains, causing swelling, thus plastics that absorb relatively larger amounts of water tend to change dimension.

When water is absorbed by the plastic, hydrolysis (chemical decomposition by water) can occur. Hydrolysis damages the polymer chains, reduces molecular weight, and decreases mechanical properties. Longer exposure at elevated temperature and loads increases hydrolysis. Cellulosic esters, casein-formaldehydes, polyamides, unsaturated polyesters, epoxies, and poly*ester* urethane elastomers (millable thermoplastics as well as thermosetting

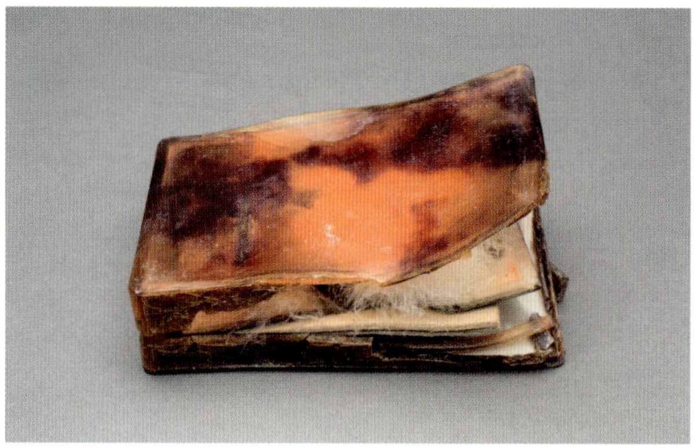

FIGURE 4.33
Effect of humidity on cellulose acetate.

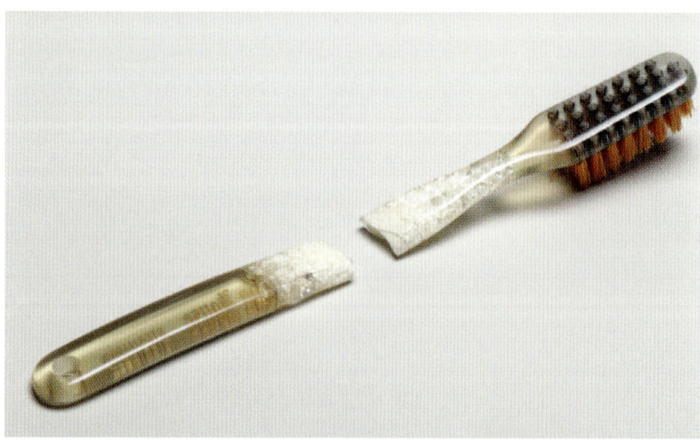

FIGURE 4.34
Effect of humidity on cellulose nitrate.

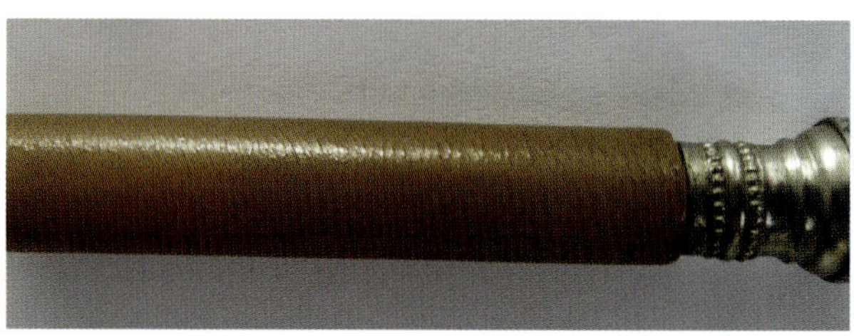

FIGURE 4.35
Effect of humidity on casein-formaldehyde.

types) are among the plastics most affected by changes in relative humidity. Results include crazing, warping, change in dimensions, and delamination. Crazing and warping have been noticed in the earlier cellulosic plastics such as cellulose acetate (fig. 4.33) and cellulose nitrate (fig. 4.34), but most often in casein-formaldehyde (fig. 4.35).

Various tests can be performed to assess the sensitivity of a plastic to water: percentage of moisture at 23°C and 50% relative humidity, water absorption at saturation, and percentage of water absorption for twenty-four hours. For example, absorption data may be obtained by immersing a plastic specimen in 23°C water for twenty-four hours or longer (ASTM D570, ISO 62); specimens are then removed from the water, dried, and weighed. The increase in weight is the original weight compared to the test weight expressed as percentage gained.

Casein-formaldehyde, cellulose acetate, and polyamides (nylons) have the greatest water absorption of all the plastics (see below). Nylon can absorb up to 10% of its own weight in water, and this affects the rigidity of its linear strength. However, water absorption and loss of strength also happen with polyurethane elastomers and silicone rubbers.

One example of the effect of water absorption is the deterioration of glass fiber-reinforced plastic (fiberglass) sculptures (figs. 4.36a, b). At aging, under environmental circumstances (light, ultraviolet radiation, change in temperature, and change in relative humidity), microcracks and cracks appear at the surface of the polyester, through which moisture can enter. Moisture will assemble in the voids between polymer chains and in any voids between the glass fiber and the polyester that may be left as a result of improper manufacturing. Under outdoor conditions over the years, this water can also freeze and thaw, causing damage not only to the polyester itself but also to the metal or wood armature that supports the polyester construction.

FIGURE 4.36
Joep van Lieshout (Dutch, b. 1963), *Mobile Home for Kröller-Müller*, 1995. Glass fiber–reinforced polyester, polyurethane foam, multiplex, wood, glass, rubber, metal, 325 × 1025 × 750 cm. Otterlo, Kröller-Müller Museum, acquired with support from the Mondriaan Foundation, KM 127.508 (a), with a detail showing damage to the exterior caused by water absorption (b).

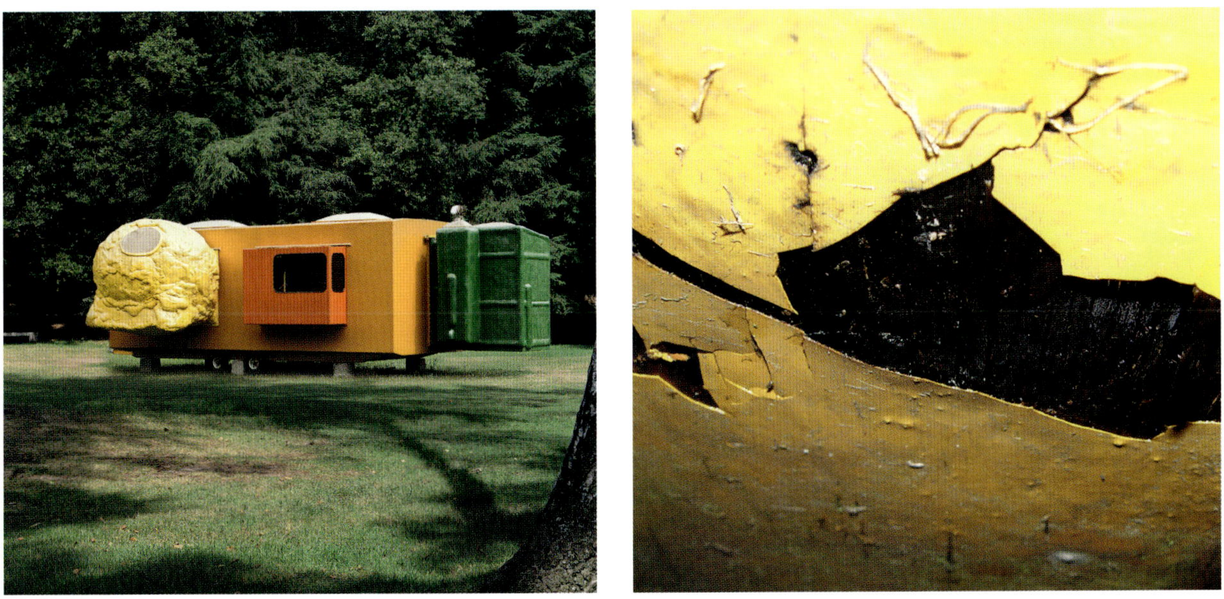

a. b.

Figure 4.37 shows the relative water absorption of selected plastics and elastomers and which are most affected by water absorption and desorption. The most commonly used test for water absorption for plastics is ASTM D570, whereby the specimens are immersed in distilled water for twenty-four hours at 23°C. Water absorption data are useful when objects are exposed to outdoor (weathering) conditions—or in cases where objects submerged in water, such as the basketballs in a PMMA tank in a work by the artist Jeff Koons (fig. 4.38), must be conserved.

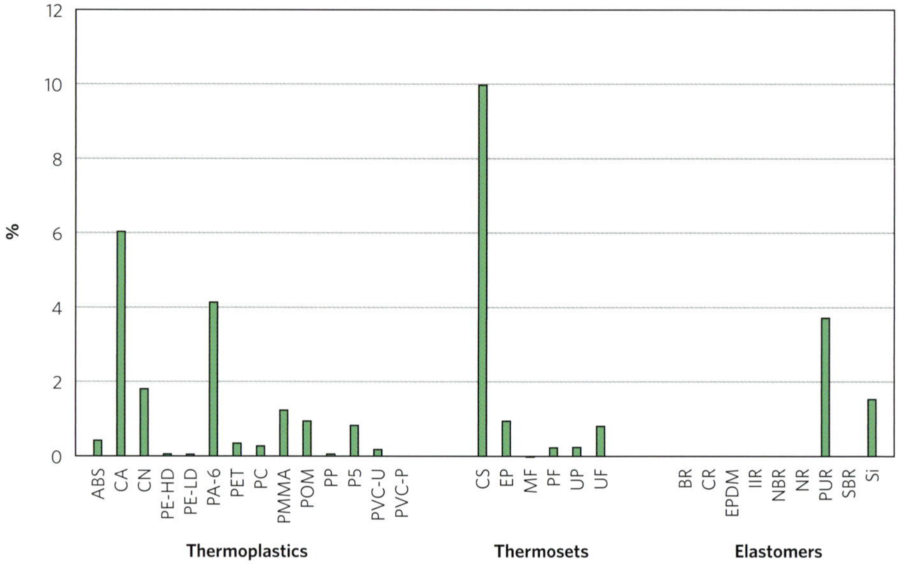

FIGURE 4.37
Average water absorption of plastics and elastomers, determined using ASTM D570.

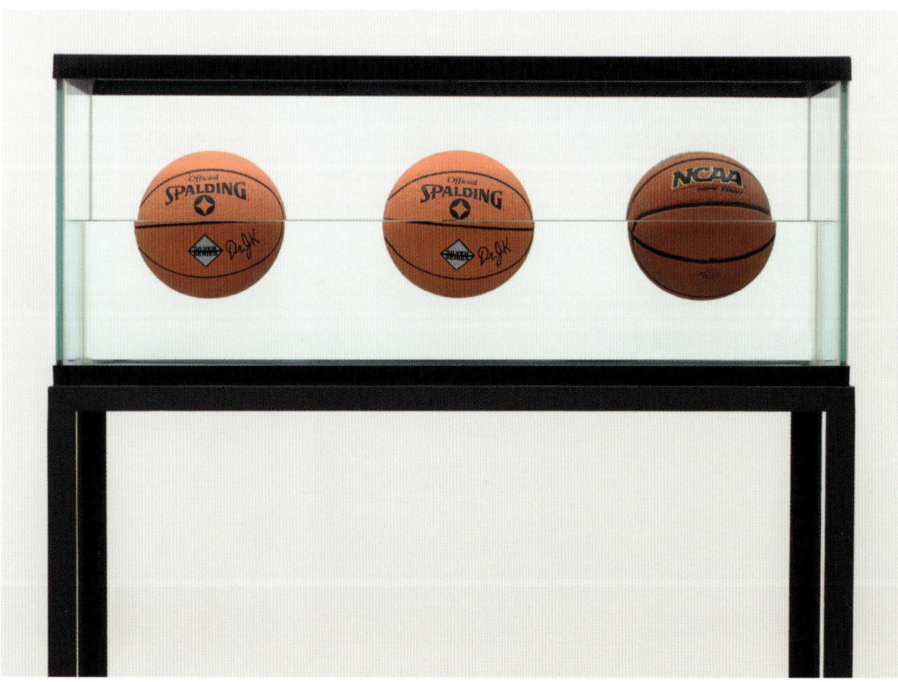

FIGURE 4.38
Jeff Koons (American, b. 1955), *Three Ball 50/50 Tank (Two Dr. J. Silver Series, One Wilson Supershot)*, 1985. Glass, painted steel, distilled water, plastic, three basketballs, 154 × 123.9 × 33.6 cm. New York, The Museum of Modern Art, Gift of Werner and Elaine Dannheisser, 226.1991.a–f.

4.3 THERMAL PROPERTIES

Plastics are sensitive to changes in temperature; the mechanical and physical properties of plastics cannot be considered without attention to the temperature at which the values were derived (Shah 2013). Understanding the behavior of plastics during temperature changes is essential for conserving plastics.

Five thermal properties influence plastics' behavior: glass transition temperature, melting point, softening point, coefficient of linear thermal expansion, and low-temperature performance. Data on these thermal properties are given in the fact sheets for the selected plastics in part III.

Glass Transition Temperature and Melting Point

The glass transition temperature (T_g) of a thermoplastic or elastomer is the temperature at which the polymer shifts from hard and brittle to rubbery and flexible. As discussed in chapter 1, plastics are essentially a long chain of entangled molecules. The extent of their entanglement varies with the length and shape of the polymer chain. When the plastic is heated, the polymer chains reach a temperature at which segments of the entangled chains can move. Above the glass transition temperature, a plastic behaves like a ductile solid (think of cooked spaghetti) or a highly viscous liquid.

At temperatures below T_g, molecular motion in an amorphous region of thermoplastics is restricted to molecular vibrations, but the polymer chains cannot rotate or move in space (like uncooked spaghetti), therefore all polymers are glassy below their T_g. In this state, the molecules are frozen in place, and the material is hard, rigid, and brittle, like a crystalline solid. The glass transition temperature may range from −125°C in the case of very flexible molecules, such as silicone rubber, up to 300°C for stiff high-performance plastics.

Being highly cross-linked, the polymer chains of thermosets do not have the ability to move at elevated temperatures; therefore thermosets always behave like rigid, hard, glassy plastics (Lower 2017).

The melting point (T_m) is the temperature at which a material goes from a solid state to liquid. Semicrystalline plastics such as polyethylene and polypropylene have a precise melting point, whereas amorphous plastics do not.

Glass transition temperature and melting temperature are not the same (fig. 4.39). While most thermosets are amorphous, thermoplastics may be amorphous or semicrystalline, meaning they display crystalline regions, called crystallites, within an amorphous matrix.

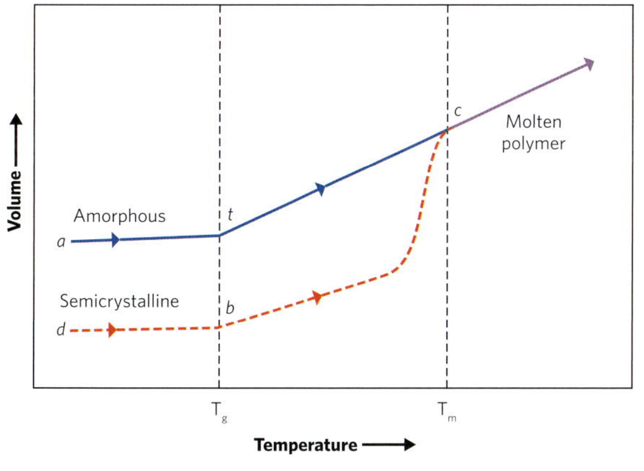

FIGURE 4.39
Volume-temperature diagram for amorphous and semicrystalline polymers, showing volume increasing with temperature.

Semicrystalline plastics such as polyethylene and polypropylene exhibit their most useful properties between T_g and T_m. Normally, raising the temperature will increase the mobility of polymer chains. If chains cannot rotate or move easily due to the size of the side chains, then a higher T_g is needed to induce the polymer chains to move and bring the polymer to a more flexible state.

Table 4.7 shows examples that illustrate variations in T_g caused by the molecular chain formations.

Table 4.7. Variations in T_g caused by molecular chain formation

Molecular chain formation	Polymer	Glass transition temperature (T_g)
Chain stiffness Stiffening groups (such as aromatic rings) reduce the flexibility of the polymer chain and increase the value of T_g.	Polyethylene adipate	T_g = −70°C
	Polyethylene terephthalate	T_g = 69°C
Intermolecular forces Stronger intermolecular forces lead to a higher T_g. Polyvinyl chloride has stronger intermolecular forces than polypropylene because of the dipole-dipole forces from the carbon-chlorine bond.	Polypropylene	T_g = −20°C
	Polyvinyl chloride	T_g = 81°C
Pendant groups		
Bulky pendant groups, such as an aromatic ring, can catch on neighboring chains like a fish hook and restrict rotational freedom. This increases T_g.	Polypropylene	T_g = −20°C
	Polystyrene	T_g = 81°C
Flexible pendant groups such as aliphatic chains tend to limit how closely chains can pack. This increases rotational motion and lowers T_g.	Polymethyl methacrylate	T_g = 105°C
	Polybutyl methacrylate	T_g = 20°C
Cross-linking Cross-links between chains restrict rotational motion and increase T_g.	Polystyrene	T_g = 81°C
	High-impact polystyrene	T_g = 100°C
Plasticizers Plasticizers weaken the intermolecular forces and decrease T_g.	Unplasticized polyvinyl chloride	T_g = 83°C
	Plasticized polyvinyl chloride	T_g = −43°C

As the size of side chains in polymer chains increases Tg, adding plasticizers at plastic compounding can decrease Tg to below room temperature. One example is rigid polyvinyl chloride (PVC) water pipes that have a Tg of 83°C and are suitable for cold water but unsuitable for hot water. Adding plasticizer to a rigid plastic PVC compounding can decrease Tg to −43 °C, bringing the plastic to a highly rubbery state, rendering it less rigid and brittle, and ideal for garden hoses, for example. However, a plasticized PVC can become stiff and brittle in winter.

Vicat Softening Point and Heat Deflection Temperature

As most plastics do not have precise melting points that can be measured, tests to observe the behavior of plastics at elevated temperatures can measure their softening point (fig. 4.40). Two standard methods are available: Vicat softening point and heat distortion temperature.

The Vicat softening point is the temperature of a plastic at which a specific needle penetrates the softened plastic; the measurement is standardized in ASTM D1525. Heat distortion or heat deflection temperature is the highest temperature at which a plastic or elastomer can withstand a specific load without deforming; the measurement is standardized in ASTM D648.

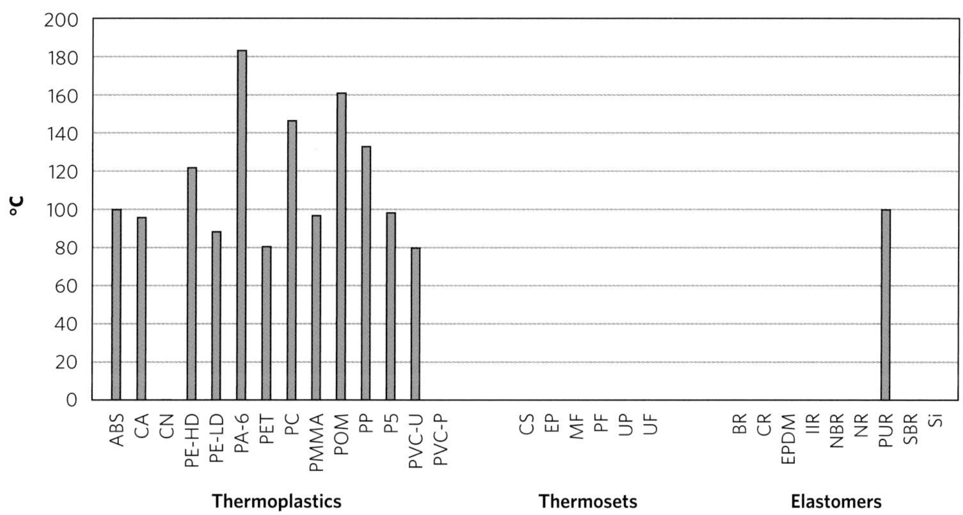

FIGURE 4.40
Average Vicat softening point of selected thermoplastics, determined using ASTM D1525.

Coefficient of Linear Thermal Expansion

The coefficient of linear thermal expansion (CLTE) of a plastic is the amount that a material will change in dimension with a change in temperature, measured according to ASTM D696 and expressed as a ratio per degree Celsius. All materials expand at higher temperatures, though plastics expand considerably more than metals, for example.

CLTE, also referred to as (α), is calculated as $\alpha = \Delta L / (L_o \times \Delta T)$, where ΔL is the change in length of the sample, L_o is the original length of the sample, and ΔT is the temperature change during the test.

For example, consider a polypropylene strip for which the CLTE (α) = 10.5×10^{-5} / °C (see fact sheet 13.16). If a one-meter-long strip is heated from −20°C to 50°C, then $\Delta L = 10.5 \times 10^{-5} \times 1 \times 70 = 735 \times 10^{-5}$ m = 7.35 mm.

Thus, if a one-meter polypropylene rod is heated from −20°C to 50°C, it will increase in length by 7.35 mm.

Low-Temperature Performance

At low temperatures, the mobility of polymer chains is greatly reduced and plastics tend to become rigid and brittle. Amorphous polymers show brittle failure at temperatures well below glass transition temperature, while for semicrystalline polymers, toughness is dependent on the degree of crystallinity. When storing plastics at cold temperatures to prevent degradation, care must be taken in handling after storage—some plastics can become very brittle, while others can become deformed (Shashoua 2004). Standard test methods for brittleness are defined in ASTM D746.

4.4 OPTICAL PROPERTIES

Optical properties of plastics are closely related to their molecular structure. Plastics have various optical properties, including refractive index, transparency, birefringence, haze, gloss, color, and yellowness index.

Refractive Index

Light entering a transparent plastic will undergo a change; part of the light is reflected while the other part is refracted. The index of refraction is expressed as Nd and calculated as the angle of incidence (i) divided by the angle of refraction (r) (ASTM D542, ISO 489). Refractive index plays an important role when adhering transparent plastics; adhesive and plastic should ideally have the same refractive index to provide an invisible join (van Oosten and Laganà 2011).

Transparency

A plastic is considered transparent if things can be seen through it clearly and if it absorbs little or no light in the visible spectrum—that is, light transmission is almost 100%. A plastic can be called transparent if a 1–2 cm thick sheet of it is transparent (ASTM D1003).

Plastics through which objects can be seen, but not clearly, are considered translucent, while plastics are considered opaque if you cannot see through them (fig. 4.41). To compare the translucency, transparency, or opacity of different thermoplastics and thermosets, one has to use samples with the same thickness.

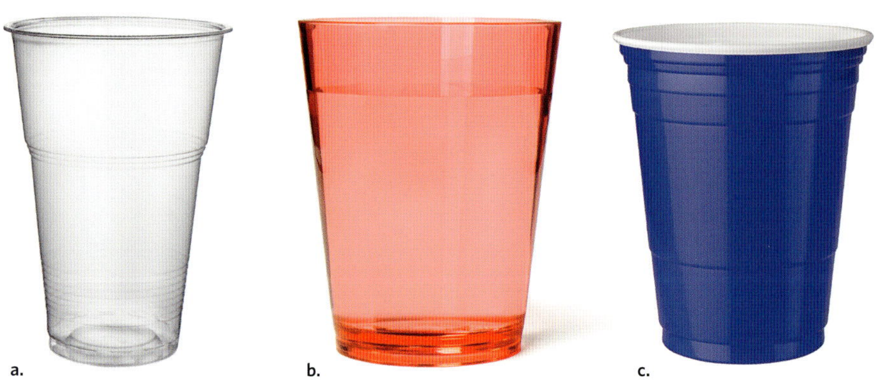

a. b. c.

FIGURE 4.41
Examples of translucent (a), transparent (b), and opaque (c) polystyrene.

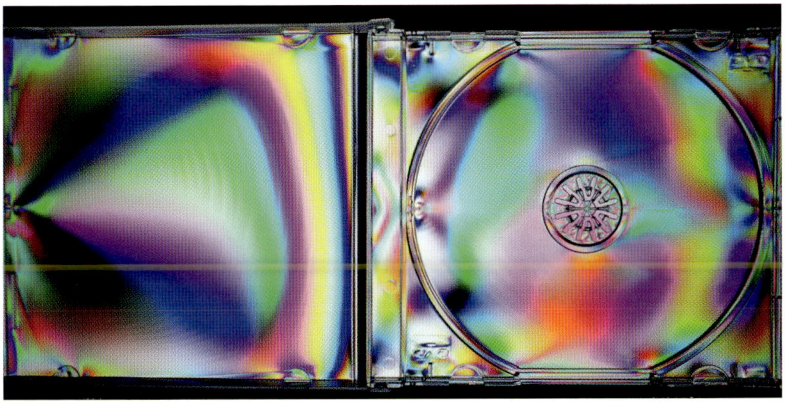

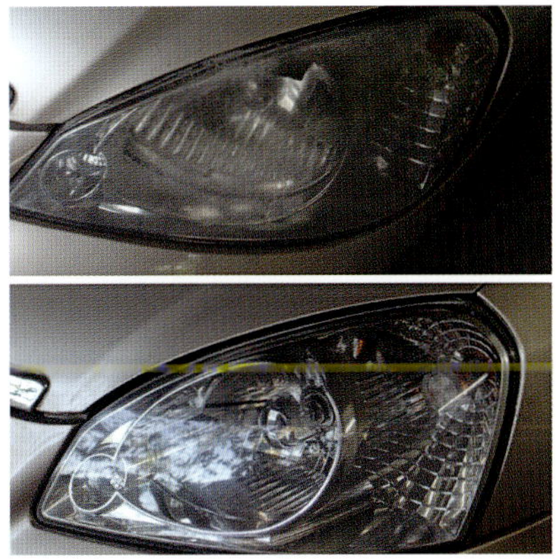

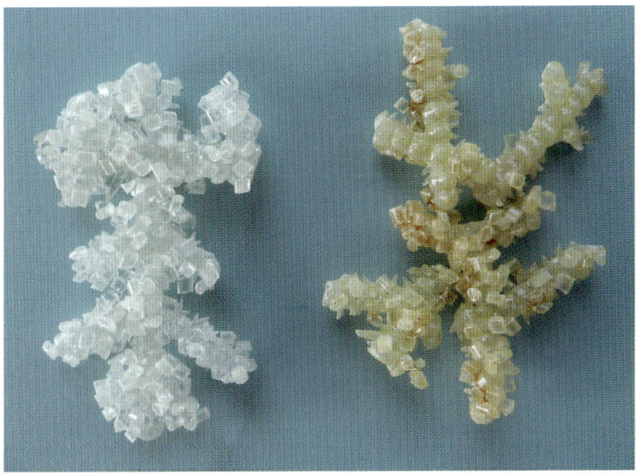

FIGURE 4.42
Example of stresses in plastic. Polarized light is used to show birefringence (areas of highest stress), where the darker color is most concentrated.

FIGURE 4.43
Transparent plastic with and without haze.

FIGURE 4.44
Example of yellowing of polystyrene.

Birefringence

The refractive index of a plastic differs based on the makeup of its polymer chains at the molecular level. Polarized light is used to achieve double refraction or birefringence effects, which serve as a way to discern stress in transparent plastics, seen as colored patterns. The highest stress is in the areas of concentrated darker color (fig. 4.42). Stresses can be imparted due to production processes, environmental circumstances, or various forces encountered in use.

Haze

Haze in a transparent plastic is observed as a milky, cloudy appearance; it may be caused by degradation of the plastic due to weathering, processing that produced a poor surface texture, or abrasion during use (fig. 4.43). Haze and luminous transmittance can be measured using a spectrometer, which measures the ratio of transmitted light to incident light (ASTM D1003).

Gloss

Gloss is the relative luminous reflectance factor of a plastic; it can be the result of the molding process or the surface texture. Gloss can be affected by weathering and surface abrasion. Decrease in gloss can be measured using a gloss meter. The gloss meter directs light onto the sample at incidence angles of 20, 45, and 60 degrees, and the light that reflects off the surface is collected and measured by a photosensitive device (ASTM D2457, ASTM D523).

Color

Three components are used to describe color: *value* or *lightness* is the range from dark to light; black is one extreme of the value range and white is the other extreme, while gray lies in the middle. *Chroma* or *saturation* represents the intensity of the color. *Hues* are distinguished between red, blue, yellow, and green. Two methods of measuring color are used: the human eye and spectrophotometer measurements (ASTM D6290).

Yellowness Index

Photooxidation produces yellowing of plastics (fig. 4.44) and is caused by light—especially UV radiation—and the presence of oxygen. Yellowing can be measured by spectrophotometry (according to ASTM E313), and the data used to calculate yellowness index, or the color change from white to yellow. The addition of light stabilizers (containing UV absorbers, heat stabilizers, and light absorbers) to plastics can inhibit yellowing.

Summary

The properties of plastics and elastomers can be divided into mechanical, physical, thermal, and optical categories:

- Mechanical properties are the reaction of plastics and elastomers to applied forces; they are measured by testing machines. Mechanical properties include tensile strength (yield or ultimate), tensile modulus (Young's modulus), tensile elongation (elongation at break), impact strength, compressive strength, compression set, flexural strength, shear strength, tear strength, torsional strength, and hardness.
- Physical properties are properties that are inherent to plastics and elastomers; they can be measured without applying force. Physical properties include density, viscoelasticity, creep, shrinkage, molded-in stress, permeability, solubility, stress cracking, surface tension, and water absorption.
- Thermal properties are properties that can be measured as the reaction of plastics or elastomers to the application of heat. Thermal properties include glass transition temperature, melting point, Vicat softening point, heat deflection temperature, coefficient of linear thermal expansion (CLTE), and low-temperature performance.
- Optical properties are the response of plastics or elastomers to visible light and UV radiation. Optical properties include refractive index, transparency, birefringence, haze, gloss, color, and yellowness index.

Suggested Reading

David R. Burfield, "Polymer Glass Transition Temperatures," *Journal of Chemical Education* 64, no. 10 (October 1987): 875, https://doi.org/10.1021/ed064p875.

Anthony Davis and David Sims, *Weathering of Polymers* (London: Applied Science Publishers, 1983).

John Lopez, "Microhardness Testing of Plastics: Literature Review." *Polymer Testing* 12, no. 5 (1993): 437–58, https://doi.org/10.1016/0142-9418(93)90016-I.

Chapter 5 | Manufacturing of Plastics and Elastomers

Industrial manufacturers of plastics tend to think of plastics as resins, and production data sheets often do not clearly distinguish between polymers, resins, and plastics. Originally *resin* referred to the natural exudative secretions of plants and trees, such as rosin, amber, and mastic, and to the resin-like substances of animal origin, such as shellac. Early synthetic prepolymers resembled natural resins (Knapczyk and Simon 1992). The polymer industry—which creates polymer granules, pellets, sheets, and powders, the primary raw materials of plastic making—often uses the words *polymers* and *resins* interchangeably to refer to unblended, synthetic polymeric materials and prepolymers (Knapczyk and Simon 1992).

In contrast, the terms *polymers* and *plastics* are not interchangeable. All plastics are made from polymers, but not all polymers are plastics. A polymer is a collection of long-chain molecules, and few polymers are processed in the pure state, while plastics are polymers that are compounded with additives to give them special properties, as discussed in section 5.1.

Plastics and elastomer manufacturing are important in the world economy, and a variety of methods are available for both shaping and compounding thermoplastics and thermoplastic elastomers and setting the shape once formed. The processing of thermosets and natural and synthetic rubbers includes chemical cross-linking (curing) in addition to setting and shaping. These processes are explored in section 5.2.

In addition, today developments in the plastics manufacturing processes include plastics recycling and new technologies such as additive manufacturing (also called 3D printing) are commonly used for modeling and prototyping, as well as production applications. These topics are covered at the end of the chapter.

5.1 ADDITIVES

Additives are organic and inorganic substances that are mixed with the polymers when plastics and elastomers are compounded, prior to manufacturing, to change the material's properties in ways that make it better suited for a given application, for example to inhibit degradation and extend a product's service life, enhance its strength, or both.

Additives for Plastics

Polymers produced as pellets, granules, or powders are seldom used in their pure form; normally during processing, ingredients are added that give them special properties. Making plastic products out of polymers involves reorganizing the molecules of the polymer into new configurations without affecting their color, causing them to stick to the mold, or without having any negative effect on the finished object; additives help avert such unfavorable outcomes.

Table 5.1. Selected plastics' resistance to chemical aging

Polymer	Hydrolysis	Thermal oxidation	Photochemical oxidation
Polypropylene (PP)	++++	-	-
Polyvinyl chloride (PVC)	++++	-	+
Polyethylene (PE)	++++	+	+
Polystyrene (PS)	++++	++	+
Polyoxymethylene (POM, acetal)	++	+	+
Polyamide-6 and -6,6 (PA-6, PA-6,6, nylon)	++	++	++
Polycarbonate (PC)	+++	++	+
Polyethylene terephthalate (PET, polyester)	+++	+++	+++
Polymethyl methacrylate (PMMA)	+++	+++	++++
Silicone rubber (Si)	+++	++++	++++

- = Poorly or not at all resistant
+ = Not very resistant
++ = Moderately resistant
+++ = Resistant
++++ = Very resistant

Additives are also used to improve processing ability, reduce material costs, lessen shrinkage, enhance surface finish, improve thermal and electrical properties, prevent degradation during fabrication and service life, provide color, and improve mechanical properties such as modulus, hardness, abrasion resistance, and toughness (see chap. 4).

Anti-aging agents | Chemical aging may result from hydrolysis, oxidation, and/or photooxidation and is caused by exposure to light and UV radiation, air (oxygen and ozone), temperature, and/or humidity during plastics processing and service life (table 5.1). Chemical aging may lead to degradation or cross-linking of polymers, causing loss of impact strength, loss of elongation, surface cracks, and discoloration. Physical aging may occur without chemical aging as a result of long-term stress. For a plastic to stand the test of time, its properties should remain stable and change as little as possible with use or exposure. To inhibit or prevent aging, heat stabilizers, UV absorbers, and antioxidants can be added, usually in small quantities, to ensure the plastic's properties remain stable.

Antistatic agents | Plastics are generally insulating, meaning their surfaces prevent transfer of heat, electricity, or sound. Consequently, plastics tend to accumulate static charges on their surfaces, which can negatively impact processing procedures as well as attract dust when on display or in storage. Antistatic agents such as carbon black and fatty acid esters are often added during compounding to eliminate static charges.

Blowing agents | To produce a cellular structure in plastic, manufacturers utilize a foaming process using blowing agents or foaming aids. Physical blowing agents are gases that do not react chemically during the foaming process. Examples include pentane, isopentane, and liquid CO_2, all inert to the polymer composing the matrix. Chemical blowing agents are chemicals that may either play a part in a reaction or simply decompose, releasing gaseous chemicals in the process. For polyurethane foams, isocyanate and water produce the blowing agent CO_2, whereas for other thermoplastic and elastomeric foams, mostly azodicarbonamides (nitrogen-based) materials are used, whereby N_2 is formed as the blowing agent.

Cross-linking agents | Peroxide cross-linking agents such as di-tert-butyl peroxide are used to connect the polymer chains of a thermoplastic. Polyethylene, a thermoplastic, can be changed into a cross-linked, thermosetting, closed-cell foam using peroxide treatment in an extruder at high temperatures.

Colorants | For practical, aesthetic, or artistic reasons, most consumer plastic products are colored. One advantage plastics have over other materials is that color can be easily incorporated throughout a molded article, whereas the color of metals and ceramics, for example, is usually applied as a surface coating. Coloring compounds can be used to add color either as insoluble pigments or as soluble dyes (Charvat 2003). The most frequently used pigments for plastics are titanium dioxide and carbon black, as well as phthalocyanines and ultramarines.

Fillers | Reinforcement fillers are added to polymers to thicken the plastic, improve strength and flow properties, and decrease cost. Fillers can consist of large-scale fillers and nanocomposites. Examples of large-scale fillers, which can make up from 3% to 40% of the material, are kaolin, calcium carbonate, talc, wood flour, sand, and quartz. Fumed silica is a very fine filler; this submicroscopic silica is added to achieve thixotropy, a state in which the material is gel-like at rest but becomes fluid when agitated. Glass, in the form of microballoons, is a relatively cheap filler that is often used to improve physical properties. Nanocomposites are fillers on a submicron scale. They were developed in the 1980s for engineering plastics, but their use was extended to commodity plastics beginning in 1990.

Fibers | To enhance the strength of a plastic, glass in the form of fibers, chopped strands, rovings, glass fiber mats, or glass balls can be added to plastics during the shaping process. Other fibrous materials such as nylon fibers (aramid) and carbon fibers are also used.

Flame retardants | To prevent flammable plastics from easily catching fire, flame retardants are often added during processing. In the early days of the plastics industry, toxic chlorinated cycloaliphatic and brominated aromatic compounds were used, but these have now been replaced by bromine- and chlorine-free fire retardants.

Heat stabilizers, UV absorbers, and antioxidants | Heat stabilizers such as copper salts and phosphoric acid esters prevent thermal oxidation reactions when plastics are processed at high temperatures. Antioxidants and UV absorbers prevent photooxidation when plastics are exposed to light and UV radiation. Antioxidants (e.g., hindered phenols and tertiary amines) and UV absorbers (e.g., benzophenones and benzotriazole) can be effective in concentrations as low as a few parts per million.

Lubricants | Lubricants and antislip agents are types of processing aids; they can be added at compounding to alter the physical properties of polymers to prevent failures during processing. External lubricants like mineral oil that have low compatibility with the polymer are often used to reduce friction between the material and the processing machine. They are also applied as mold release agents inside the mold during processing to help extract the object easily and prevent damage. A negative side effect of external lubricants is their exudation to the surface shortly after processing.

Internal lubricants, or antislip agents, are modified esters of long-chain fatty acids that have high compatibility with the polymer. They are added at processing to improve the flowability of the melt of thermoplastics such as polyethylene and polyvinyl chloride. A negative side effect is their blooming on the surface as the polymer ages.

Plasticizers | *External plasticization* is the term used for adding plasticizers to improve the flexibility of the polymer end product and the processability of plastics. The term *internal plasticization* describes copolymerization of the parent monomer with comonomers that provide flexible chain units.

There are thousands of different types of external plasticizers—all of them liquids with high boiling points and low molar mass. About 80% are used for polyvinyl chloride; the most widely used is di(2-ethylhexyl) phthalate (DEHP), also called dioctyl phthalate (DOP). This plasticizer is also used for polyurethanes, unsaturated polyesters, and phenol-formaldehydes. Organic esters of phosphoric acid are used as plasticizers for cellulose acetate. A negative side effect of external plasticizers is their migration to the surface as the polymer ages.

Additives for Elastomers

When natural or synthetic rubber is compounded, sulfur or inorganic or organic accelerators are added for vulcanization. Fillers are added to all elastomers to strengthen products and lower their cost. Plasticizers are added to disperse the fillers during blending, antioxidants and antiozonants increase the service life of the final product, and colorants and odorants are added for aesthetic reasons. Blowing agents are added in the production of foam rubber, and mold release agents are used to make it easier to extract the product from the mold.

Anti-aging agents | Organic materials such as amines and phenolics are used as antioxidants and can inhibit natural and synthetic rubber degradation caused by oxygen and ozone. Amines are known to tint rubber, which is an unwanted side effect, while phenolics do not. Waxes, though sometimes used as antiozonants, can have the negative effect of blooming on the surface (Agrawal et al. 2005).

Colorants and odorants | Inorganic colorants such as ultramarine green, chromic green, and chromic yellow were used in the past, but they have been replaced by organic colorants for two reasons: inorganic metals can degrade the rubber, and organic colorants have greater coloring strength. Odorants are used to mask the distinct and often unpleasant smell of rubber.

Curing agents/activators | Vulcanization agents for natural and synthetic rubbers comprise sulfur and organic peroxides. For faster and more efficient curing, organic accelerators such as aniline and mercaptobenzothiazole are added to activate the sulfur during vulcanization. They have replaced traditional accelerators such as lead oxide, magnesium oxide, and calcium oxide. Zinc oxide (an inorganic activator) and stearic acid are still in use.

Fillers | During compounding, two types of fillers can be added. Fillers such as carbon black and silicon dioxide enhance properties like tear strength, tensile strength, and abrasion resistance, while fillers such as calcium carbonate, barium sulfate, and kaolin (hydroxy aluminum silicates) lower the price of rubber products but provide less reinforcement.

Plasticizers | Plasticizers are added to soften the rubber formulation during processing. They can also be added at compounding to get a better dispersion of the fillers. Plasticizers comprise mineral oils and esters of phthalic acid.

5.2 PROCESSING PLASTICS AND ELASTOMERS

In general, processing can be described as shaping plastics and elastomers into a structure. For thermoplastics and thermoplastic elastomers, the process involves heating and cooling, while thermosets, natural rubber, and synthetic rubbers undergo heating, cooling, and curing (chemical reactions, cross-linking/vulcanization). Table 5.2 provides a snapshot of the various processes and which types of polymers they are used for.

Processing Thermoplastics and Thermoplastic Elastomers

Polymer industries produce polymers in powders, pellets, and granules that are then further processed into products using machines. Plastics processing can be divided according to the type of machine or process used and whether the plastic is a thermoplastic or a thermoset. As discussed in chapters 1 and 2, thermoplastics are polymers with the ability to soften and flow when heated and to become hard again upon cooling. This cycle can be repeated multiple times with the application of such heat-fabrication techniques as extrusion and injection molding. Types of thermoplastic elastomers that have hard block units and soft block units in their polymeric chains do not have to be cured at all (see chap. 3). The soft blocks act as the rubbery part, and the hard blocks act as a thermoplastic. Thermoplastic elastomers can be manufactured using the same processes as thermoplastics and can be repeatedly processed, and therefore recycled.

Processing Thermosets and Rubbers (Natural and Synthetic)

At processing, thermosets, natural rubber, and synthetic rubbers have to be shaped, molded, and cured using both heat and pressure. Forming natural rubber and synthetic rubber compounded mixtures into the desired shape relies on processes also used for thermosetting plastics.

The starting material for processing natural rubber is rubber sheets manufactured at rubber plantations from the exudate of rubber trees, whereas synthetic rubber and thermoplastic elastomers, as monomers, are derived from the petrochemical industry.

The natural rubber production process involves stages: mastication (softening by mechanical shearing), blending, and vulcanization or curing at molding.

Mastication | At mastication, the first step in processing natural rubber, the thick sheets of rubber from rubber plants are shredded and softened by shearing in a powerful mixer. (The Banbury mixer is the one most commonly used in the rubber industry.) Processing synthetic rubber differs somewhat in that the synthetic rubbery substance is polymerized from monomers derived from petroleum, and therefore it does not need to be masticated first before being mixed with additives.

Table 5.2. Production processes used for different polymer types

Process	Thermoplastics	Thermosets	Rubbers (natural and synthetic)	Thermoplastic elastomers
Extrusion processes				
Extrusion	X	X	X	X
Blown-film extrusion	X			
Calendering	X		X	X
Cast-film and cast-sheet extrusion	X			
Extrusion coating	X		X	X
Molding processes				
Injection molding	X	X	X	X
Blow molding	X			X
Dipping	X		X	
Rotational molding	X			X
Thermo- and vacuum forming	X			X
Heating/cooling and curing (cross-linking) processes				
Casting		X	X	
Compression and transfer molding		X	X	
Foaming	X	X	X	X
Laminating		X		
Reinforcing		X	X	
Pultrusion		X		

Blending | To achieve the desired range of properties in the finished products, the next processing step is to add fillers and pigments, as well as vulcanizing and processing agents, to the softened rubbers, and to then mix this compounded rubber in another blender. Vulcanizing agents, or curatives, are active chemicals that bring about the cross-linking of the long-chain rubber polymer. There are three types of vulcanization:

- Sulfur vulcanization takes place under heat and pressure at processing in the mold, where sulfur reacts with the double bonds in the rubber polymer chains.
- Cold vulcanization uses sulfur chloride rather than sulfur; as a result, no heat is required.

- A third type of vulcanization involves the use of peroxides. The types of thermoplastic elastomers that have no double bonds in their polymer chains, such as silicone rubber, cannot be vulcanized using sulfur but must be cured with peroxides.

Molding and curing | After mastication and blending, the rubber is processed, fixed in its mold, and then vulcanized (cured) at a temperature of about 150°C.

Processing Techniques

At processing, two main groups can be distinguished: those that need external support for the shaping of products (molding processes) and those that do not (extrusion processes). Within each group, many different types of processes are used to convert polymers into plastic and elastomeric products (Brydson 1999; Lokensgard 2004; *Ullmann's* 2016).

Extrusion-based processes | Extrusion is a continuous processing method. It is most frequently used for thermoplastics and rubbers and to some degree for thermosets. The following sections describe the common types of extrusion.

Extrusion | The basic type of extrusion is used to make both finished objects or components, or to form smaller particles that are then reprocessed to create the finished product. Most extrusion processes make use of a die. Typically, compounded plastic pellets are fed into an extruder in which the plastic is heated (fig. 5.1). A continuously turning screw forces the molten mass through a die at the end of the extruder to shape the finished product, which is then left to cool. The shape of the die determines the shape of the end product: pipes, bars, or sheets (fig. 5.2). Extruded objects may have parallel marks along the long axis from the material being forced through a shaped die.

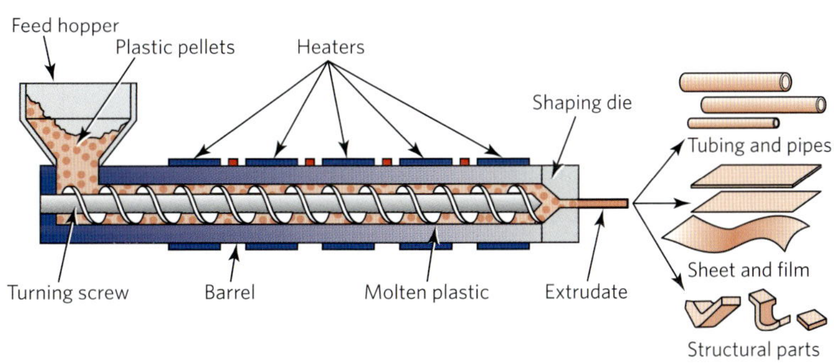

FIGURE 5.1
Extrusion molding schematic.

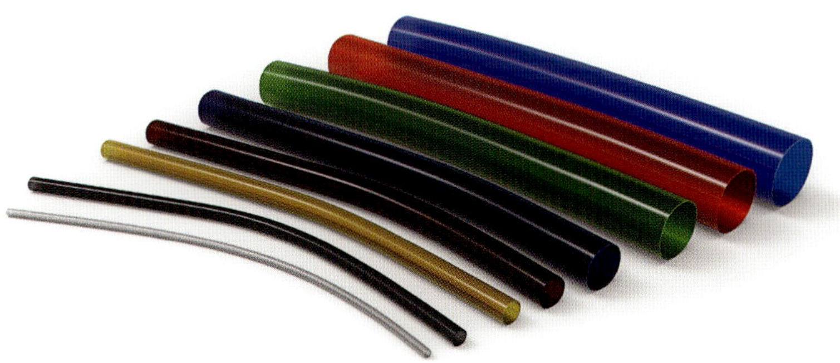

FIGURE 5.2
Tubes made by extrusion.

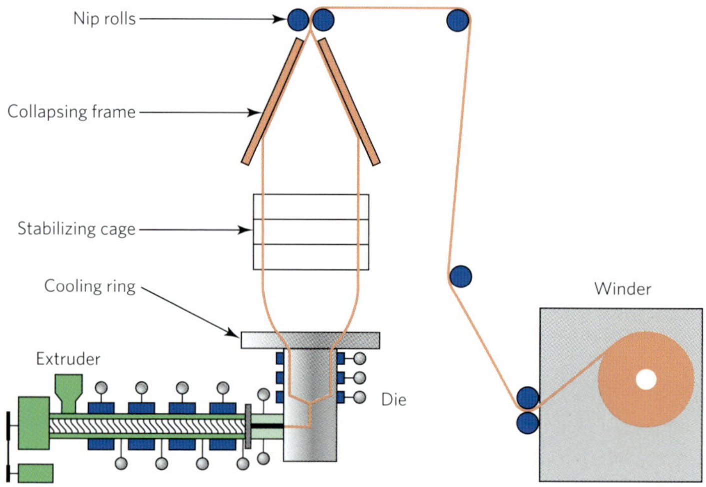

a.

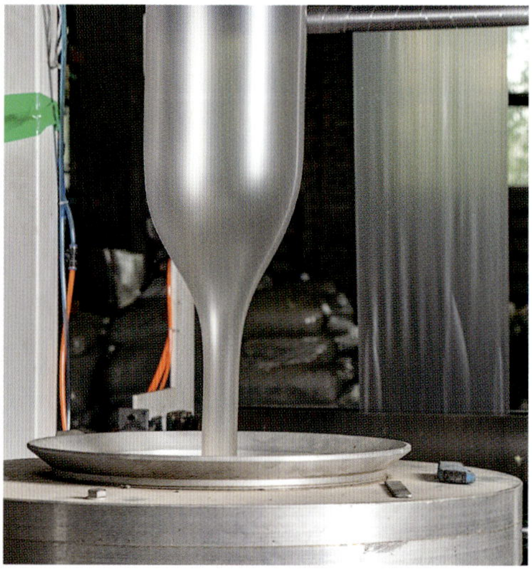

b.

FIGURE 5.3
Blown-film extrusion schematic (a) and equipment (b).

Blown-film extrusion | In film-blowing extrusion, a molten polymer such as polyethylene or polypropylene is extruded upward from a circular die to form a tubular film. Blowing air inflates the film tubing, which is then cooled by air from a cooling ring. The resulting tubing is then collapsed, flattened by rollers, and finally taken up by a winder (figs. 5.3a, b). The resulting tube is used for the manufacture of polyethylene bags, among other things. The emerging thin-walled tube can also be slit to obtain a single-sheet film that is then wound up into rolls.

Coextrusion, whereby up to three polymers are extruded, forms multilayered films in which the layers can have different properties, for example, outer layers for strength and an inner layer that is impermeable to oxygen or other gases. Multilayer barrier films can be used by conservators to create low-oxygen (anoxic) environments to encapsulate natural rubber objects (see fig. 4.30).

Calendering | Calendering is another process for manufacturing plastic films and sheets. During calendering, the heated plastic is moved between several heated rollers to homogenize the compounded plastic and to achieve a distinct thickness for the end product.

The plastic film is cooled on the last set of rollers, and the final roller winds up the produced film (fig. 5.4). Calendering is suitable for thermoplastics such as polyvinyl chloride as well as for elastomers (fig. 5.5).

Cast-film and cast-sheet extrusion | In the cast-film and cast-sheet extrusion process, the molten polymer is forced through a broad-slit die, and the resulting film or sheet solidifies as it cools. The cast-film extrusion method is used to produce cellulose acetate, polyamide, and polyester films (fig. 5.6).

FIGURE 5.4
Calendering schematic.

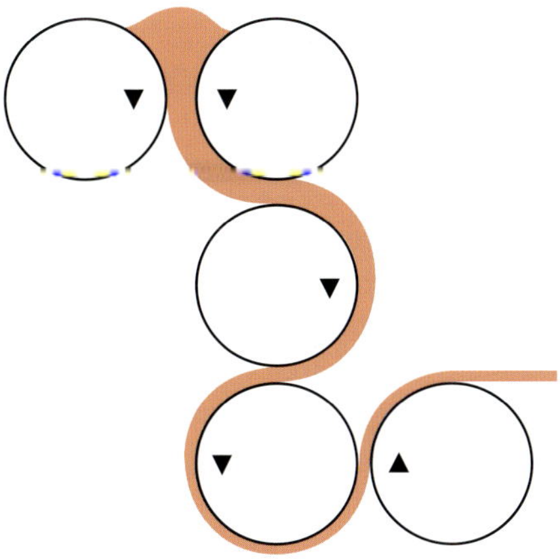

FIGURE 5.5
Calendered polyvinyl chloride sheet.

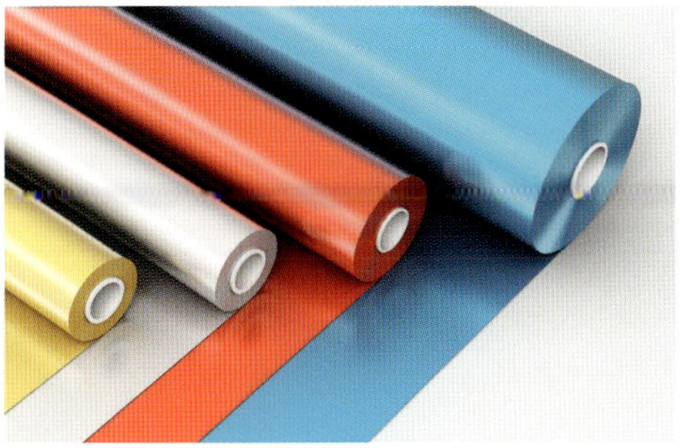

FIGURE 5.6
Cast-film extrusion schematic.

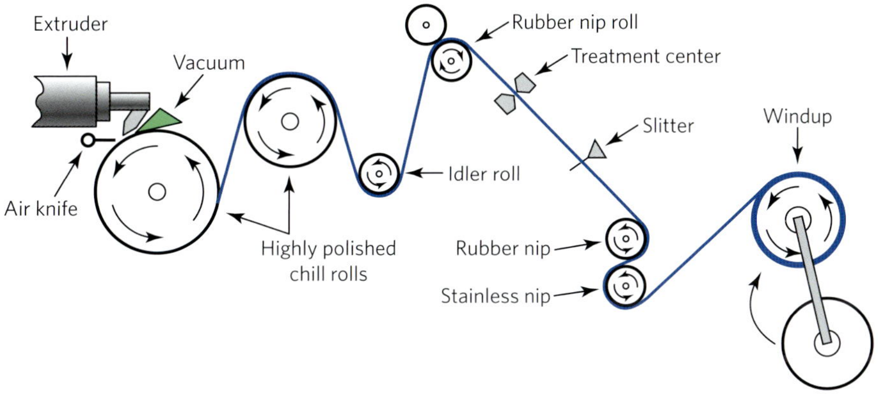

Foil, Film, or Sheet?

- *Foil* means thin materials of the type used for wrapping and packaging, generally in the order of 25–50 μm.
- *Film* means products produced by the cast-film extrusion process, including materials 70–200 μm thick and those used for semirigid packaging up to 380 μm.
- *Sheet* means materials in excess of 380 μm, which are typically produced by slicing blocks of material.

FIGURE 5.7
Transfer coating schematic.

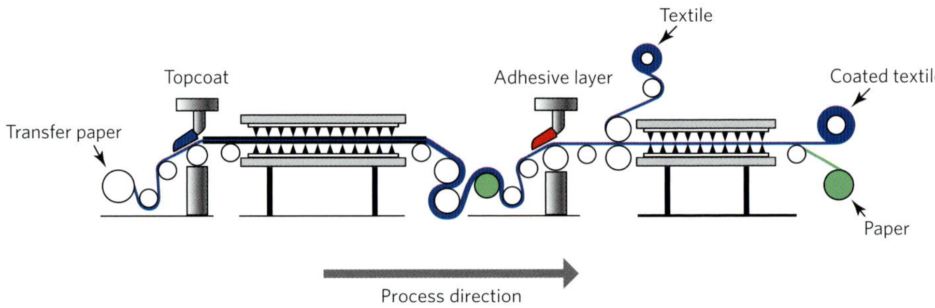

Extrusion coating | Coating is a process whereby a thermoplastic (polyethylene, polyvinyl chloride, or thermoplastic polyurethane) is applied onto a substrate such as paper, textile, wood, or wire to alter the surface of that material or to produce a composite material. This process is called direct coating and includes spray coating and roll coating.

Transfer coating is a similar process in which a layer of the polymer is applied first on a transfer paper, which is often embossed to provide texture to the finished product. Then subsequent layers of the same or another polymer are applied on top of the first, followed by an adhesive layer. Finally, a textile or paper substrate or backing is applied to the adhesive. After heating and drying, the transfer paper is released (fig. 5.7).

Direct or transfer coating onto a textile backing with polyvinyl chloride or thermoplastic urethane is used to make synthetic leather. Other applications include more utilitarian purposes such as adding a layer of plastic or rubber to the grip of metal forceps.

Molding processes | The following processes make use of a mold to form objects.

Injection molding | In the injection molding process, molten plastic is pushed through a heated chamber into a mold by a reciprocating screw (fig. 5.8). Injection molding is similar to extrusion, but the screw stops turning to allow the plastic objects to cool and harden in the mold. As soon as each object has cooled, it is ejected from the machine, and a new cycle starts: the screw moves backward, allowing new material to enter, then starts turning again. Depending on the complexity of the molded product, cycles can allow several injections per minute, making injection molding a suitable process for mass production.

Both thermosets and thermoplastics can be injection molded. When molding thermosets the mold is heated, and when processing thermoplastics the mold is cooled (fig. 5.9).

FIGURE 5.8
Injection molding schematic.

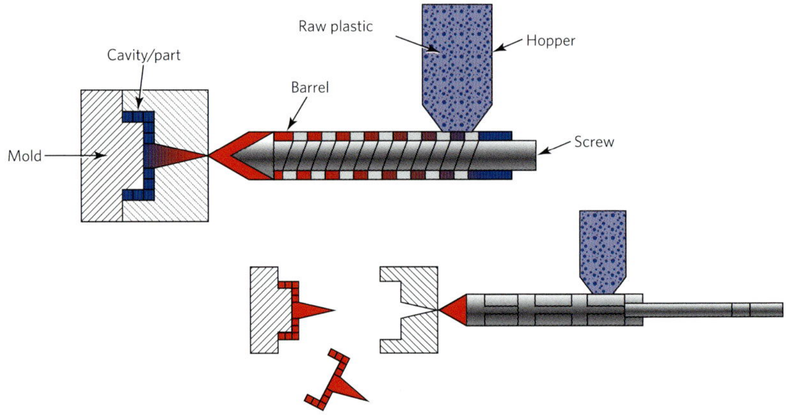

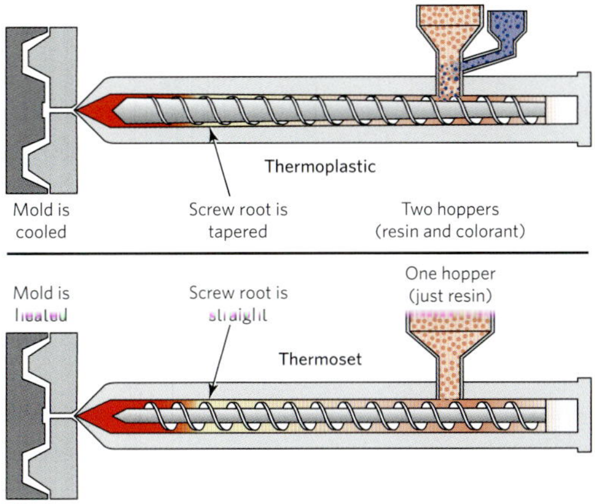

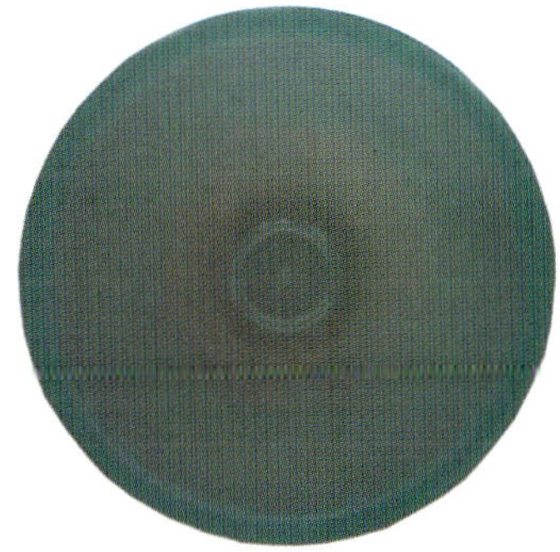

FIGURE 5.9
Schematic comparing injection molding of thermoplastics and thermosets. Red indicates molten material and blue, cooled.

FIGURE 5.10
Injection molding sprue on the bottom of a polyethylene beaker.

Injection-molded objects are recognized by the waste piece, or sprue, typically found on the bottom of the object (fig. 5.10).

Blow molding | Blow molding is an injection molding technique for hollow plastic objects, such as plastic bottles. Using that example, the molten plastic comes out of a heated extruder through a die in the form of a tube and is introduced into the machine for the blow molding step. Compressed air is used to expand the tube to conform to the interior of a cold mold (fig. 5.11). Blow molding is commonly used to make bottles out of various thermoplastics, including polyethylene, polypropylene, and polyethylene terephthalate (fig. 5.12).

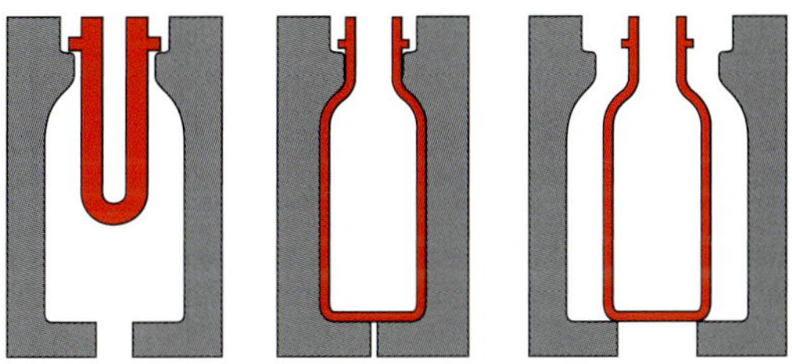

FIGURE 5.11
Schematic showing process of blow molding bottles: Injection-molded preformed tube is inserted into heated mold (a). Tube is blown out to conform to mold (b). After cooling, mold is removed (c).

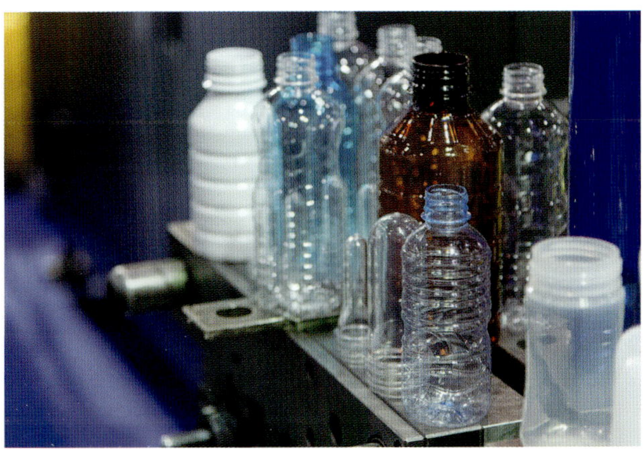

FIGURE 5.12
Blow-molded bottles.

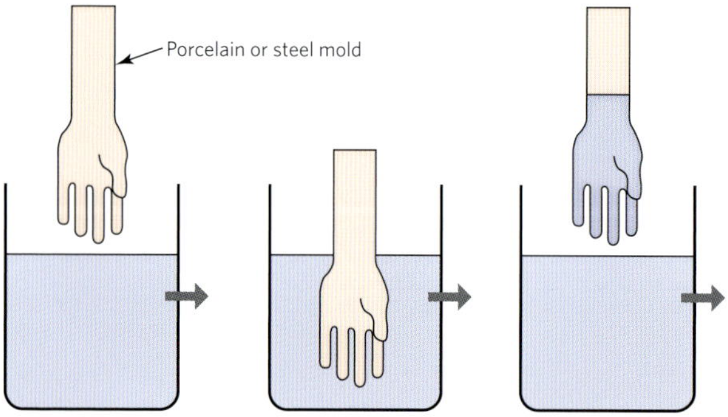

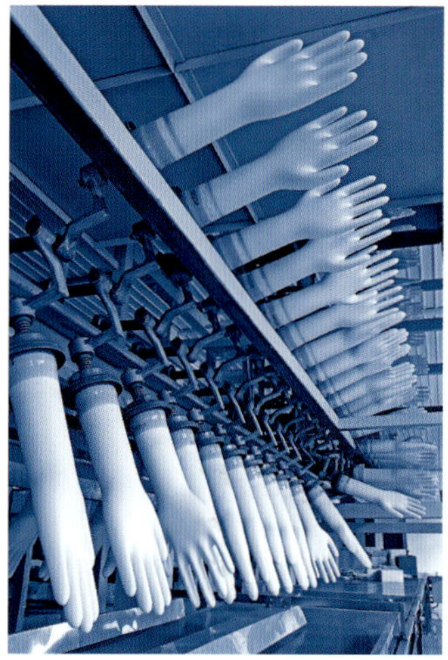

FIGURE 5.13
Dipping schematic and photograph showing the process of forming nitrile gloves.

Dipping | Dipping, or dip molding, is a process that involves coating a ceramic or metal mold with a liquified plastic or rubber compound like plasticized polyvinyl chloride, latex, or nitrile rubber, after which the product is cooled and hardened/cured, in the case of latex or nitrile rubber. Depending on the material used, dipping can take place at room temperature or at elevated temperatures. To achieve the desired thickness of a product, layers can be built up through successive dips, as needed (figs. 5.13). Products made by dipping include prophylactics and latex, nitrile, and PVC gloves.

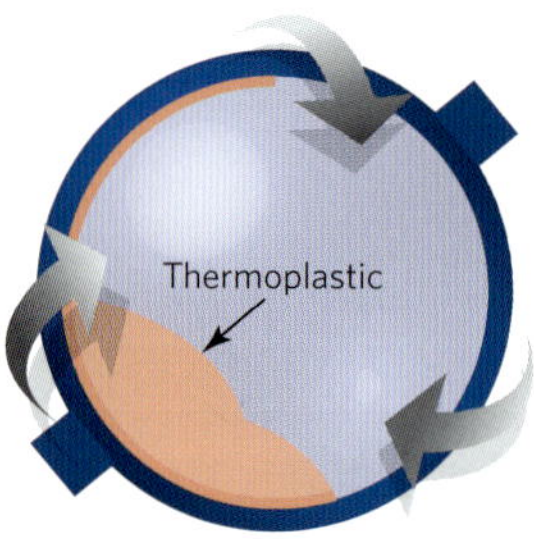

FIGURE 5.14
Rotational molding schematic.

Rotational molding | Rotational molding is a technique used for producing hollow items. Fluid or powdered thermoplastic is introduced into a heated mold, which is then gently rotated both vertically and horizontally—no pressure is involved. During rotation, the fluid plastic forms a thin film inside the mold. After cooling, the mold is opened and the product removed (fig. 5.14). Dolls' heads were once made by rotational molding (fig. 5.15). Rotationally molded objects are hollow with no seams along the sides.

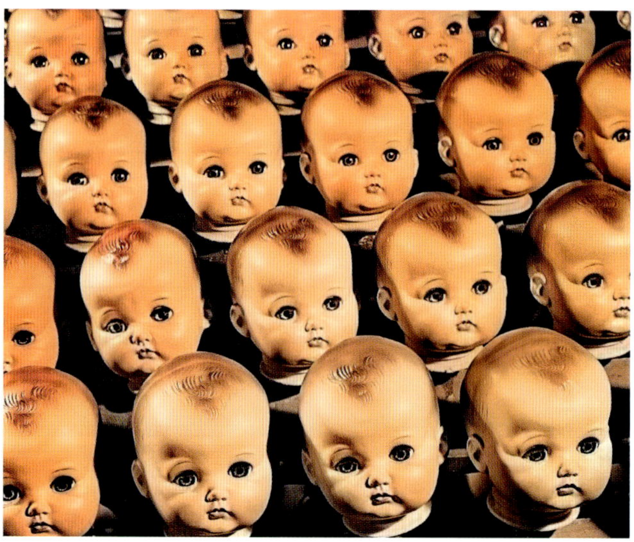

FIGURE 5.15
Dolls' heads made by rotational molding.

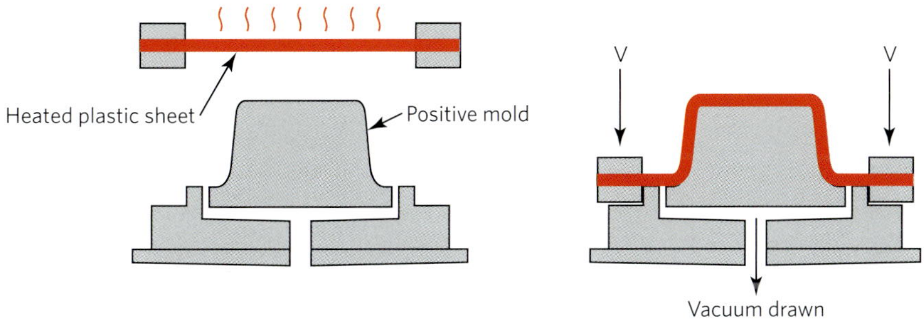

FIGURE 5.16
Schematic showing process of thermoforming (vacuum forming) thermoplastics.

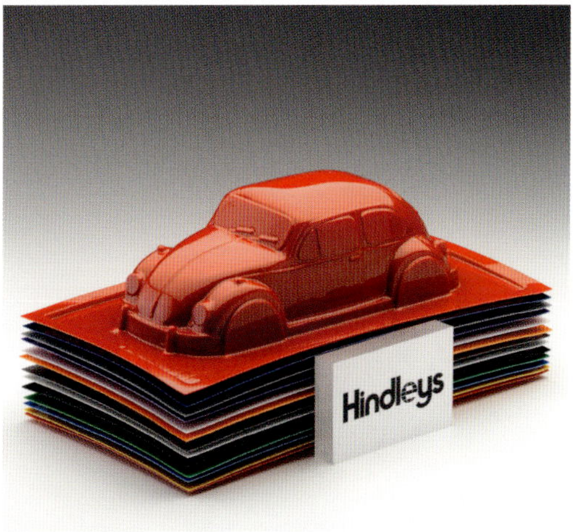

FIGURE 5.17
Vacuum-formed model of a toy car available in various colors.

Thermo- and vacuum forming | Thermoforming involves heating an extruded sheet of plastic (e.g., polystyrene, copolymers of styrene, or polyvinyl chloride) and forming it into a mold (fig. 5.16). The most commonly used thermoforming technique is vacuum forming. A vacuum is created between the heated sheet and the mold. The vacuum forms the product, which is then cooled and removed from the mold. Refrigerated food containers such as margarine and yogurt tubs, restaurant trays, and cold-drink cups are formed from polystyrene in this manner (fig. 5.17). Vacuum-formed sheets of acrylonitrile-styrene-acrylate and high-impact polystyrene have been used by artists such as Marcel Broodthaers to create works of art (van Wegen 1999). Vacuum-formed and other thermoformed objects do not have seams or sprues.

Curing processes | The following techniques for thermosets and natural and synthetic rubbers entail heating, cross-linking/vulcanization, and cooling to process and form the material.

Casting | The simplest method of processing plastics is casting, in which reactive liquid thermosetting plastic is poured into a mold. The liquid plastic, at room temperature, is poured into a mold made of metal or silicone rubber. The liquid plastic contains an initiator that is mixed into the polymeric material before it is poured into the mold. Once the mold is filled, the process of cross-linking takes place through the action of the initiator on the polymer, and the plastic hardens.

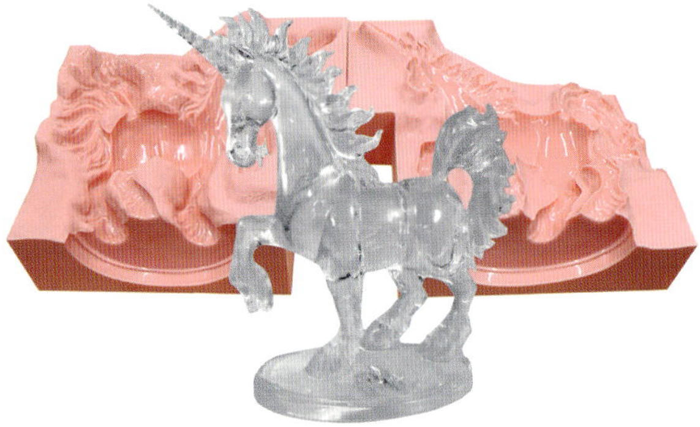

FIGURE 5.18
A cast thermoset figurine of a unicorn.

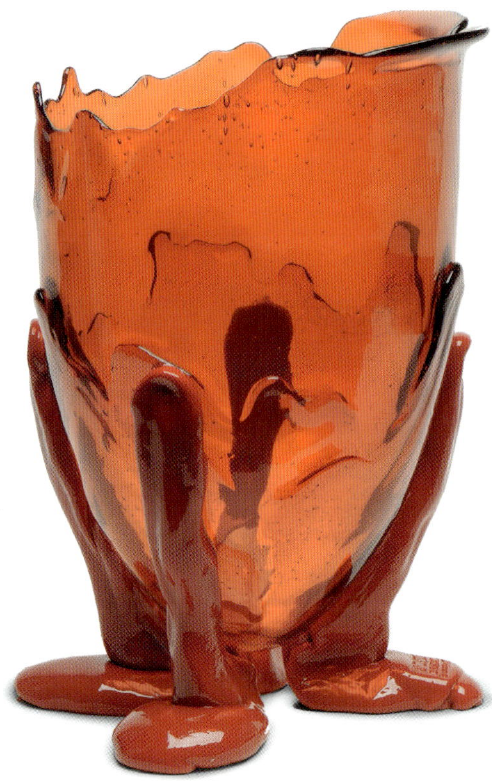

FIGURE 5.19
Gaetano Pesce (Italian, b. 1939), *Vase, Fish Design Collection*, 1995, manufactured by Corgi Design, Italy. Soft resin, various dimensions.

Examples of cast plastics are unsaturated polyesters, epoxy type plastics, and polyurethanes. While cast plastics are widely used for industrial purposes for medium-size and large products, hobbyists and artists also create small cast plastic objects such as jewelry and toys (fig. 5.18) and larger objects such as the vases and chairs of Gaetano Pesce (fig. 5.19).

Compression and transfer molding | In compression molding, thermosetting plastic products are molded using heat and pressure. Where injection molding uses a screw to transport the plastic into the mold, compression molding makes use of a plunger (fig. 5.20). Typically, in this process powder or granules of a phenolic or melamine plastic are forced

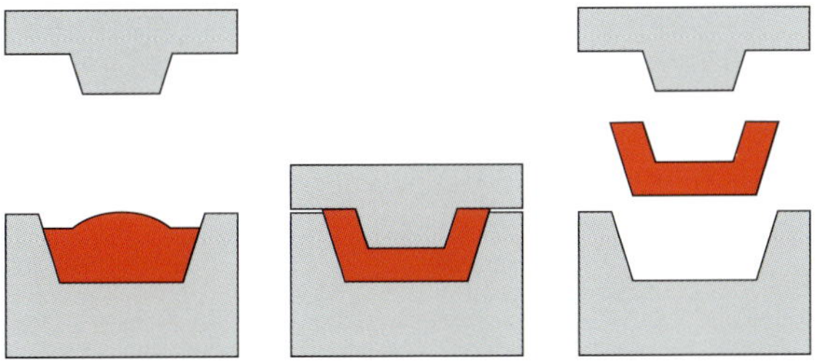

FIGURE 5.20
Compression molding schematic. Plastic (powder or granules) is poured into heated mold (a). Mold is closed and pressure is applied (b). After cooling, mold is opened and object removed (c).

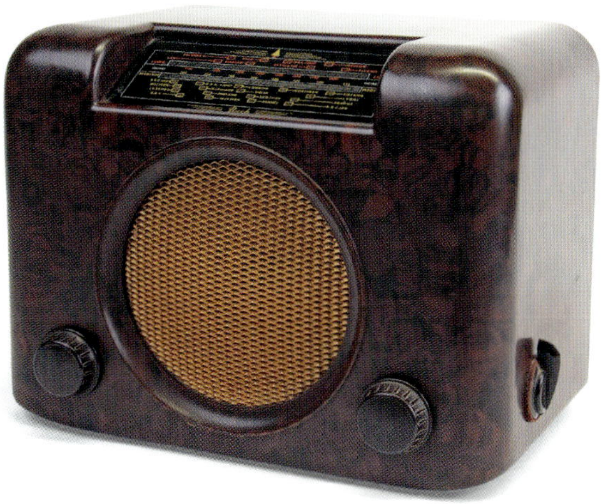

FIGURE 5.21
Bush Type DAC 90A radio, 1950, manufactured by Bush, United Kingdom. Compression-molded phenol-formaldehyde, 22.5 × 31 × 19 cm. Bournemouth, Arts University, Museum of Design in Plastics, AIBDC: 005896.

into a heated mold, the mold is closed, pressure is applied, and the powder is cured (cross-linked) by the heat and pressure. Transfer molding is similar, but the plastic is preheated to achieve a viscous or liquid state before introduced into the mold.

After the mold is cooled, the shaped product is removed from the mold with ejector pins. Sometimes a compression-molded object can be recognized by the marks of the pins. More often they are recognized by the mold seams, although not infrequently these have been buffed away. Compression molding is most often used for small objects but is also useful for large or odd-size thermosetting products (fig. 5.21).

Foaming | Foam is a polymer matrix with either closed or open cell structures, and foaming is the production process by which it is made. A foam is produced by introducing a blowing agent into a liquid mixture of monomers. A cell-structured matrix is created at polymerization. The blowing agents can be air or a specific gas formed as a reaction product of the polymerization of the liquid monomers. Foams can be made from thermoplastics that harden by cooling, and thermosets and elastomers that harden by chemical reaction (cross-linking).

When air is used as the blowing agent, it is frothed into the liquid polymer, much as it is when making whipped cream. Foam rubber and urea-formaldehyde foam are created by frothing; it produces irregularly formed air bubbles, hence irregular cells.

When gas is used as the blowing agent, it is either a physical or a chemical agent. Physical blowing agents include volatile gases such as pentane, which is used to manufacture expanded polystyrene foam (fig. 5.22). Polystyrene foam has closed cells and is used for protective packaging or to provide buoyancy (e.g., in surfboards). They are also sometimes

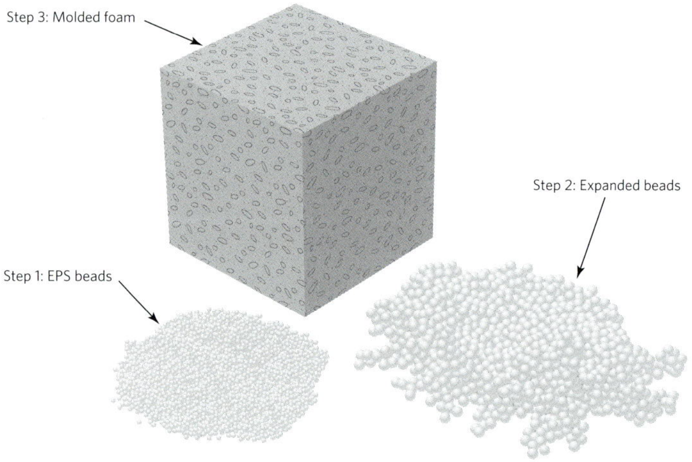

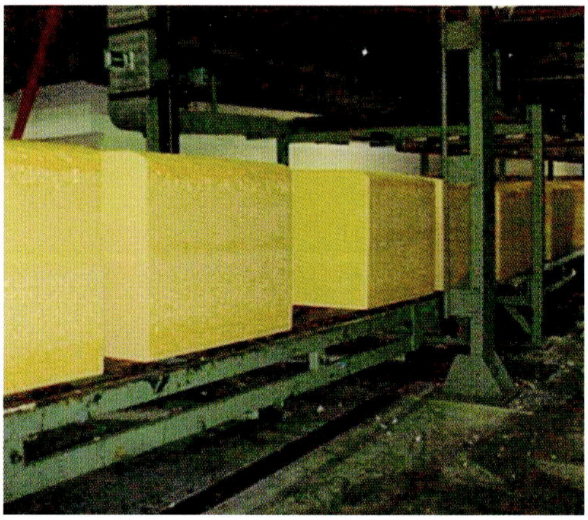

FIGURE 5.22
Stages of expanded polystyrene (EPS), or Styrofoam.

FIGURE 5.23
Slab stocks of polyurethane foam.

used as support material inside works of art, such as those by the artist Jean Dubuffet (see fig. 2.14).

Chemical blowing agents come from a chemical reaction of organic compounds that gives off gases, much like the CO_2 released by yeast that causes bread dough to rise. Foams produced using chemical blowing agents include those made from polyethylene, polypropylene, and polyvinyl chloride, and are used in insulation and consumer goods.

Chemical blowing agents can also come from the reaction of the thermosetting polymer itself during polymerization. An example is the reaction of the isocyanate groups of polyurethane with water-forming amines, which releases CO_2, the blowing agent that produces the cells in polyurethane foam. Polyurethane foams can be rigid (closed cells) or flexible (open cells), depending on the formulation of the prepolymer. Foamed products can be made into slab stock blocks by pouring (fig. 5.23) or formed by reaction injection molding.

Laminating | In the lamination process, large sheets of paper or textile impregnated with a thermosetting prepolymer are stacked up and then compacted in a heated press. This causes polymerization (cross-linking) to take place, forming the layered, laminated sheets (fig. 5.24). The liquid prepolymers used in lamination include unsaturated polyesters and epoxies, along with a catalyst.

Laminated products include structural materials used in the construction of homes, furniture, and aircraft.

FIGURE 5.24
Schematic of the manufacture of laminated sheets.

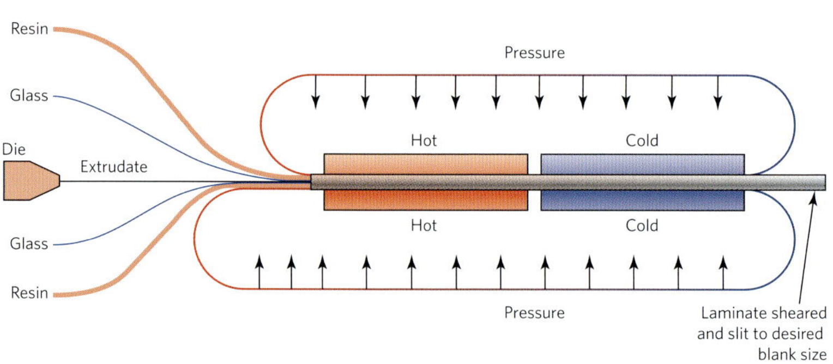

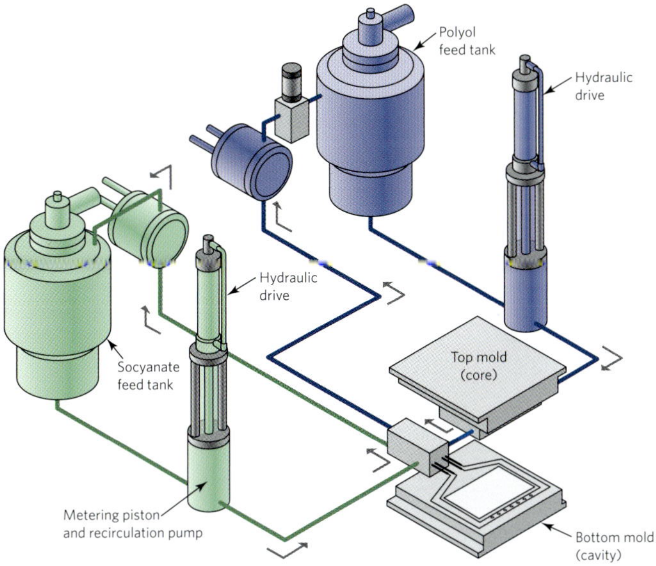

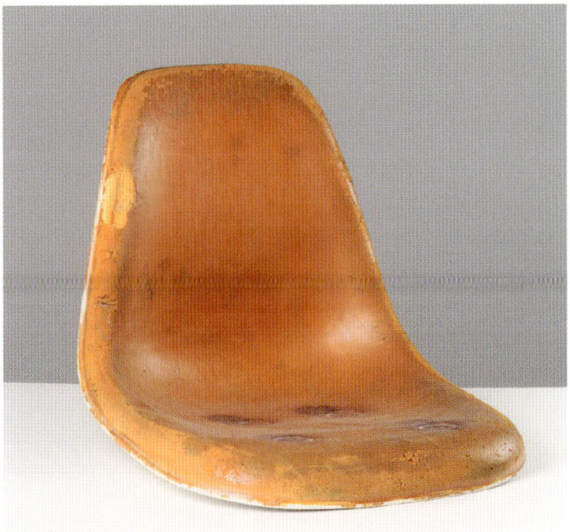

FIGURE 5.25
Reaction injection molding schematic.

FIGURE 5.26
Charles Eames (American, 1907–1978) and Ray Eames (American, 1916–1988), experimental reaction injection-molded seat shell, 1968. Fiberglass, urethane foam, rubber, 40 × 48 × 45 cm. Venice, California, The Eames Office.

Reaction injection molding | In the reaction injection molding process, two or more reactants (liquid monomeric compounds) are mixed and forced into a mold that has the form of the final product; the reaction (polymerization) then takes place (fig. 5.25). This process, typically used for thermosetting polyurethanes, results in rigid, semirigid, or flexible structural foamed products (see "Foaming," above) such as automobile bumpers and chairs (fig. 5.26).

Reinforcing | The manufacture of glass fiber–reinforced plastics involves pouring thermosetting reactive liquid plastic, such as epoxy or unsaturated polyesters, into a mold. The mold (or form) is first covered with a layer of a gelcoat, an unsaturated polyester with excellent weathering properties. On top of the gelcoat layer, reinforcing material is positioned (see "Fibers" in section 5.1). The mixture of thermosetting liquid plastic, combined with an initiator or catalyst, is then applied on top of the reinforcement material using either a brush (hand layup) or sprayer (spray layup). The layers are then lightly compressed by using a roller to remove trapped air (fig. 5.27). Polymerization, which creates a three-dimensional cross-linked structure, occurs at room temperature. Once it has hardened, the object can be removed from the mold. Hand layup and spray layup are low-cost tooling techniques that are often used for objects of various sizes, including bathtubs, furniture, and boats.

FIGURE 5.27
Glass fiber–reinforced plastic schematic.

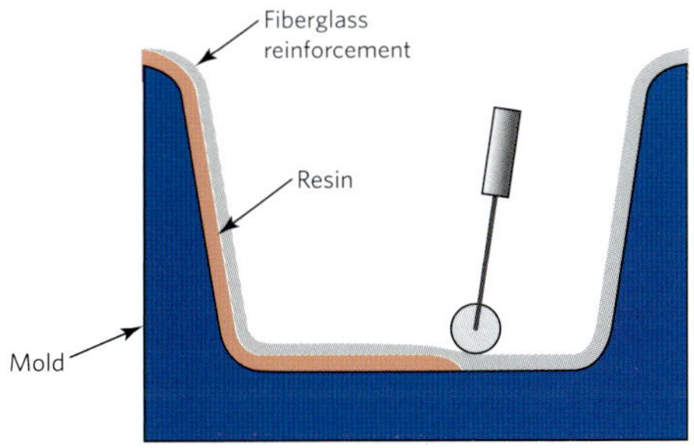

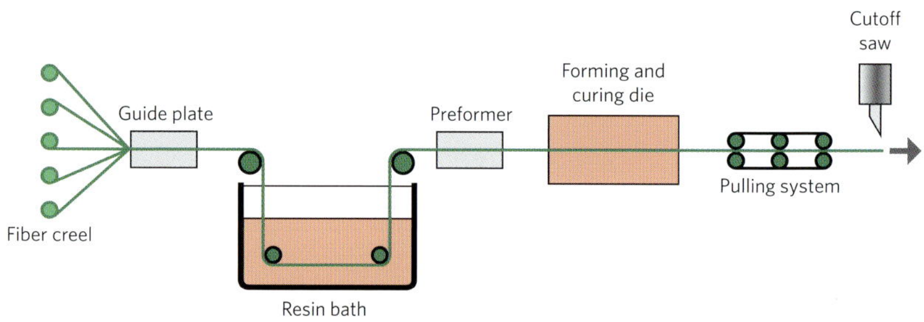

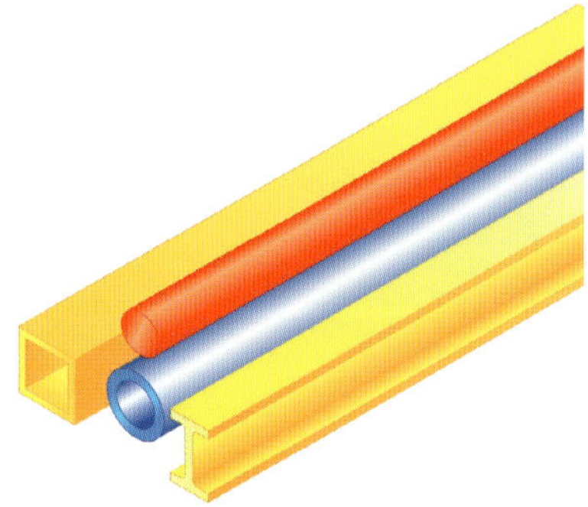

FIGURE 5.28
Pultrusion schematic.

FIGURE 5.29
Pultruded glass fiber structural sections for construction.

Pultrusion | Pultrusion (a combination of *pull* and *extrusion*) is the process of making reinforced materials by coating glass fibers, or carbon or textile threads, by pulling them through a thermosetting liquid plastic and then extruding the threads through a die to shape and cure the product (fig. 5.28).

Unsaturated polyesters and epoxies are the primary plastics used in pultrusion. Carbon threads and nylon threads are used for strength in tennis rackets, ski poles, and fishing rods, while pultruded glass fiber rods and other structural parts are used in construction (fig. 5.29).

5.3 ADDITIVE MANUFACTURING

Additive manufacturing, also known as 3D printing and rapid prototyping, is, according to its original ASTM standard, the "process of joining materials to make objects from 3D model data, usually layer upon layer" (ASTM F3413).[1]

The term *additive manufacturing* refers to the technology's method of creating 3D objects by building up layer upon layer of a chosen material, as opposed to subtractive manufacturing methodology. It makes use of a computer, 3D-modeling software (computer-aided design, or CAD), manufacturing equipment (3D printer), and the layering material. After a CAD drawing is created (or a model is scanned in three dimensions to produce a CAD file), the 3D printer reads the file and lays down successive layers of liquid, powder, or sheet material to manufacture a physical 3D object. An advantage of additive manufacturing over more conventional manufacturing technologies is the ability to produce very complex shapes.

The technology for printing physical 3D objects from digital data began in the 1980s with stereolithography, which was patented in 1986. Other technologies are fused deposition modeling (patented in 1989) and selective laser sintering (introduced in the 1990s). Material jetting technology was developed in the early 2000s.

The initial additive manufacturing machines focused on the production of visualization models and prototypes. Today, additive manufacturing is being used to fabricate end-use products in industrial design, architecture, engineering, and construction; educational tools; consumer products like jewelry, clothing, and footwear; and materials and equipment for

1. Additive manufacturing includes many technologies. To avoid confusion, the standard ISO/ASTM 52900:15 was published in 2015 to replace the original (ASTM F2792). It standardizes all terminology and classifies each of the different additive manufacturing technologies.

the automotive, aerospace, dental, and medical industries. The development of additive manufacturing and printing materials is evolving so rapidly and the objects being produced are so wide-ranging that it is difficult to keep up.

3D-printed objects will only become increasingly common in our daily lives and in museum collections. How these objects will behave in the future and how we are to conserve them have yet to be established. Moreover, manufacturers and designers are slow to provide information on material composition—knowledge that would help those charged with care of plastic 3D objects to better understand properties and aging behavior. For the time being, however, it is important to understand as best we can the technology and material as they now stand.

Additive manufacturing comprises seven different broad categories: vat polymerization, powder bed fusion, material jetting, binder jetting, material extrusion, sheet lamination, and directed energy deposition. The first five of these are relevant to this book and are discussed below. Some of these categories encompass several printing technologies.

Vat Polymerization

The three forms of vat polymerization are stereolithography (SLA), digital light processing (DLP), and masked stereolithography (MSLA). The primary difference between these technologies is the light source they use to cure the photopolymer (as well as a mask, in the case of MSLA). While these three printing techniques share the same principles, each has its advantages and disadvantages: DLP and MSLA are faster than SLA, and SLA gives smoother surface structures than the others. As SLA is the oldest of the three and therefore the best known of the techniques, we'll concentrate on that.

SLA makes use of a bath filled with a liquid photopolymer (resin) that is cured by a high-powered point laser to generate objects (fig. 5.30). Once the printing is complete, the objects must be rinsed in a solvent (isopropyl alcohol) to remove any uncured resin from their surface. Visible building layers are evident on the surface of unfinished objects or parts, but postproduction processes such as painting, varnishing, or sandblasting can create a variety of finished effects, from high glosses to coarse textures (fig. 5.31). SLA creates parts with no weakness in any direction, which is not the case with some other techniques (see "Material Jetting," below).

FIGURE 5.30
Diagram of a stereolithography printer.

FIGURE 5.31
3D-printed geometric knots.

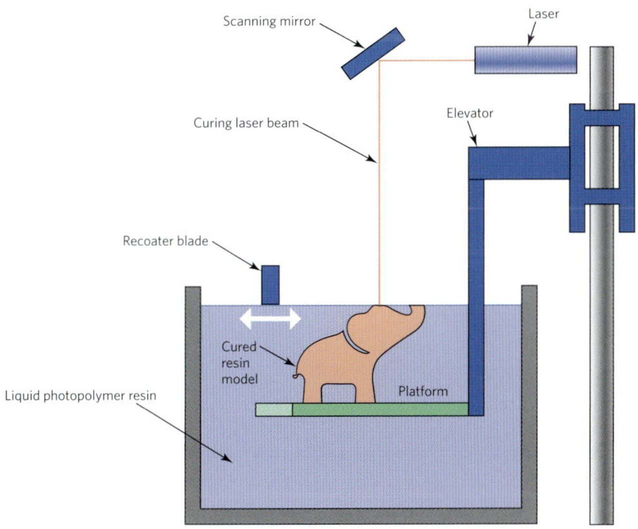

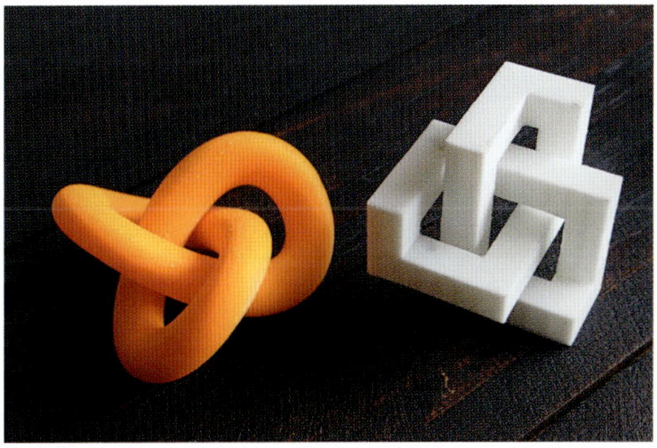

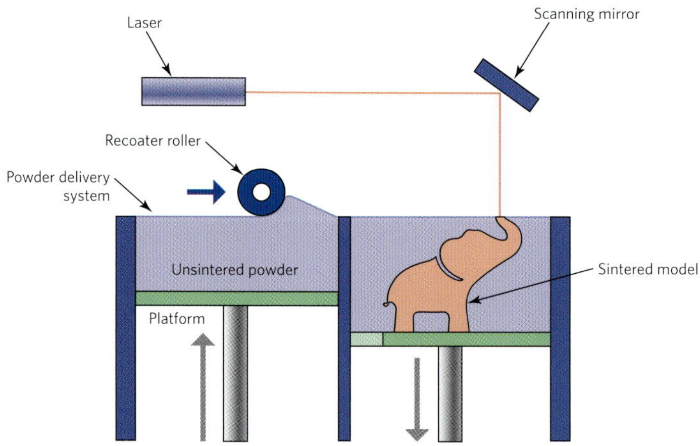

FIGURE 5.32
Diagram of a selective laser sintering device.

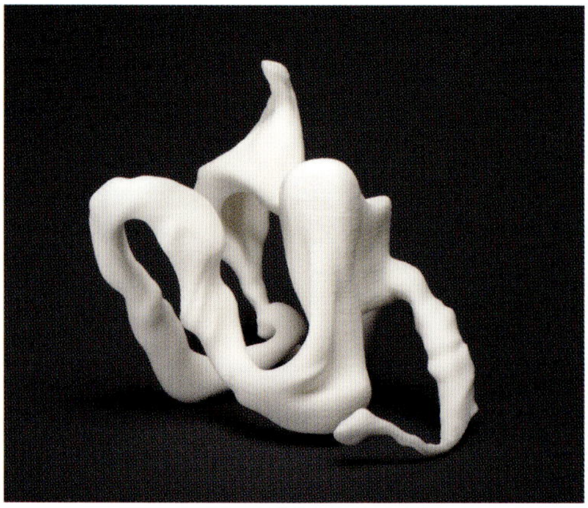

FIGURE 5.33
Marcel Wanders (Dutch, b. 1963), Airborne Snotty Vase: Influenza, 2001. Manufactured by Cappellini S.p.A., Italy. Polyamide, using selective laser sintering digital scanning and printing, 15 × 15 × 15 cm. Art Institute of Chicago, purchased with funds provided by the Architecture & Design Society, 2007.61.

An abundance of materials are on the market for SLA. However, manufacturers provide information about their composition only sparingly, using brief descriptors such as "photopolymer resin," "standard," "castable," "transparent," "high-temperature," or "polypropylene like," "ABS like," and "rubber like."

Powder Bed Fusion

Powder bed fusion is a manufacturing technique that uses a heat source, usually a laser or electron beams, to fuse small particles (powder) into a mass that has a desired 3D shape. Figure 5.32 shows one such device that uses a CO_2 laser as the heat source. Four types of powder bed fusion exist; selective laser sintering (SLS) is the most relevant for plastics.[2]

In SLS, the platform on which the object is positioned lowers by a single layer of thickness per fusing (sintering) pass (Deckard 1989). The laser sintering process is repeated until the object is complete. It is then removed from the platform and excess powder is blown off. In contrast to stereolithography technology, where support material is needed, in SLS the build is surrounded and supported by unsintered material. SLS objects can be sandblasted or coated to smooth their surfaces.

2. Two others, selective laser melting and electron beam melting, are used for metal parts. Multijet fusion is a laser sintering technique used to make ultra-thin layered products.

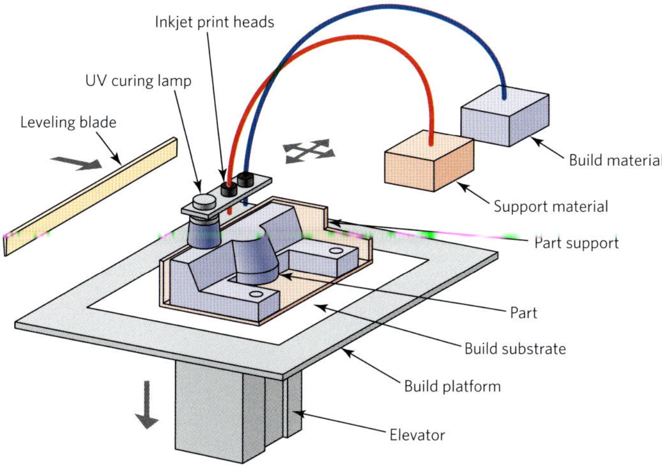

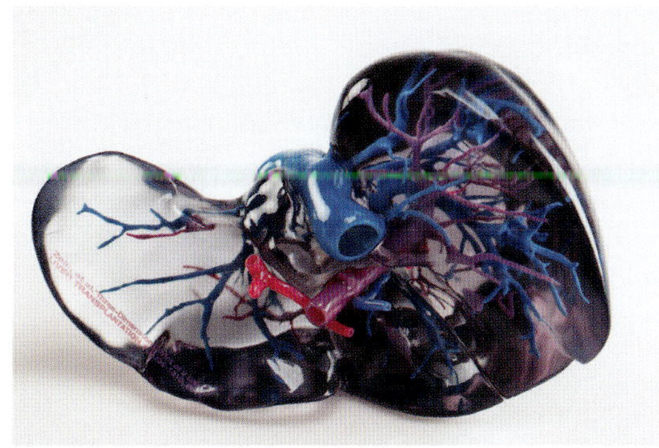

FIGURE 5.34
Diagram of a material jetting printer.

FIGURE 5.35
Scale model of a liver made using material jetting.

Polymer materials that can be used for SLS include polyamide (nylon-12), glass-filled polyamide, polyurethane elastomer, and polystyrene. Marcel Wanders's nylon Airborne Snotty Vases were made using SLS (fig. 5.33).

Material Jetting

Like stereolithography, material jetting is based on the principle of photopolymerization. It works in a manner similar to that of a standard inkjet printer. The material jetting process involves selectively depositing droplets of photopolymers on a build plate. Once the layer of droplets cures (following exposure to light), the next layer is added. As with SLS, objects are built up one layer at a time until the part is complete (fig. 5.34). Because of their extremely thin layers, the resulting parts are very precise and feature exceptionally smooth surfaces. Objects made with material jetting require support, which is provided during the building process in the form of a dissolvable material that is removed when the process is complete.

In the medical sector, this method can be used to create, for example, accurate anatomical scale models with different colors used to show the various parts of an organ (fig. 5.35).

Material jetting technology also allows different materials—with diverse physical and mechanical properties—to be incorporated in the same object. For instance, a single object might contain material with properties ranging from rubber-like pliability to ABS-like rigidity. Materials used for material jetting are photopolymers (available in standard, transparent, and high-temperature types) with names such as Vero, Vero Clear, and Vero Ultra Clear (the latter simulates polymethyl methacrylate).

Binder Jetting

Binder jetting is an additive manufacturing technology similar to stereolithography. Whereas SLS uses a laser to sinter powder, binder jetting utilizes a print head to deposit droplets of a binder as it moves over a surface of powdered gypsum or sandstone. The droplets bind the powder particles together, thereby producing each new layer of the object (fig. 5.36).

Once a layer has been printed, a new layer of powder is laid down over the just-printed layer, and the process repeats until the object is completed. After the object is complete and the binder has dried, excess powder can be removed from the surface. The loose powder provides adequate support for the object during production. As a final step to enhance

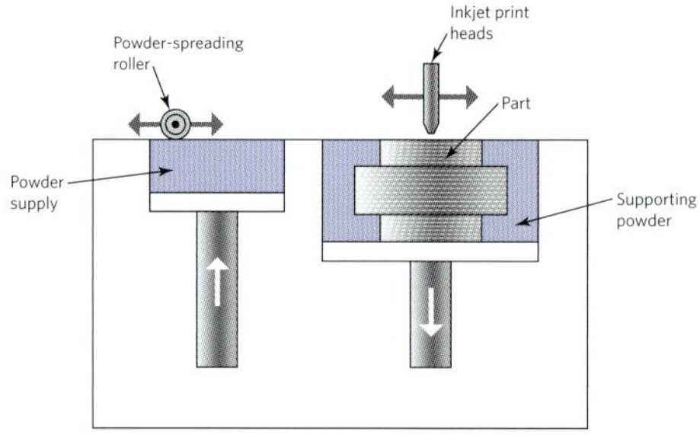

FIGURE 5.36
Diagram of a binder jetting printer.

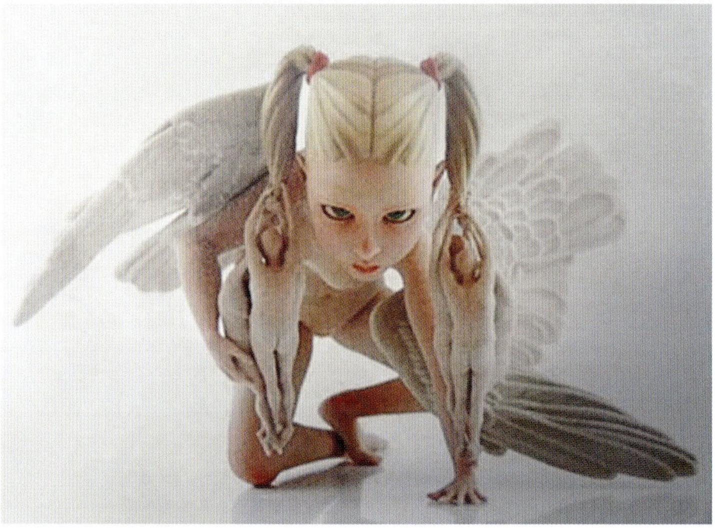

FIGURE 5.37
Eric van Straaten (Dutch, b. 1969), *Fallen Angel*, 2011. Multicolor composite 3D print using binder jetting.

strength and impart a desirable finish, various infiltrating agents such as wax, cyanoacrylate, or an epoxy material can be applied to the object (fig. 5.37).

Fused Deposition Modeling

Fused deposition modeling (FDM) is a form of material extrusion that uses 3D computer modeling. The process involves introducing filaments of solid thermoplastic build material and support material to heated nozzles, which lay down molten filaments of these materials in cross sections to establish a desired shape (figs. 5.38a, b). As the molten material cools, it solidifies to form an object. This process is repeated and new layers are added until the object or part is completed. Depending on the shape and complexity of the object, support structures may need to be added, for example, if an object has overhanging parts.

The layer adhesion mechanism of FDM makes parts created using this method anisotropic (weaker in one direction), and layer lines are visible on the surface. FDM is the most inexpensive additive manufacturing technology and the one most used by artists and designers (fig. 5.39).

A wide range of thermoplastic materials (in filament form) is available for FDM, including polylactic acid, nylon-6, nylon-12, acrylonitrile-butadiene-styrene (ABS), acrylonitrile-styrene-acrylic, thermoplastic polyurethane, polycarbonate (PC), PC-biodegradable blend, and PC-ABS blend.

FIGURE 5.38
Fused deposition modeling schematic (a), and equipment (b).

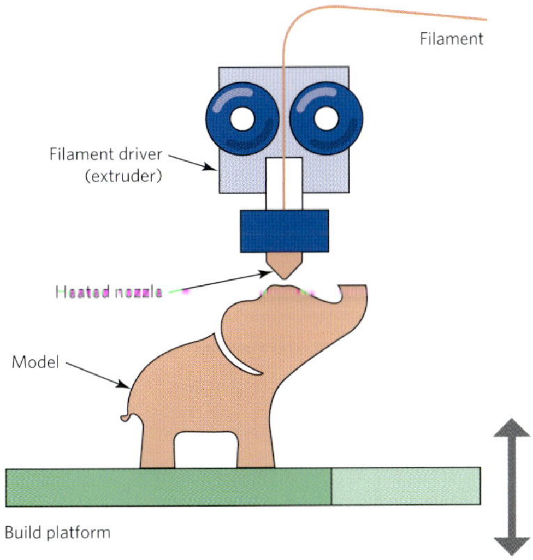

a.

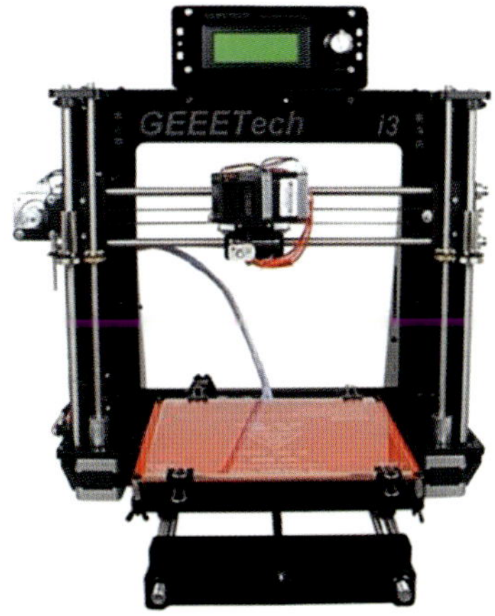

b.

5.4 PLASTICS RECYCLING

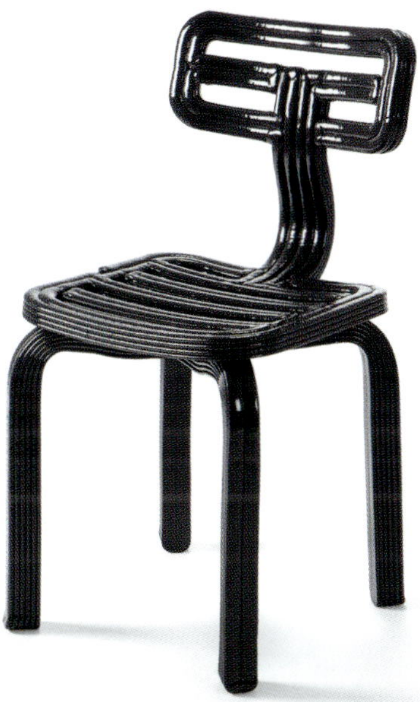

FIGURE 5.39
Dirk van der Kooij (Dutch, b. 1983), Chubby Chair, 2012. Recycled plastic (96% reclaimed synthetics) using fused deposition modeling, 80 × 43 × 49 cm. Weil am Rhein, Vitra Design Museum, MNL-1035.

Because plastics have become so much a part of our everyday lives since they were first introduced, consumption has increased tremendously. Until the beginning of the 1970s, consumers were unaware of the impact plastics have on our environment. Most plastic ends up as garbage, and it does not degrade very quickly; it takes hundreds of years to disappear. In the meantime, it is filling up landfills, where, rather than breaking down completely, it disintegrates into minute pieces that end up in rivers, lakes, and oceans as a "soup"' of microplastics.

Plastic pollution can be addressed by reducing plastics use, reusing plastics, and recycling plastics. Reducing plastics pollution can be achieved by consuming less plastic, producing less fossil-based plastic, and using biobased or biodegradable plastics instead. Reuse can be accomplished by using manufacturing scrap. However, this reused plastic, because it is not virgin material, is less durable and thus has a shorter service life.

Recycling is effectively restricted to thermoplastic products, which can be remelted and reprocessed. Feedstock for recycling comes from plastics collected from recycling containers, facilities where bottles are returned for deposits, and curbside pickup bins. Like those made from reused plastics, recycled products may be lower quality and have a limited life span; uses include plant pots, packaging, and plastic grocery bags. PET bottles can be recycled into polyester fiberfill or fibers for the textile industry. For better quality, remelted plastics are mixed with virgin plastics to enhance durability.

To make identification of to-be-recycled plastic easier and thus improve the quality of recycled products—because recycled products should comprise a single type of polymer—a system of triangular symbols called Resin Identification Codes is used to distinguish the different types of plastics (table 5.3).

Table 5.3. The Resin Identification Code system

Plastic name	RIC symbol	Original uses	Recycled uses
Polyethylene terephthalate (PET or PETE, polyester)	1 PETE	Carbonated beverage bottles, food and condiment jars, oven-ready and microwavable meal trays, textiles, carpet, straps, films	Carpet fiber, polar fleece, insulating fiberfill, tote bags, straps, containers for food and beverages, film and sheet
High-density polyethylene (PE-HD)	2 HDPE	Milk jugs, bottles for shampoo and household cleaners, trash bags, shopping bags, cereal box liners, pipe and conduit, wire and cable covering	Bottles for shampoo and household cleaners, plastic lumber, floor tiles, buckets, crates, film and sheet, recycling bins
Polyvinyl chloride (PVC)	3 PVC	Food trays, shrink-wrap, cling film, bottles (for cleaners, shampoo, cooking oil, etc.), blister packs, hinged carryout food containers (clamshells), pipe, siding, window frames, fencing, decking, railings	Decking, mud flaps, flooring, cables, carpet backing, traffic cones, film and sheet
Low-density polyethylene (PE-LD)	4 LDPE	Squeezable bottles, bags (for bread, dry cleaning, shopping, etc.), tote bags, shrink wrap, toys, wire and cable covering	Trash bins and liners, shipping envelopes, paneling, plastic lumber, floor tile, film and sheet
Polypropylene (PP)	5 PP	Margarine tubs, microwavable meal trays, yogurt containers, bottle caps, medicine bottles, durable consumer and automobile parts, carpeting	Automobile signal lights, battery cables, battery cases, ice scrapers, garden rakes, storage bins, pallets, sheeting
Polystyrene (PS)	6 PS	Foam food trays and egg cartons, disposable tableware (plates, cups, cutlery), hinged carryout food containers (clamshells), compact-disc cases, packing material, toys	Electric switch plates, egg cartons, foam packing material, carryout containers, rulers, plastic decorative molding
Other	7 OTHER	Large water bottles, sunglasses, DVDs, computer cases, nylon products, packaging	Plastic lumber, bottles
Acrylonitrile-butadiene-styrene (ABS)	09 ABS	Monitor/TV cases, coffee makers, cell phones, calculators, most computer plastic, Lego bricks	FDM 3D-printed parts

Summary

- Compounding plastics and elastomers at processing involves mixing and/or blending basic polymer with additives that alter the characteristics of a plastic or elastomer. Additives can include anti-aging agents, antistatic agents, blowing agents, cross-linking agents, curing agents and activators, colorants, fillers, fibers, flame retardants, heat-light stabilizers, UV absorbers, lubricants, odorants, and plasticizers.
- A wide variety of methods for manufacturing plastics, rubbers, and elastomers are used:
  - Processing techniques that make use of a die are called extrusion processes. Used for thermoplastics and thermoplastic elastomers, these include blown-film extrusion; calendering, for producing films or sheets; cast-film and cast-sheet extrusion; and coating layers of plastic onto wires, fabric, or paper.
  - Processing techniques that make use of a mold, used for thermoplastics, thermosets, and thermoplastic elastomers, include injection molding; blow molding; dipping; foaming; rotational molding; and thermo- and vacuum forming.
  - Processes that involve heating and cross-linking (curing) and cooling, used for thermosetting plastics and natural and synthetic rubbers, include casting, compression and transfer molding, foaming, laminating, reaction injection molding, reinforcing, and pultrusion.
- Additive manufacturing (rapid prototyping, 3D printing), is a manufacturing process that has come about since the advent of the personal computer. Types of additive manufacturing processing used with plastics and elastomers include stereolithography and material jetting, powder bed fusion, selective laser sintering, binder jetting, and fused deposition modeling.
- Recycling thermoplastic products is a technique whereby thermoplastic products are remelted and processed into new products. To improve the quality of recycled products, a single type of thermoplastic should be used. The system of Resin Identification Codes, used to distinguish the various types of plastics, was developed to facilitate that end.

Suggested Reading

All3DP, 3D Printing Technology Guide: The Types of 3D Printing Technology (February 7, 2020), https://all3dp.com/1/types-of-3d-printers-3d-printing-technology.

British Plastics Federation, Plastipedia: Plastics Additives, https://www.bpf.co.uk/plastipedia/additives/Default.aspx.

Manas Chanda, *Plastics Technology Handbook*, 5th ed. (Boca Raton, FL: CRC Press, 2018).

Robert A. Charvat, ed., *Coloring of Plastics: Fundamentals*, 2nd ed. (New York: Wiley-Interscience, 2003).

Formlabs, Guide to Manufacturing Processes for Plastics (August 29, 2018), https://formlabs.com/eu/blog/guide-to-manufacturing-processes-for-plastics/.

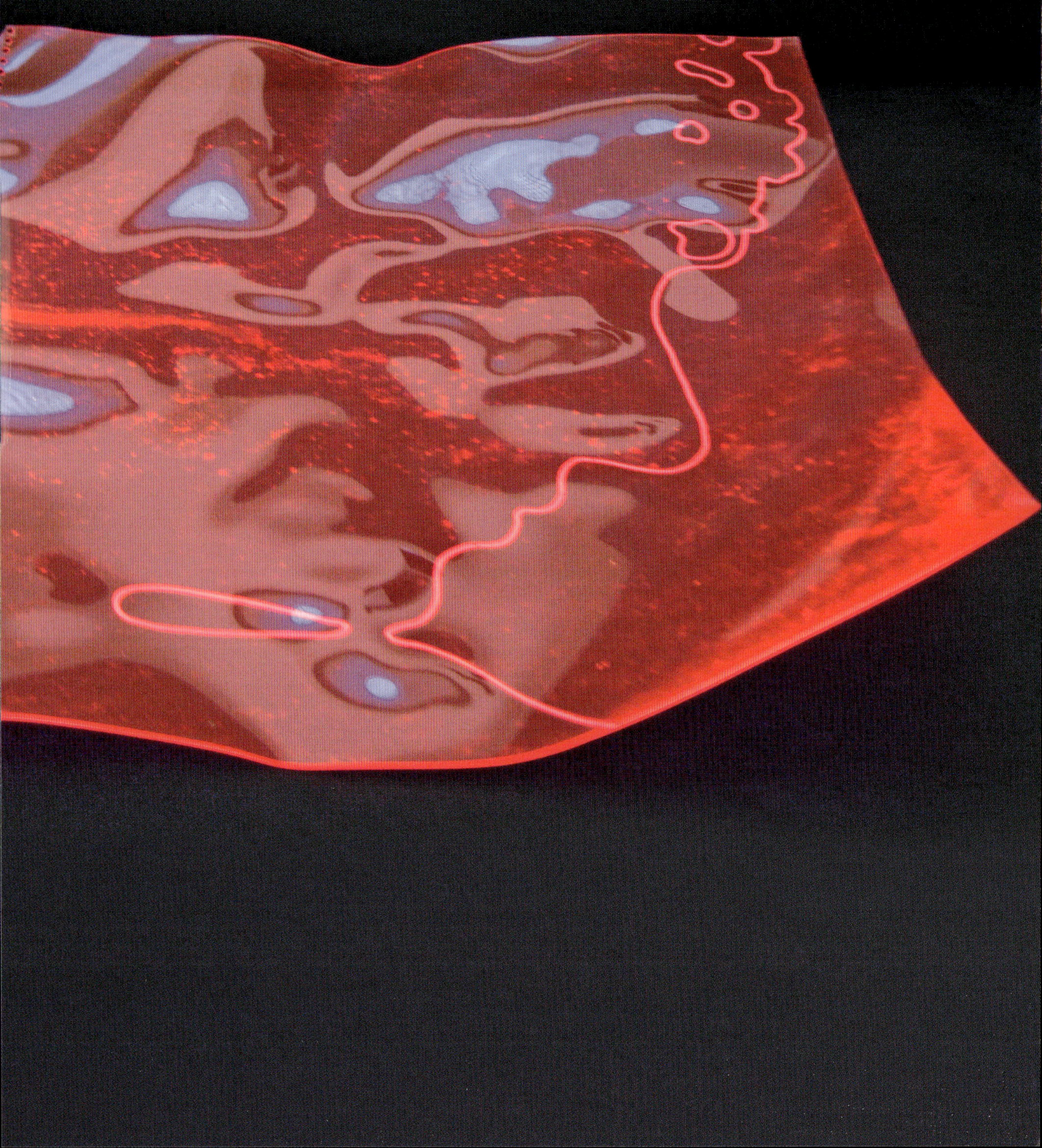

Part II

Conservation Case Studies

Understanding the physical properties of plastics and elastomers, as well as interpreting the corresponding values, is necessary for the conservation of works of art and design made of plastics and elastomers. However, although we can use values described in chapter 4 to quantify many attributes of plastics—stiffness and flexibility, strength and weakness, hardness and softness—how these physical properties affect the behavior of plastics and elastomers over time is not yet well known.

The case studies in this section provide examples of how some plastics and elastomers used in works of art are conserved, and highlights the relevant physical properties in the context of these works of art. Values for the materials are given at the end of each case study.

Chapter 6 | Repairing Polyvinyl Chloride Film

Lydia Beerkens

Henk Peeters (1925–2013) was a member of the Dutch artist collective called the Nul Group (1960–65), which followed the ideas of the international Zero movement. (Pörschmann and Schavemaker 2015). The Nul artists—Peeters, Armando, Jan Schoonhoven, and Jan Henderikse—officially presented themselves in dark suits and ties, rather than as artists had in the past, with easels, paints, and brushes. As a declaration against painting with oils and other traditional artists' materials, they started making serial, minimalist artworks, produced using materials from daily life and available in hardware stores. Their main concept—that their artworks were to be without any meaning—was underpinned by titling their works with only the year of production and a sequential number for the year.

Henk Peeters was fascinated by soft, tactile materials such as cotton pads, feathers, hair extensions, voile, polyurethane foam, synthetic fur, and white, transparent, or silver plastic films. He mounted combinations of these materials onto traditional wood painting stretchers. Peeters replaced painters' canvas with white polyvinyl chloride (PVC) film, stretching it with tacks onto the back of the stretcher bars. He found he could glue soft pads onto the film, stitch a square with synthetic wool through it, or burn a grid of holes through the white plastic surface. These so-called Pyrographies became his signature works over time (Rodrigo and Beerkens 2005).

In this chapter, two different works by Henk Peeters are presented to show that the same material (plasticized PVC) may need different conservation treatments due to the degree of degradation of the plastic and the way it is used in an artwork.

6.1 CASE STUDY A: DESCRIPTION OF THE ARTWORK

The artwork *60-21* (100 × 80 cm), made in 1960, is one of Peeters's early Pyrography works. It represents the technique of "drawing with fire," in which Peeters made brownish-black burn stains, with holes sometimes accidentally occurring. This randomness that led to the actual artwork was key to Peeters's work. For *60-21*, he tacked a semimatte white film material over a regular wood painting stretcher with one horizontal crossbar and made a total of fifty-six burn holes in eight vertical rows, each with seven horizontal positions.

During an extensive artist interview in 2003 (Beerkens 2012), he explained the technique: to draw the burn stains, the stretcher with the white film was positioned facedown across two ropes stretched between the backs of two chairs. Peeters then would position a lit candle underneath it and carefully draw the stains in the planned rows. With a close

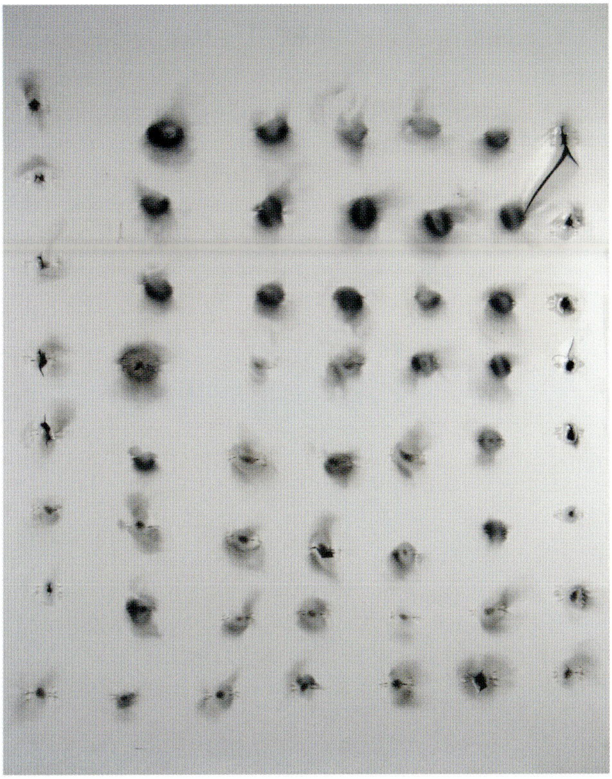

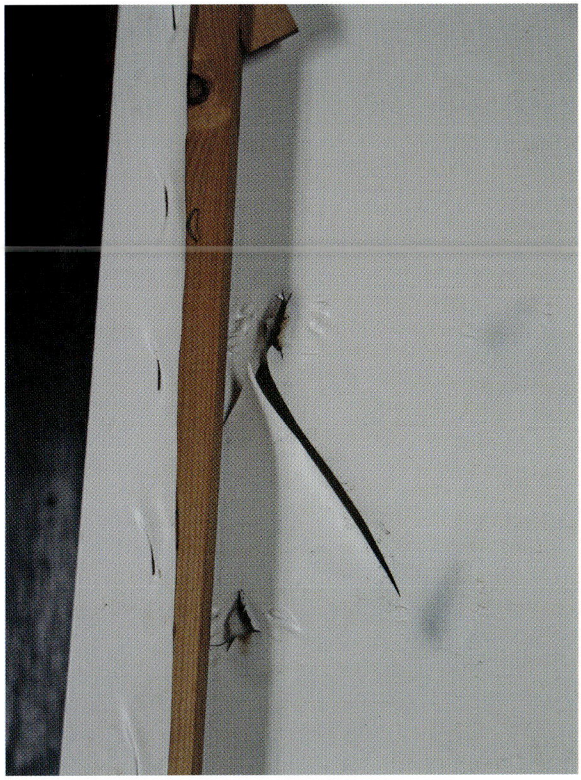

FIGURE 6.1
Henk Peeters (Dutch, 1925–2013), *60-21*, 1960. Semimatte white film material and wood, 100 × 80 cm. Note tear in upper right corner. An old tape repair was also previously made (invisibly) in the upper left quadrant.

FIGURE 6.2
Detail of tear from back.

view of the backside of the film, Peeters could see the plastic film starting to soften and wrinkle from the heat of the flame, and when the brown stain developed, he could stop the process by removing the candle. Some of the stains would be burned away, producing a hole. Peeters said that several attempts were needed to get a good resulting piece, and warned about the risk of inadvertently starting a fire.[1]

Material Condition

Peeters preferred to use PVC film, as it stretched the best due to the plasticizers in the film. Stretching plasticized PVC film over a common painting stretcher gives a good tension in the film as well as a smooth, flat, white surface.

In 2007, Peeters presented *60-21* for conservation. The work had dropped from its nail on the wall and the impact had caused a tear. The tear ran from the stain in the top right corner diagonally to stain number six of the second row (figs. 6.1, 6.2).

Prior to treatment, the condition of the artwork was assessed. Close inspection of the surface showed a visible difference between the untouched PVC film and the smoke stains. The areas around the burn holes varied: in degree of gloss; in texture, from strong wrinkling to the formation of tiny air pockets; and in color, from light brown to dusty charcoal black in the center. All of these characteristics of (intentional) damage depended on the distance to the center of the flame during the making of the Pyrography. Each spot could be technically

1. Filmed interview with Henk Peeters (unpublished), December 12, 2003.

regarded as a chart for the different stages of very fast deterioration under the extreme temperature exposure of the flame. The decay in and around the burn holes is in fact also the aesthetic "designed" aspect in Peeters's fire-drawing. The stages of decay run from unchanged white PVC film to slightly melted (glossy), more melted (warping), discolored (brown), and burnt (black) to nothing left (the holes at the center of some of the burn marks).

The tear, predictably, ran from one weak spot to the next, from one burnt hole to another. As the artwork was almost fifty years old when the damage occurred, much of the plasticizer—which normally provides good stretch properties—had migrated out of the PVC, rendering it stiff, thin, and hence very brittle. New, unaged PVC film, with its full stretch properties, would have absorbed the sudden shock when the work fell.

Conservation Treatment

Peeters, declaring himself too old to restore his own works, turned to a conservator for the treatment of this damage, with the general advice to apply regular cellophane tape onto the back over the torn area. In fact, during the initial condition assessment, an old repair was found on the left side between the second row and the third row of burn stains that had been executed with Sellotape, probably by Peeters.

The making technique was investigated. To enable the conservators to better understand the production process, Peeters provided a larger piece of PVC film from his studio as test material. From this, a 30 × 30 cm mock-up was made,[2] including the burn stains produced using the artist's indications and warnings (Beerkens 2012) (figs. 6.3a, b).

A first attempt to damage the new film of the mock-up and cause a "natural" tear failed. The undeteriorated PVC film was too strong and flexible, thus cuts had to be made with a scalpel knife. A variety of mending pads, glues, and pressure-sensitive tapes were tested: a selected mix from the conservation studio; regular office supplies, especially all kinds of tape; double-sided adhesives; and sticky films from the graphics industry. The mending performance and the matching of flexibility with the PVC foil was tested with Japanese paper, Beva 371 film, nonwoven polyester, sticky acrylic glues such as Lascaux 498 HV, Gudy tapes used in paper conservation, various transparent pressure-sensitive tapes (from Nitto, 3M, and Tesa), and unconventional materials such as Kliniderm and Tegaderm (second-skin-type medical dressings).

Many materials performed surprisingly well; however, most did not seamlessly follow the small changes in the overall tension of the PVC film, varying slightly by day and with differences in temperature. Further, any additional material, tape, or patch will, by doubling the layers, influence the local stretch properties of the PVC film and may become visible on the front by forming a "shade" in the shape of the tape. The best performing tapes on the mock-up were Tesa 4024 and 3M 471, but both were judged too sticky and too flexible to mend the full length of the tear in the artwork without bulging.

In the end, Neschen filmoplast P conservation tape was used. It allowed for more adjustments during the repair of the tear, including some retouching in white on the front side.

Result

Although the tear is not immediately noticeable and the artwork can be displayed again, as a conservation treatment of an artwork, the repair of the torn PVC film did not achieve a perfect or invisible final result.

2. The mock-up was produced by Stephanie Gagné, a conservator of modern art, who participated in this research while interning at the studio for several months in 2007.

FIGURE 6.3
Mock-up made in 2007 with new PVC film from the artist's studio, including burn holes, inflicted tears, and test repair materials, front (a) and back (b).

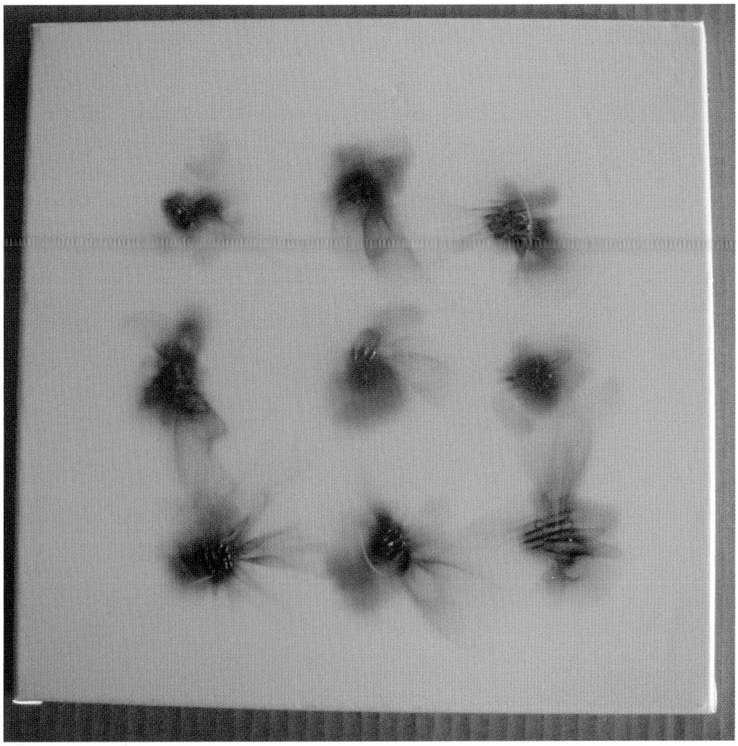

a.

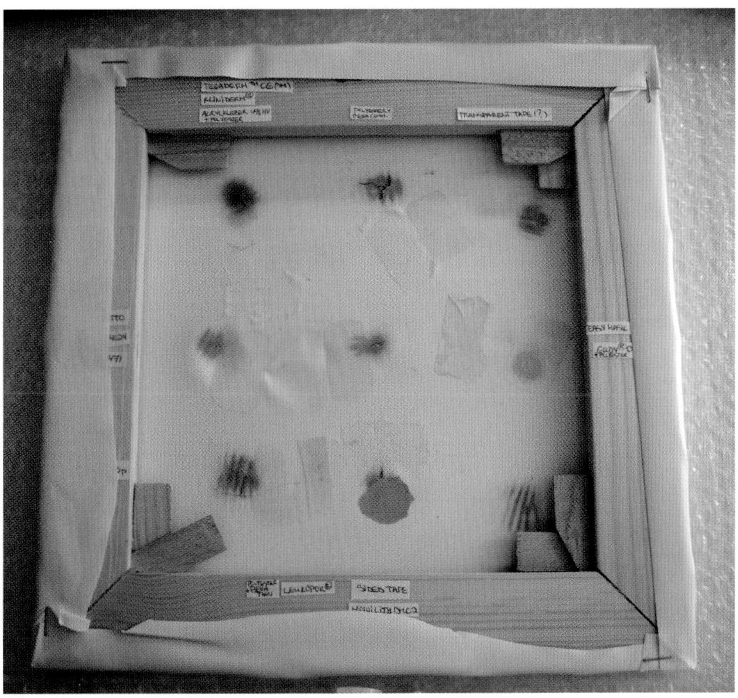

b.

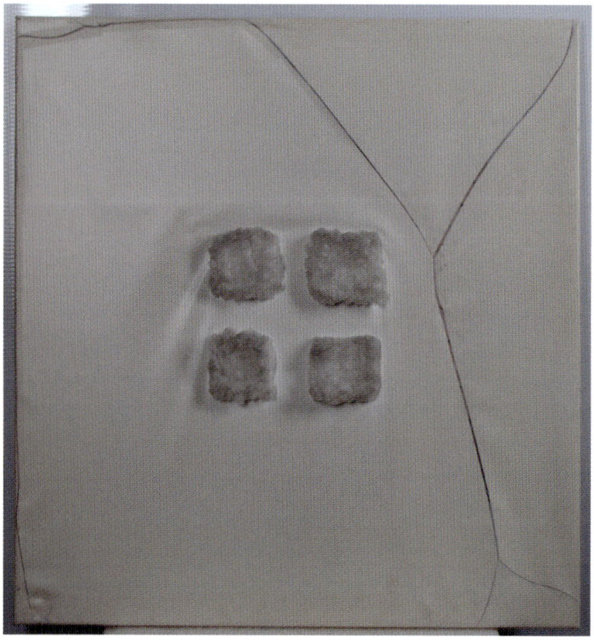

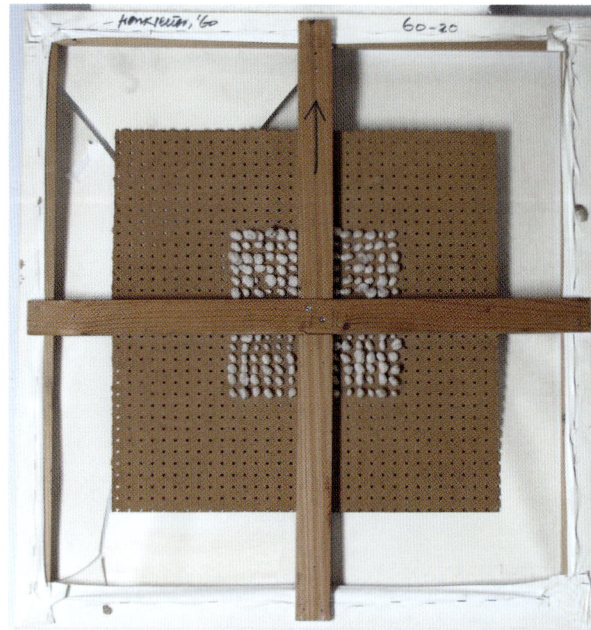

FIGURE 6.4
Henk Peeters (Dutch, 1925–2013), *60-20*, 1960. White PVC film and wood, 79 × 74 cm. Note the long tears in the aged and brittle PVC film.

FIGURE 6.5
View of the back, with the wood crossbars and the perforated cardboard plate into which the tufts of fluffy nylon are locked in four squares.

6.2 CASE STUDY B: DESCRIPTION OF THE ARTWORK

Henk Peeters made *60-20* (79 × 74 cm) in 1960. The piece consists of a large sheet of white PVC film stretched and tacked over a standard wood painting stretcher with double crossbars on the back. In the center, four equal squares of fluffy white, wool-like material are positioned, standing out against the semimatte white plastic surface. The work has an intriguingly soft look that almost invites one to touch and experience its tactile properties. Although detailed material information is not available, Peeters made *60-20* right before the Pyrography *60-21*, discussed in section 6.1 above, and may well have used PVC film from the same stock or roll.

The front of the artwork looks as if the wool-like squares are adhered onto the stretched film (fig. 6.4). The reverse side of the artwork, however, unveils the artist's complex technique (fig. 6.5). Peeters created *60-20* in several steps. Beginning with a wood stretcher with crossbars, he attached a pegboard (hardboard) plate with screws from the back. Next, the film was stretched over the stretcher, similar to the way canvas would be stretched for a painting, attached with tacks on the back. Then the separate tufts were pulled through from the front to the back using a hook or bent wire. When each tuft extended 1 cm through the board, it was unhooked, and the tuft end was adhered flat onto the board plate with what seems to be wallpaper adhesive. Four squares have six rows of 6 tufts each, 36 tufts per square, for a total of 144 tufts in the artwork.

Along with the tufts being pulled through the pegboard on the back, the white PVC film was pulled close to the intersection of the crossbars. This increased the tension in the white film, and the slight concavity of the PVC in the center of the artwork draws the viewer's attention toward the wool-like squares on the plane.

Material Condition

Over six decades, the plasticized PVC film had completely lost its flexibility. The PVC film was identified as plasticized PVC, and the fluffy white, wool-like material as nylon (polyamide).[3]

Over the years, the nylon had darkened and grayed through aging; it also had collected dust, which was stuck to its fine fibers. The consequential discoloration had changed the original intent: the absence of color in the artwork's originally all-white materials was the typical Zero aspect the artist was pursuing. Although the full transparent Perspex showcase that was added later functioned to physically protect the artwork, preventing mechanical damage while locking it off from the air, it could not prevent the two plastic materials from aging.

Aside from normal aging, the artwork did not change substantially until 2019. Then its handling in preparation for a loan caused the PVC film to suddenly break in long, sharp tears. As the reverse plate of the Perspex box is screwed directly into the wood crossbars of the artwork, loosening the screws to dismantle the showcase (90 × 86 × 8 cm) may have caused a small shift in the stretcher, the keys in the corners, and the crossbars—in effect, giving a different direction to the permanent tension in the film. (The film had also been "damaged" originally, during the making in 1960, when Peeters punched pinholes for the fluffy nylon fibers.)

Signs of decay included brittleness, resulting from sixty years of daylight and stress, especially over the corners of the stretcher and near the center, where the four squares with the original needle holes are located. Additionally, dismantling the Perspex box released an unpleasant chemical smell, which indicated that plasticizers had been migrating out of the PVC over the years. This loss of plasticizer made the PVC more rigid and brittle.

Repairing the tears with pressure-sensitive tapes from the back, as was done in the conservation treatment of the Pyrography *60-21* described above, was not an option. Merely touching the PVC film near the large tears caused it to break into many smaller pieces, and the PVC had lost any of the ability to sustain strain provided by its original property of stretch.

Conservation Treatment

Once damaged, the artwork could no longer be displayed, and there was no option for repair of the PVC film, which exhibited many tears and was disintegrating completely. Because the PVC film composed two-thirds of the visible artwork, the piece was technically considered a total loss.

From his practice, we know that Henk Peeters often restored his own works by replacing plastic materials. He also had a full practice in the 1990s in remaking his works from the 1960s and 1970s. In interviews, Peeters explained that any person can make a Zero artwork with these simple materials—that is the essence of Zero. His signature, however, would make it his work (Beerkens 2007, 2012).

For a conservator, the decision to remake or reconstruct a major section of an artwork is a very complex one. In this case, it required the full agreement of all the stakeholders: the owner of the work; the representative of Peeters's ideas, his widow, Truus Nienhuis; and the approval of the heirs of the oeuvre of Henk Peeters.[4] The collaboration with the family

3. The PVC film, as well as the wool-like material, was analyzed by Suzan de Groot, Cultural Heritage Agency of the Netherlands (RCE) Amsterdam, using Fourier-transform infrared (FTIR) spectroscopy and a Mitutoyo 102-217 micrometer.

4. Conversation with Truus Nienhuis (1929–2019) during a visit to the artist's studio, ca. February–April 2019. At that time she also gave permission for a reconstructive conservation in the presence of two of Henk Peeters's sons.

also enabled the vital use of relatively new white PVC film from the stock material in the artist's studio, making a partial reconstruction possible.

Partial Reconstruction

A first test was done on a 1:1 mock-up of the artwork: trying to stretch a film over the damaged PVC, with four windows to allow the fluffy nylon squares to stay visible. This did not yield a satisfactory effect, as the new film with the four windows proved difficult to stretch in such a way that the exact position of the squares was matched. Also, the PVC film did not "lean back" behind the squares as the old film had.

The remaining option was to dismantle each of the 144 tufts, stretch the new PVC film from the stock roll in the artist's studio over the original, broken PVC, tack it onto the back, and then reposition each nylon tuft the same way the artist had, by pulling them through newly made holes in the new PVC, one by one (figs. 6.6a, b).

With this conservation treatment, the look and feel of the original stretched PVC film has been reproduced in the very same way with similar effect. The single issue for the conservator was the very fine dust that was electrostatically stuck to the original fluffy nylon material, which resulted in discoloration that made the squares stand out against the bright white of the relatively new PVC film (fig. 6.7).

Result

The final result is remarkably good. Optically, the artwork functions in the same way as the original, and it represents the artist's intent. Repositioned in the protective Perspex case, the artwork is conserved to the satisfaction of both the owner, a friend of Henk Peeters, and the artist's family. There is a slight discrepancy between the replacement PVC film—newer and very white—and the still-original fluffy nylon squares. Nylon is known for darkening and yellowing over the years, but on the other hand the aged look of the nylon squares gives the artwork an authentic look that reflects its age. An inspection of the reverse side will still reveal both the original (torn, broken, yellowed) PVC film, the way the new layer of PVC film is positioned over it, and how the nylon is pulled through both old and new PVC layers.

The decisions on the intervention were based on previous statements by Henk Peeters about conservation options for his works, and his own practice of remakes, which to him were in accordance with the central idea of the Nul movement: applying materials in artworks with no intrinsic meaning (Beerkens 2007). The final option for a possible treatment was discussed with the family, representing Peeters's artistic heritage. Without their collaboration, this unconventional treatment would not have been an option.[5]

5. Our thanks to Maria van Elk, artist and old friend of Henk Peeters; Nikki van Basten, conservator of modern art, for collaboration during the treatment; and to the family of Henk Peeters, in particular Truus Nienhuis.

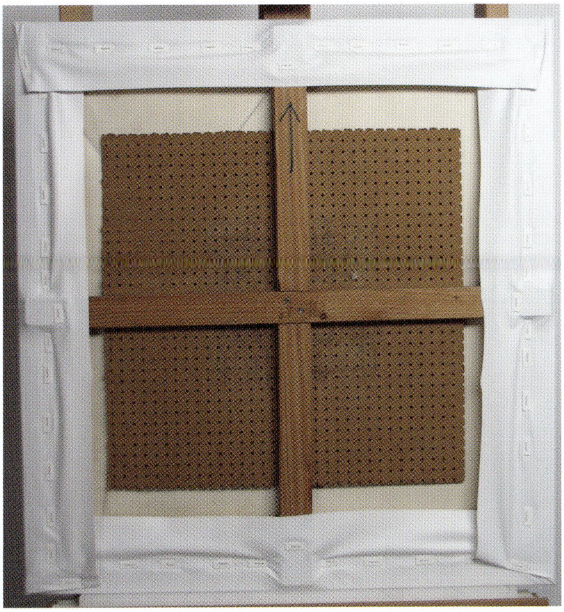

a.

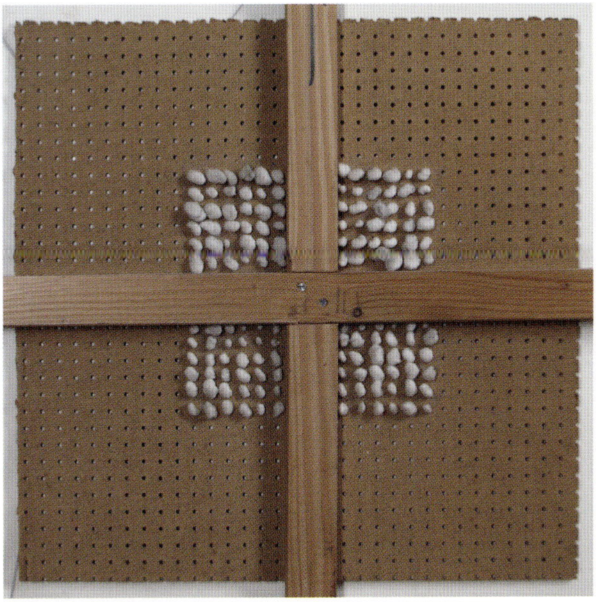

b.

FIGURE 6.6
The back with new PVC over the artwork, before placement of nylon tufts (a). Detail of back with nylon tufts back in place (b).

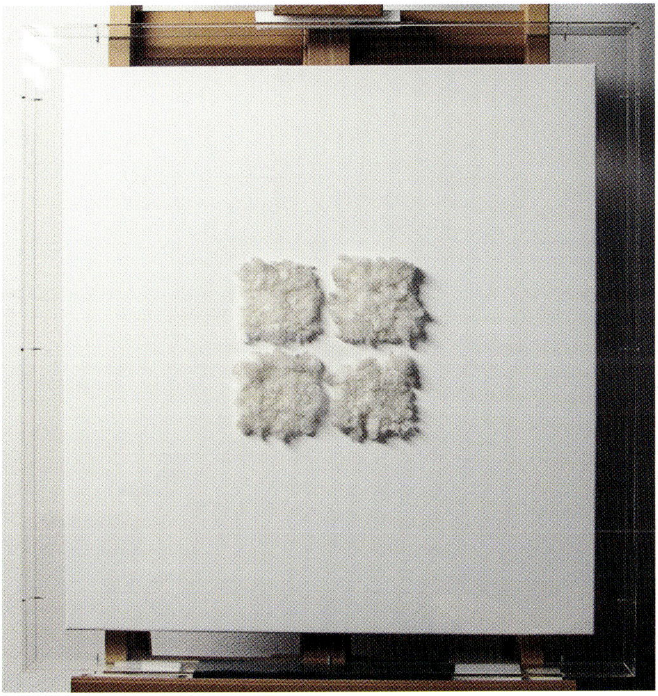

FIGURE 6.7
60-20 after reconstructive conservation treatment with new PVC film, remounted into the Perspex showcase.

Material: Plasticized Polyvinyl Chloride Film (PVC-P)

Density

Density (specific weight) can range from 1.37 to 1.43 g/cm^3, which means that PVC-P film will not float on water. Using these density values, the weight of a piece of PVC film can be calculated.

Elongation

Elongation values are 3.3%–430%, meaning that, depending on the amount of plasticizer it contains, a PVC-P film can be stretched between 3.3% and 430%. The loss of plasticizers in aged PVC-P film decreases the stretch capability, causing the once-flexible PVC-P to become brittle. This is obvious in case study B, where the PVC-P film became brittle after sixty years, and its thickness, when measured with a micrometer, decreased from 372 to 122 microns (μm). Thin, brittle, aged PVC-P film has a much lower impact strength—so low that just the touch of a finger will break the film.

Tensile strength

Tensile strength is important for a material that will be stretched or under tension. The tensile strength of a PVC film can vary from 7.1 to 68.9 MPa. Compared to other thermoplastics, PVC-P film is not a strong material (see fig. 4.7).

Tear strength

Tear strength is not available because this property is mostly available for elastomers and not for thermoplastics such as PVC. However, as PVC is sometimes highly plasticized, elongation can be measured (see fig. 4.11).

Impact strength

Impact strength is the force at which an impact force (e.g., a blow with a hammer) makes a hole or dent in a plastic. For PVC-P, values are around 0.33 J/cm. For comparison, polycarbonate (PC) has values between 6 and 7 J/cm (see fig. 4.14).

Data

Source: fact sheet 13.23. Data show minimum and maximum values due to differences in processing and the amount of plasticizer added by the manufacturer.

Material used in treatment (case study A)

Neschen filmoplast P conservation tape (self-adhesive, transparent paper-mending tape).
https://www.neschen.de/product/filmoplast-p-90/#pdetails

Chapter 7 | Consolidating Polypropylene Fibers

Thea B. van Oosten and Anna Laganà

The Dutch artist Wil Fruytier (1915–2007) is considered one of the pioneers of modern art wall tapestries. Most of her textile work was done between 1950 and 1980, at a time when she and other textile artists were pushing the art world to consider their work as expressions of conceptual art, in appearance and technique and, maybe most importantly, in that the concept of the artwork is central.

Fruytier came to international attention in 1961, when her works were shown at the Venice Biennale. She has since become known for her tapestries made of both natural materials and polypropylene rope. Because of the monumental size of her works, they are usually exhibited in large spaces such as entrances to universities, hospitals, and commercial banks, as well as in museums (Megens 1991).

7.1 DESCRIPTION OF THE ARTWORK

The wall tapestry *Fête I* (355 × 250 cm), made in 1969 by Wil Fruytier, had been in the art collection of ING Bank in the Netherlands (fig. 7.1) and was exhibited in the hall of a bank for decades. In 2005, it was donated to the Netherlands Institute for Cultural Heritage to investigate the possibility of conserving it.

Fête I was constructed using black, white, yellow, orange, and blue ropes made of polypropylene (PP). Each rope consists of PP slit-film fibers that are twisted together. The technique used by the artist to make this object is tapestry weave: a specific, handwoven textile construction based on plain weave, which is the most basic of the three fundamental textile weaves. Warp and weft are aligned in a crisscross pattern. The artist used white rope for the warp and black, yellow, orange, blue, and white PP ropes for the weft.

Materials and Technique

Polypropylene is a semicrystalline thermoplastic polymer obtained by addition polymerization of the monomer propylene. PP can vary in molecular chain structures (stereoregularity; see chap. 1). The isotactic form of PP, in which the methyl side groups are positioned on the same side of the polymer chain, is currently the most produced (van Oosten et al. 2008). This type of polypropylene, developed in the early 1960s using Ziegler-Natta catalysts, proved to be very stable and was a successful product (Brydson 1999). Over the years, the use of even more efficient catalysts led to increased production of polypropylene, and it has become one of the most consumed plastics in the world (Elias 2003; Lemaire, Arnaud, and Lacoste 1988).

FIGURE 7.1
Wil Fruytier (Dutch, 1915–2007), *Fête I*, 1969. Polypropylene ropes in different colors (PP slit-film fibers twisted together), 355 × 250 cm. Amersfoort, Netherlands Institute for Cultural Heritage.

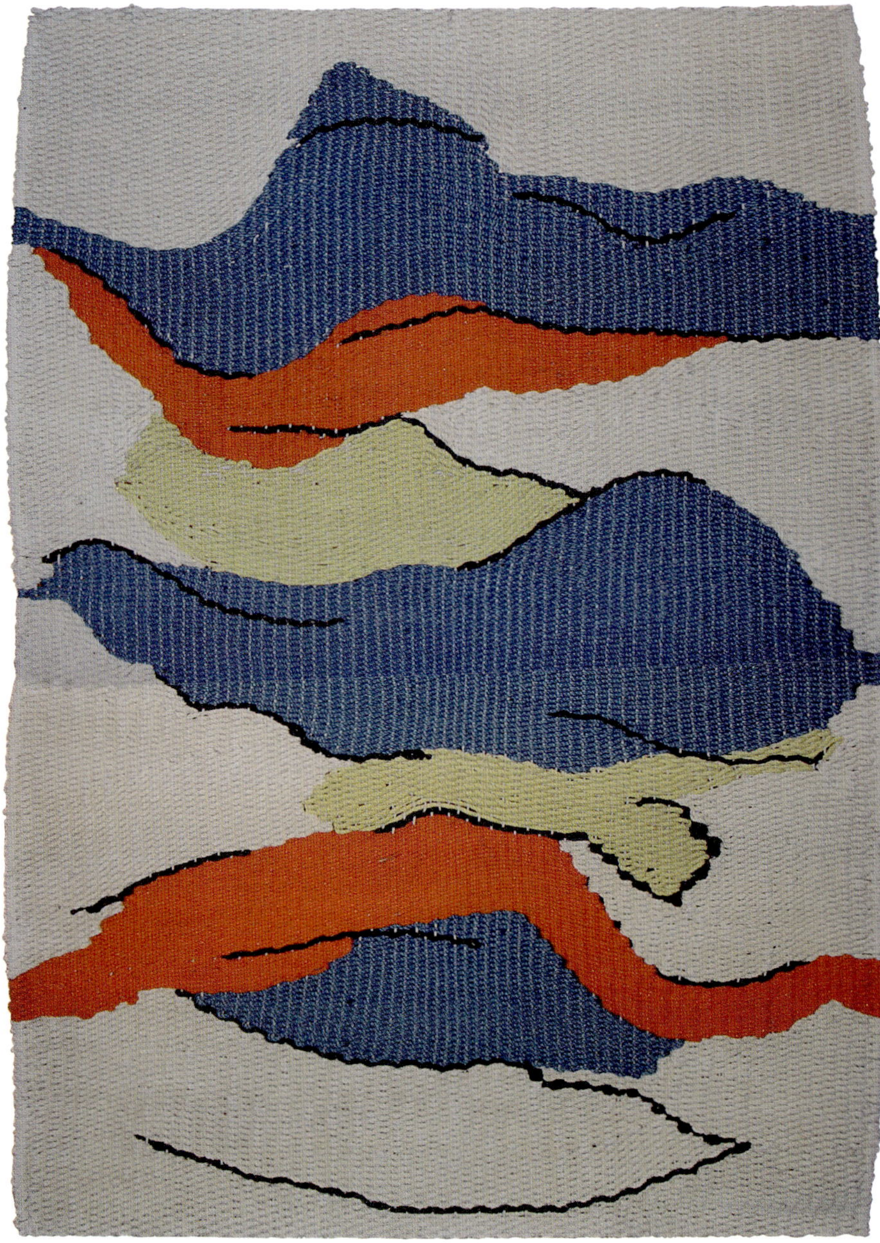

Polypropylene is used to produce a wide range of products using a variety of manufacturing processes, from rigid outdoor furniture and car parts to flexible fibers used to make diapers, rope, woven and nonwoven fabrics, and reinforcement materials for other plastics. Products can also differ in their properties: some are clear, others completely opaque; some are heat resistant and used for food containers, while others melt easily and are used for heat-sealable food packaging.

Most of the world's synthetic carpets are made from polypropylene fibers, produced by melt spinning of the molten polymer through tiny round holes in a spinning device called a spinneret; the material is then stretched. The slit-film method, in which the polypropylene is extruded through a small slit, producing flat strips, was used from the beginning of PP production. The greatest use for the slit-film tape yarns was in carpet backing and industrial sacking. Polypropylene can now be found in monofilament (split or slit film in one form or another), multifilament, and staple forms (Cook 1993).

Polypropylene fibers have many desirable properties: they are the lightest of all synthetic fibers (0.91 g/cm^3), have the lowest moisture absorption of all fibers, have low surface energy, have the same tensile strength wet or dry, are not penetrable to dirt, can be easily cleaned, do not rot, and are not prone to microorganisms or molds. In addition, PP fibers are highly resistant to both acids and alkalis and most organic solvents.

Material Degradation

The usefulness of polypropylene depends on how long the material can retain its properties during a prolonged service life. For instance, under mild conditions, nonstabilized PP will retain its properties for long periods of time; however, when exposed to environmental circumstances that affect its stability, polypropylene can deteriorate (Lemaire, Arnaud, and Lacoste 1988).

The major causes of degradation of polypropylene are photooxidation (caused by light, especially UV radiation), oxygen, and heat, and some pigments accelerate their degrading effects. Degradation leads to loss of strength and flexibility, and ultimately to powdering (van Oosten, Joosten, and Megens 2006; van Oosten et al. 2008). Therefore polypropylene artworks displayed in public buildings under uncontrolled climate conditions degrade relatively quickly, and this is a challenge to conservators.

To retard oxidation and prolong service life, antioxidants and light stabilizers are normally added to polypropylene. These can be incorporated by any of the conventional thermoplastic compounding techniques (Al-Malaika 1999; see chap. 5). Davis and Sims (1983, 121–23) have reported that some pigments are UV absorbers, and a concentration of 0.5% of various organic and inorganic pigments along with channel black and furnace black can restrict degradation of PP to surface oxidation, which slows the overall rate of deterioration. Degradation can also be inhibited by using photostabilizers either separately or in synergistic combinations at conservation treatments (van Oosten et al. 2008).

Material Condition

The degradation of *Fête I* was detected in 2005, at which point it had been exposed to light and UV radiation for more than thirty years, resulting in degraded and very brittle PP fibers. A detailed examination and condition report were completed two years later (van Oosten and Haider 2007). Another examination was performed in 2011, which showed that the most degraded fibers of the tapestry were the white weft fibers (fig. 7.2a).

The white rope in both the warp and weft contains titanium dioxide (TiO_2), a pigment that accelerates degradation (Davis and Sims 1983, 122). The white warp fibers, however,

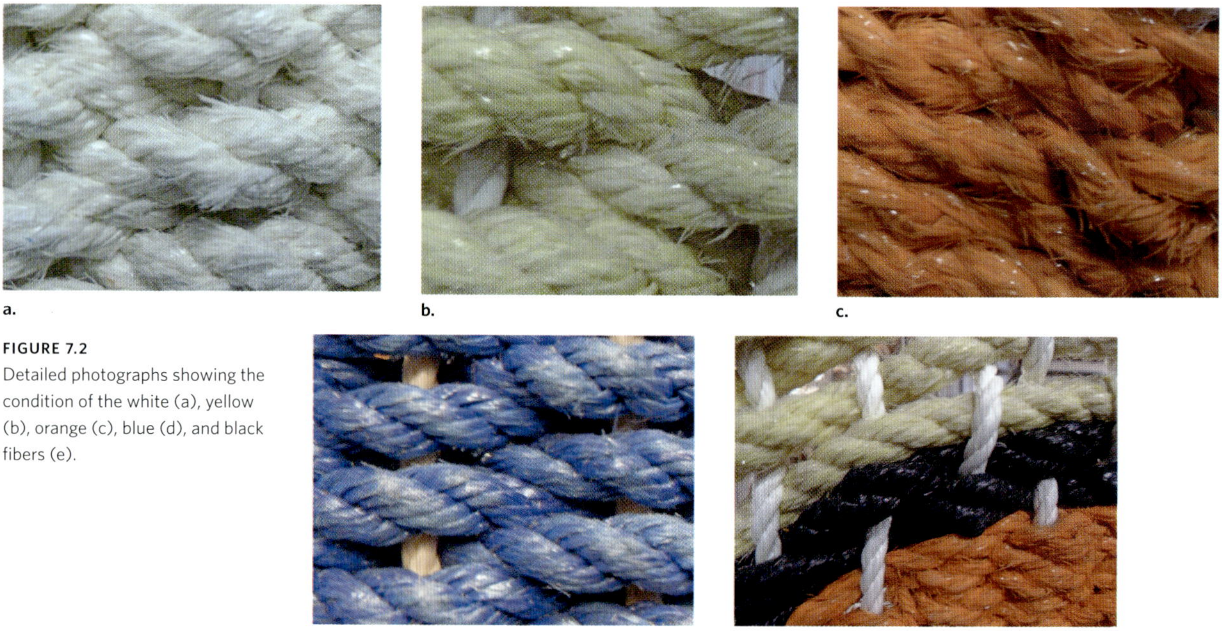

FIGURE 7.2
Detailed photographs showing the condition of the white (a), yellow (b), orange (c), blue (d), and black fibers (e).

were in better condition, as they were protected from UV radiation and light by the weft. The ropes on the reverse side of the tapestry were likewise in better condition, since they were also not exposed to light and UV radiation.

The yellow, orange, and blue weft fibers (figs. 7.2b–d) showed some degree of degradation, while the black fibers were in good condition (fig. 7.2e); as mentioned, this is explained by the presence of black pigment (carbon black in this case), which has an inhibitive effect on the degradation of plastics (Davis and Sims 1983, 123, 201–2).

7.2 TREATMENT STUDY

Because *Fête I* was supposed to remain exposed under nonmuseum conditions, conservation treatment to preserve the work from light and UV radiation was considered necessary. A conservation treatment able to consolidate and protect the polypropylene from further degradation was developed. In a previous study, the use of light-stabilizing systems proved to inhibit the degradation of polypropylene (Allen, Chirinis-Padron, and Henman 1985). Therefore this treatment study focused on identifying a protective treatment consisting of a consolidant and light-stabilizer systems and evaluating its applicability to the PP ropes of the artwork.[1]

Determining Surface Tension

Polypropylene, as a result of its hydrophobic property, is very inert and has generally quite a low surface energy (30 mNm^{-1}) (see chap. 4), which makes coating this material difficult. To improve adherence and printability, surface treatment with corona, atmospheric plasma, or UV radiation is commonly used in the plastics industry to print ink on polypropylene or

1. This study is largely based on previously published research (van Oosten et al. 2008).

ethylene bags (van Oosten et al. 2008). These treatments create temporarily active polar groups (e.g., hydroxyl, carbonyl, and carboxyl groups) on the surface of polypropylene, enhancing the surface tension, which improves wettability and provides adherence of the light-stabilizing coatings.

To determine the surface tension of the PP ropes, the need for surface treatment, and the applicability of a consolidating and protective layer, measurements were carried out. To measure the surface tension of the polypropylene, an ink test was performed on various PP test samples using a series of blue test inks (from Tigres) composed of water, glycerol, formamide, alcohols, and ethylene-glycol. These inks have surface tensions ranging from 30 to 58 mNm^{-1} (see fig. 4.32)

This test determines the surface tension of PP film by comparing the behavior at the edge of the ink strokes. If the ink stroke applied to a PP surface contracts, forming droplets, this measurement must be repeated with the ink of the next lower value. The surface tension of the material is equal to the value of the ink at which the corresponding stroke shows good wetting for at least two seconds without demonstrating any changes.

The results of the ink test showed that seriously degraded PP fabric (fig. 7.3) has a surface tension of 52 mNm^{-1} (on both sides of the test sample), that dyed PP rope (fig. 7.4) has a surface tension of 46 mNm^{-1}, and that opaque PP sheet (fig. 7.5) has a surface tension of 30 mNm^{-1}.

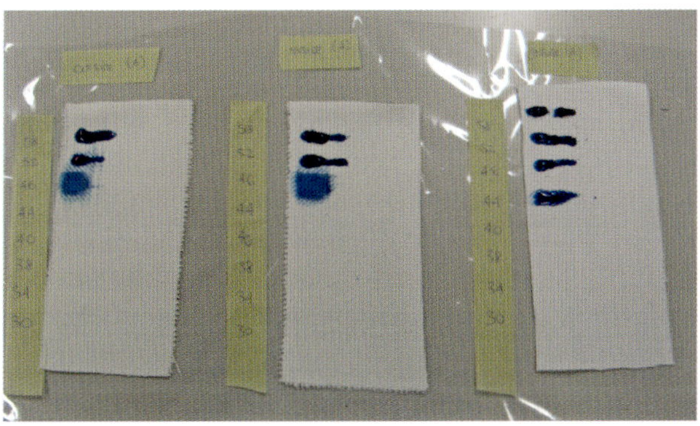

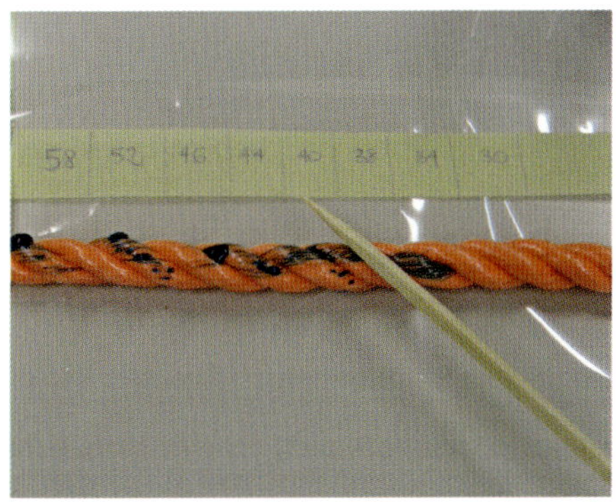

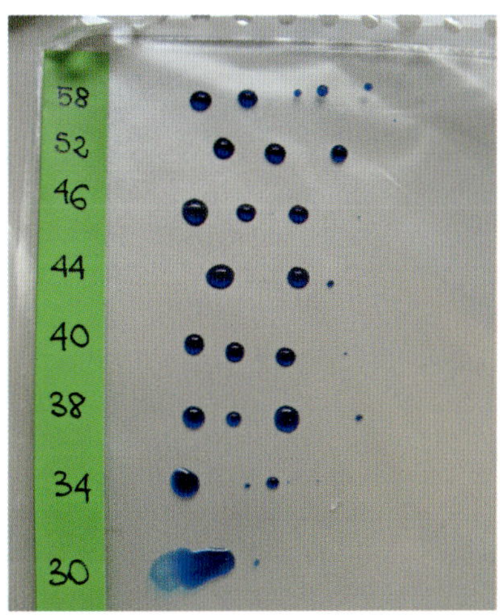

FIGURE 7.3
Test inks on degraded PP fabric showing surface tension of 52 mNm^{-1}.

FIGURE 7.4
Test inks on orange polypropylene rope showing surface tension of 46 mNm^{-1}.

FIGURE 7.5
Test inks on opaque polypropylene sheet showing surface tension of 30 mNm^{-1}.

Treatment Development

Various acrylic consolidants were considered, and Plextol D 498 was determined to be ideal, since it had a suitable surface tension of 44 mNm^{-1}, and this closely matched the surface tension of the dyed PP ropes of the tapestry, which was shown to be 46 mNm^{-1}. For an adhesive to be capable to wet out a surface, it must have a surface energy/tension as low as or lower than the material to be bonded. This meant the consolidant would adhere without requiring any surface treatment for the PP ropes, and would also be suitable for consolidating the degraded PP (surface tension = 52 mNm^{-1}). To determine the best way to inhibit the degradation of polypropylene, various UV stabilizers were added to the acrylic consolidant and tested on both PP test samples and PP mock-ups. The following light stabilizers (containing UV absorbers, heat stabilizers, and light absorbers) developed by BASF were tested:

- Tinuvin B 75, a liquid light-stabilizer system consisting of a synergistic blend of Irganox 1135, Tinuvin 571, and Tinuvin 765, soluble in organic solvents
- Tinuvin 770, a low-molecular-weight hindered-amine light stabilizer, soluble in organic solvents and methanol
- Chimassorb 2020, a hindered-amine stabilizer soluble in organic solvents, ethanol/water (95/5), and isopropanol

Treated test samples were artificially light aged using an Atlas Xenotest Alpha chamber. Aging performance was studied by the increase of oxidation products (carbonyl, hydroxyl groups) using Fourier-transform infrared (FTIR) spectroscopy. The efficacy of the consolidants in adhering to the surface of the polypropylene and inhibiting its photooxidation was evaluated by performing crosscut tests (ASTM D3359) and burst tests (van Oosten, Joosten, and Megens 2006).

A solution composed of Plextol D 498 consolidant and Tinuvin B 75 light stabilizer (4.7% Plextol D 498, 2.4% Tinuvin B 75, 20.1% water, and 72.8% ethanol) was found to give the best result: "No brittleness was observed on the PP test samples which were consolidated/treated with two layers of this solution; moreover, once applied this system appeared to remain transparent, not interfering with the color of the PP rope" (van Oosten et al. 2008, 101–2).

In addition, various application methods including airbrush, nebulizer, and brush were tested to apply the consolidating agents on polypropylene ropes. To easily evaluate which method would give a more homogeneous application on the test samples, the solution was slighty colored by adding 0.5 ml of methylene blue solution. The treated PP rope test samples were examined visually, with magnification and with UV light. A flat, soft brush proved to be the most suitable method for homogeneously consolidating PP ropes. Very good results were also obtained with the airbrush, but the consolidation showed to be less homogeneous.

Treatment Tests

Finally, treatment tests were carried out on a small representative area of the tapestry to finalize the treatment methodology and the steps to be used on the whole tapestry.

As a result of fiber disintegration due to degradation, the tapestry was covered with loosely held dust, fixed particles in some areas, and white powder from the white PP rope, which spread onto the colored ropes. To avoid consolidating these deposits onto the PP surface—especially the white, disintegrated fibers on the colored ones—it was necessary to first clean the surfaces of the ropes. These loose deposits were easily removed with soft brushes.

Cleaning tests were then performed to remove fixed particles. Distilled water, applied with either a microfiber cloth or a polyvinyl alcohol Blitz-Fix sponge, efficiently removed them. However, while Blitz-Fix sponges and microfiber cloths were evaluated to be suitable for cleaning the ropes that were in good condition, they were not suitable for the more brittle and degraded ones. For those ropes, any mechanical action to remove dirt would increase the risk of breaking and further damaging the PP fibers. Therefore it was decided to clean the most fragile ropes by applying Japanese paper (a thin, absorbent paper) wetted with distilled water, and carefully removing it after a couple of minutes. This method allowed the absorption and removal of dirt without harming the brittle fibers.

The last step was to evaluate the consolidation treatment. Tests showed that the consolidant and light-stabilizer solution could be easily and homogeneously applied on those ropes in good condition using a flat, soft brush. However, as with the cleaning method, for the most degraded ropes it was decided the treatment method would need to be adapted. To consolidate those ropes, the brush was not directly used on the brittle fibers; instead, the consolidant and light-stabilizer solution was gently brushed (twice) onto a sheet of Japanese paper placed on the area to be treated. In this way, the solution was successfully and safely applied to the damaged ropes through the paper.

Conclusions

The results of this study showed the importance of determining the properties of the materials to develop a treatment. The ink tests showed that the surface tension of PP can vary by both type of material and condition, allowing in some cases a successful adhesion of products with comparable or lower surface tension without the need to adopt surface treatments. In fact, the consolidant and light-stabilizer solution composed of Plextol D 498 consolidant and Tinuvin B 75, with a surface tension comparable to that of the dyed PP ropes, proved to successfully consolidate them on both test samples and the degraded tapestry. In addition, in line with previous studies, the use of light stabilizers in the solution proved to be crucial in protecting polypropylene from photooxidation. The study also identified methods for cleaning and consolidating the degraded ropes safely and homogeneously.

The treatment procedure and method developed were taught to students in the Conservation and Restoration of Modern and Contemporary Art Program at the University of Amsterdam, who carried out the treatment of this tapestry in 2018.

Material: Polypropylene (PP)

Hardness

The hardness of polypropylene can vary between 85 and 100 (Rockwell R), which is comparable to high-density polyethylene (PE-HD) and polymethyl methacrylate (PMMA). It is harder than low-density polyethylene (PE-LD), which is easily scratched.

Solubility

The Hildebrand parameter for PP is 16 $(MPa)^{1/2}$, which is low, meaning that most organic solvents used in conservation, which have higher Hildebrand parameters, do not affect PP.

Surface tension

The structure of PP is nonpolar due to methyl side groups in the macromolecular chains of PP, and in general the surface tension value of PP is 20.4 mNm^{-1}.

In cases where a material has low surface tension, increasing the surface tension so that it is equal to or greater than the surface tension of the consolidant gives good results. Generally, adhering bonds on the order of about 50 mNm^{-1} is necessary.

To consolidate the PP ropes of the artwork *Fête I* and to protect the work against photooxidation by daylight and UV radiation using the Plextol-Tinuvin consolidant solution (surface tension = 44 mNm^{-1}), the surface tension of the ropes needed to be around this same value. This was the case (surface tension = 46 mNm^{-1}) due to the fact that the ropes were dyed and aged.

Data

Source: fact sheet 13.16. Data show minimum and maximum values due to differences in processing (manufacturer).

Materials used in treatment

Plextol D 498, Talas.
    https://www.talasonline.com/images/PDF/ProductInformation/PLEXTOL_B500_INFO.pdf
Series A blue test inks, Tigres.
    https://www.tigres-plasma.de/en/test-ink/test-ink-overview
Tinuvin B 75, BASF.

Chapter 8 | Flattening Cellulose Acetate Sheets

Anna Laganà, Ana Cudell, Rita Veiga, and Thea B. van Oosten

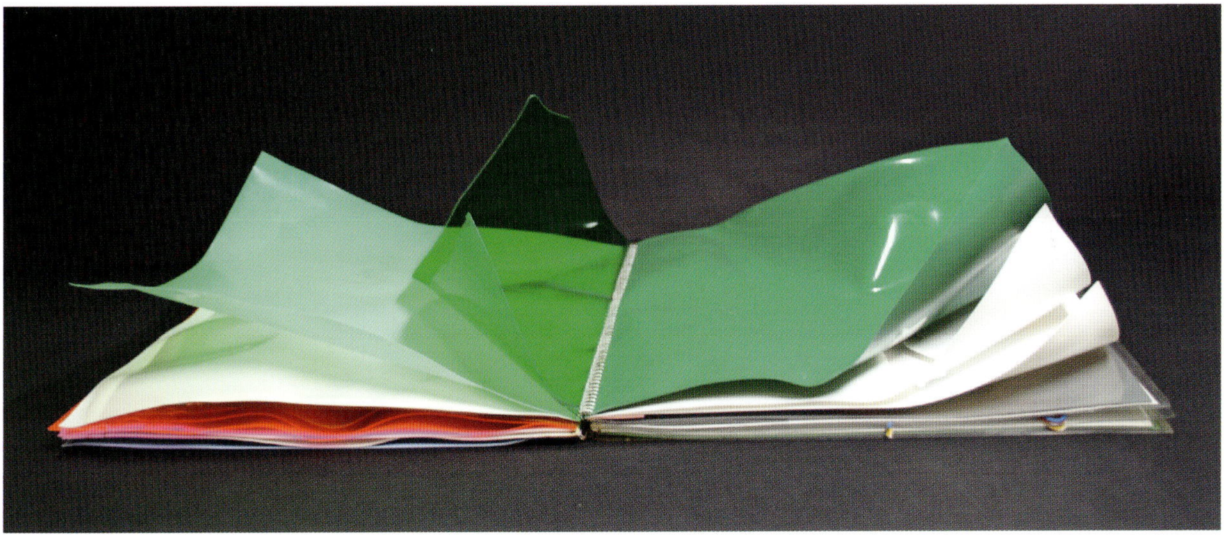

FIGURE 8.1
Lourdes Castro (Portuguese, 1930–2022), *Ombres Transparentes*, 1967. Cellulose acetate sheets and metal, 30 × 28.5 × 2 cm. Porto, Serralves Museum of Contemporary Art. The artist's book is shown before treatment, after having been on display for several months opened to the green sheets.

Lourdes Castro (1930–2022) was a Portuguese artist born in Madeira. She studied painting at the Escola de Belas Artes in Lisbon. After her studies, she moved first to Munich and then to Paris in 1958, where she lived for twenty-five years.

During her career, Castro intensively explored the themes of shadow, light, and shapes using a variety of materials. From 1964 onward, she began using plastics such as polymethyl methacrylate and cellulose acetate, especially for their transparency and lack of texture.[1]

8.1 DESCRIPTION OF THE ARTWORK

Ombres Transparentes is an artist book (30 × 28.5 × 2 cm) made by Castro in 1967 when she was living in Paris (fig. 8.1). The object belongs to a series of thirty similar artist books and is in the collection of the Serralves Museum of Contemporary Art, Porto, Portugal, where it is currently stored (Cudell et al. 2011).[2]

1. Lourdes Castro, interview with author, October 22, 2010.
2. Special thanks to Inês Venade and Marta Moreira de Almeida of the Serralves Museum for providing all facilities for the research on *Ombres Transparentes*.

FIGURE 8.2
Two cellulose acetate sheets of *Ombres Transparentes* showing warping (a, b).

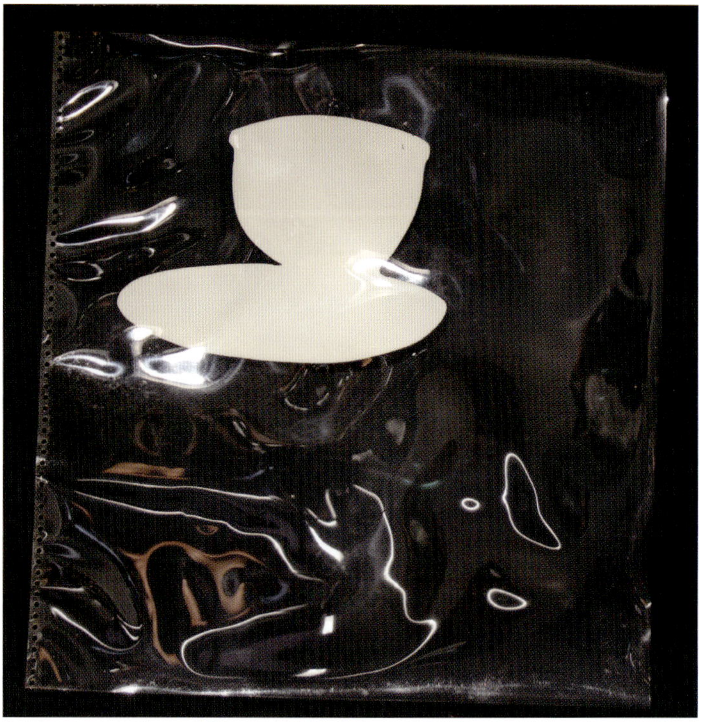

a.

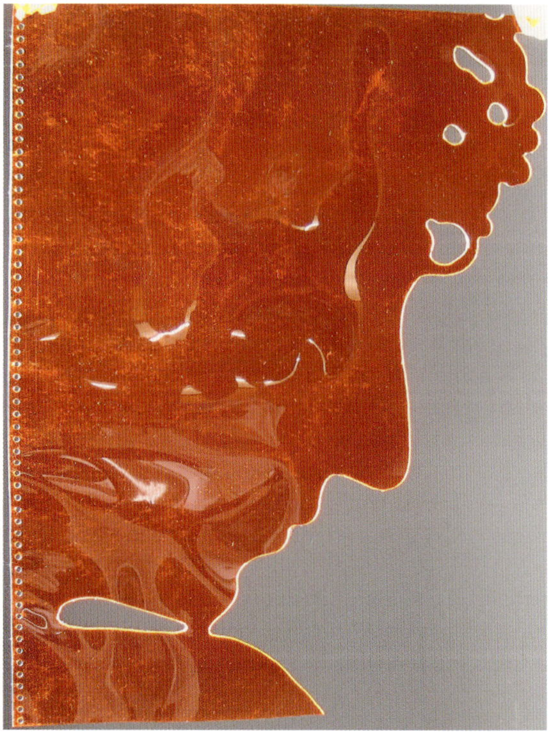

b.

Materials and Technique

The book consists of twenty-five pages made of cellulose acetate (CA) sheets—including the front and back covers—bound together with a metal spiral binding coil. The sheets have variable thicknesses (0.2, 0.3, and 0.5 mm) and their appearance also varies. The book includes both transparent and opaque sheets with matte and glossy finishes. Some are colored, some painted using the silk-screen technique, while others are cut into various shapes using a pyrography method. In an interview, Castro recalled using Rhodoïd to make the book—a French trade name of cellulose acetate that was sold in Paris in transparent and opaque sheets of different thicknesses.[3]

Material Condition

When *Ombres Transparentes* was examined in 2010, it presented several conservation issues typical of cellulose acetate objects. Some of the sheets showed warping (slight, medium, and severe) due to degradation of the CA sheets and to the migration of plasticizers (figs. 8.2a, b). The sheets that were severely warped smelled of the acetic acid produced by the cellulose acetate's deacetylation (Quye and Williamson, 1999; Shashoua 2008). White, solid, sticky plasticizer was found on some sheets' surfaces, and different types of soiling and deposits—dust, fixed particles, stains, and fingerprints—were observed on most of the sheets. Crazing, scratches, cracks, and broken parts were also observed, mostly around the holes for the spiral binding.

Considering the poor condition of the artist book and the lack of literature on established conservation treatments for cellulose acetate objects, in 2011 the Serralves Museum, together with the Catholic University of Portugal, also in Porto, established a research project to explore treatment options for cellulose acetate using *Ombres Transparentes* as a study object.[4] The project was carried out in collaboration with the Cultural Heritage Agency of the Netherlands (RCE) in Amsterdam.

This case study presents the part of the research project dedicated to the investigation into flattening methods for warped cellulose acetate sheets and the application of its findings on the sheets of *Ombres Transparentes*.

Characterization of Materials

All materials used to make *Ombres Transparentes* were analyzed to understand their chemical composition and physical properties and to formulate a plan for properly approaching their conservation and preservation. Fourier-transform infrared (FTIR) spectroscopy and pyrolysis-gas chromatography-mass spectrometry (Py-GC/MS) analyses were carried out at RCE and confirmed that all sheets were made of plasticized cellulose acetate. Transparent sheets showed a mixture of diethyl phthalate, triphenyl phosphate, and dibutyl phthalate, while the colored transparent and colored opaque ones contained a mixture of diethyl phthalate, tris(2-chloroethyl) phosphate, and triphenyl phosphate. The paint applied to some of the sheets by the artist was identified as alkyd paint (Cudell et al. 2011).

To define the glass transition temperature (T_g) of the CA sheets, one sheet was analyzed at RCE using differential scanning calorimetry (DSC). No definite T_g was observed. According to literature, cellulose acetate has a T_g of 95°C–108°C depending on plasticizer content and manufacturing method (see the "Material" box at the end of this chapter and fact sheet 13.1).

3. Lourdes Castro, interview with author, October 22, 2010.

4. An overview of the results of the research project and more details on the cleaning tests and the cleaning procedure developed can be found in Cudell et al. (2011) and Laganà (2012, 258–64).

8.2 FLATTENING THE CELLULOSE ACETATE SHEETS

Once the chemical composition and physical properties of the materials used in *Ombres Transparentes* were determined, the possibility of using heat and pressure to reshape and flatten the warped cellulose acetate sheets was investigated.

Heating and reshaping thermoplastic sheet and film material has been used in manufacturing processes since the beginning of the plastics industry. Cellulose acetate is a thermoplastic material, and when heated it softens and melts before charring or burning, enabling it to be shaped in a mold or under a press. During this process, the material is basically heated at a certain temperature until it becomes soft and pliable; then it is shaped into the desired form, using mechanical force, vacuum pressure, or air pressure. Finally, it is cooled down to allow it to solidify in its new shape. The temperature and pressure used in industry for manufacturing cellulose acetate sheets and film vary depending on the material's composition and the specific process used (Yarsley et al. 1964).

Flattening Tests

The lack of information in the conservation literature on treatments for warped plastics in museum collections led the authors to conduct a methodological study to identify viable flattening options for recovering the original appearance of the warped CA sheets. The main goal was to establish a temperature range and pressure at which the flattening could safely take place without altering the surface structure and applied paint of the CA sheets, and without pushing out plasticizers.

Factors that were considered to avoid any risk of damage during flattening were (1) the composition of the CA sheets (bulk material, plasticizer, filler, etc.), (2) their surface structure, and (3) their current condition and their physical properties (particularly glass transition temperature, softening point, and compressive strength).

Tests were carried out at RCE in Amsterdam and at the Catholic University in Porto. The materials used for the flattening tests included spare CA sheets from the original book series, which were donated by the artist specifically for this research. These spare sheets had the same characteristics present in *Ombres Transparentes*: they had thicknesses of 0.2, 0.3, and 0.5 mm; they were transparent and opaque in various colors (as well as colorless); and few were cut and painted with same paint and patterns. The spare sheets used for these tests showed slight and medium warping.

Flattening tests were also performed on other transparent 0.12 mm CA sheets (from the RCE plastics collection) that showed severe warping, had migrated solid plasticizers on the surface, and emitted a strong smell of acetic acid. These sheets were specifically included in the tests to evaluate the impact of each type of flattening treatment on severely degraded CA.

Tests were conducted using a variety of tools and equipment selected from among the ones most commonly used in conservation for flattening other materials, as well as used in the printing industry. The tools and equipment selected included the following:

- A heated spatula with a thermostatic temperature control system. The pressure was applied manually during the heat distribution.
- A lining iron with a thermostatic temperature control system (from a household shop). The pressure was applied manually during heat distribution.
- A small thermal laminator (from a household shop), which had no temperature setting and no pressure control, only an on-off button. In this laminator, the sheets pass between two heated rollers.

- A large thermal laminator (Omma jet laminator) with temperature setting but no pressure control. In this laminator, the sheets also pass between two heated rollers.
- A gilding press (Karl Krause Leipzig) with temperature setting and no pressure control.
- A low-pressure heating table (NSD 1101 from CTS) with temperature setting, electronic temperature control, and low-pressure system.
- A manual heat press (TS One from Siser) with temperature setting, electronic temperature control, and manual pressure: 300 g/cm^2 max (0.003 MPa).[5]

A variety of temperatures were tested to establish a "safe" temperature range that would flatten the warped material without causing any changes to the original surface structure.

Liquid crystal thermometers (from Hallcrest) were used to detect the exact temperature used in each test and equipment. In addition, a handheld noncontact infrared thermometer (MiniTemp, MT IR series, from Raytek) was used to measure the exact temperature on the sheets at flattening using the low-pressure table and the Siser TS One press.

Flattening tests were carried out in two ways: using the heat source directly in contact with the plastic, and protecting the sheets by inserting interleaf materials between the sheets and the heat source (the interleaves were removed after flattening). The interleaf materials tested were Melinex polyester film, Melinex RSX951 silicone-coated polyester film, and silicone release paper. To allow the vacuum to function in tests using the low-pressure heating table, a nonwoven polyester fabric was placed underneath the CA sheets with a Melinex polyester film on top.

Additional tests were performed to identify the proper interleaf material for use between the CA sheets and the rubber surface of the TS One press plates, because both the Melinex polyester and silicone-coated films often left marks. Experiments were executed by placing the CA sheets between two metal plates; however, this caused the CA sheets to become shinier after flattening. Therefore, in the last experiments using the TS One heat press, the CA sheets were placed between two 8 mm glass plates to avoid direct contact with the rubber of the press plates and thus avoid leaving marks.

Test Results

The temperature range that was shown to be effective and safe for flattening slightly and moderately warped CA sheets was between 65°C and 75°C. Above this temperature range, the surface structure of CA sheets started to change, especially at 85°C–90°C. These changes were particularly visible in sheets showing parallel lines from the manufacturing process on their surface; if the temperature was too high, the material became too soft, and these original manufacturing lines started to disappear during the flattening. In fact, according to the technical literature, a temperature of 95°C, in combination with a pressure of 13 MPa, was specifically used by the industry to smooth the surface structure of cellulose acetate sheets to remove manufacturing marks (Yarsley et al. 1964).

During testing, the higher temperatures were also seen to have an impact on the plasticizer. In some tests, the flattening carried out at 85°C caused migration of liquid plasticizer to the surface of a CA sheet.

5. Thanks go to Alfredo Vale of the printing company DESIGNAR, in Porto, Portugal, for helping with the flattening experiments on spare sheets and providing the TS One press.

The heated spatula and the lining iron had the advantage in that the temperature could be safely set and controlled. Both of these methods enabled the flattening of the cellulose acetate sheets, but the results were unsatisfactory due to the tools' size. Because the surface of these tools was smaller than that of the CA sheets, heat and pressure could not be applied equally across the surface of a sheet.

Conversely, the low-pressure heating table provided a quick and even distribution of heat across the CA sheet, and a uniform vacuum over the entire working surface, as expected (sheets were flattened with a low pressure of 45 millibars and a temperature of 67°C). However, the results were not satisfactory because the perforated surface of the table tended to leave marks on the CA surface.

Both heat laminators were shown not to be safe options. The heat was not distributed evenly, and the temperature of the rollers tended to increase rapidly, reaching temperatures high enough to cause plasticizers to migrate to the surface in some of the tests. Liquid crystal thermometer strips showed that in the small laminator, the temperature during flattening was 85°C; in the large laminator, the temperature rose to 104°C, even though it was set at 55°C.

Although the gilding press gave generally good results (CA sheets could be successfully flattened at 65°C–75°C, depending on the thickness), it was finally considered not safe because the instrument was not equipped with a temperature control system. During some tests, the temperature rose too quickly and the pressure could not be controlled.

The TS One manual heat press proved to be the best and safest flattening method for CA sheets. Warped CA sheets could be flattened without altering their surface structure and without causing the plasticizers to migrate. The press provided easy setting and control of the temperature within the range identified as safe (65°C–75°C). The different thicknesses of the sheets making up *Ombres Transparentes* required slightly different flattening times. For both unpainted and painted sheets, the following times were found to be effective and safe: 5 seconds for 0.2 mm sheets, 7 seconds for 0.3 mm sheets, and 11 seconds for 0.5 mm sheets.

The flattening procedure developed for CA sheets included the following steps:

1. Each CA sheet was placed between two 8 mm glass plates inside the TS One press, together with a liquid crystal flexible strip thermometer, to measure the exact temperature reached on the cellulose acetate (fig. 8.3).
2. The temperature of the press was set to 70°C in order to bring the cellulose acetate sheet to a temperature of 65°C.
3. The temperature was checked on two spots: on the upper glass plate, using a Raytek portable infrared thermometer, and inside the glass plates, close to a warped CA sheet, with the flexible thermometer strips.
4. The press was closed manually and the exact application time was counted with a timer: 5 seconds for 0.2 mm sheets, 7 seconds for 0.3 mm sheets, and 11 seconds for 0.5 mm thick sheets (fig. 8.4).
5. The press was then opened manually, and the flattened CA sheet was taken out and placed between two other glass plates to cool down to room temperature (fig. 8.5).

Flattening Treatment

Based on the experimental results on the spare sheets, the Siser TS One press showed to be a safe, effective, and easy method of flattening CA sheets, and it was chosen for the flattening treatment on the study object, *Ombres Transparentes*.

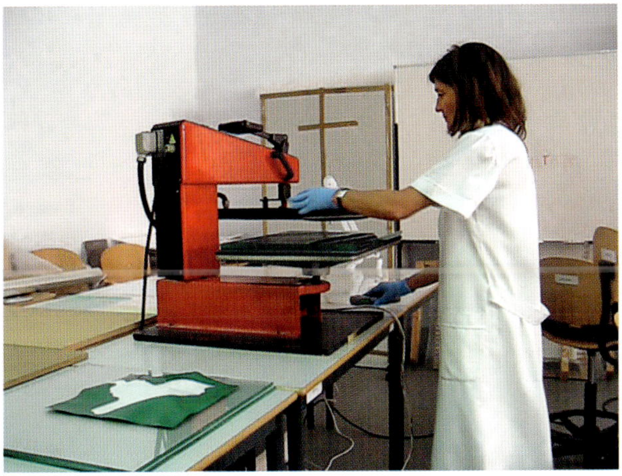

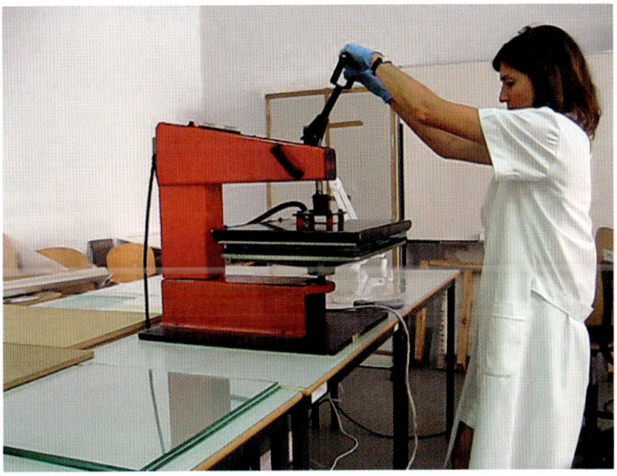

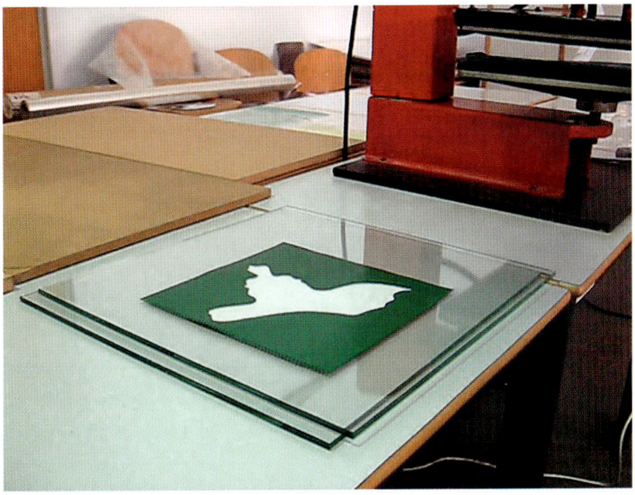

FIGURE 8.3
Opening the press to insert a warped sheet of *Ombres Transparentes* for treatment.

FIGURE 8.4
Heating and pressing the sheet.

FIGURE 8.5
The sheet placed between two glass plates to cool down after flattening.

Before flattening, the book was disassembled by removing the extremely oxidized and distorted metal spiral. Then dust and loose particles were removed from the sheets using an eyeglass (spectacle) cloth. Aqueous cleaning was preferred to solvent cleaning, as it proved to be an efficient and safe way to remove fixed particles, fingerprints, and the white, sticky, solid deposits of migrated plasticizers (Laganà 2012).

Flattening was then successfully carried out following the procedure described above (see figs. 8.3–8.5). Only the sheets showing slight and moderate warping were treated. The few severely warped sheets were considered untreatable due to their poor condition.

For some CA sheets, the procedure was repeated because it was determined to be more effective and safe to repeat the operation for a longer time than to increase the temperature. After treatment, the CA sheets were stacked with Melinex RSX951 silicone-coated polyester film as interleaf material. The sheets were placed in the storage area of the Serralves Museum at a temperature of 19°C–21°C and a relative humidity of 45%–55%.

Conclusions

All flattening tests carried out applying heat and pressure using various tools gave a good insight into the behavior of warped cellulose acetate sheets.

The experiments carried out on this study object provided the opportunity to explore several methodologies and to finally show that CA sheets with slight and moderate warping can be successfully and safely flattened (figs. 8.6a, b; 8.7a, b).

The flattening method developed using the manual heat press proved to be easy, controllable, safe, and efficient to use. In this case study, this method was tested and used on cellulose acetate, but it could also be a promising option to explore for the treatment of other warped thermoplastic sheets.

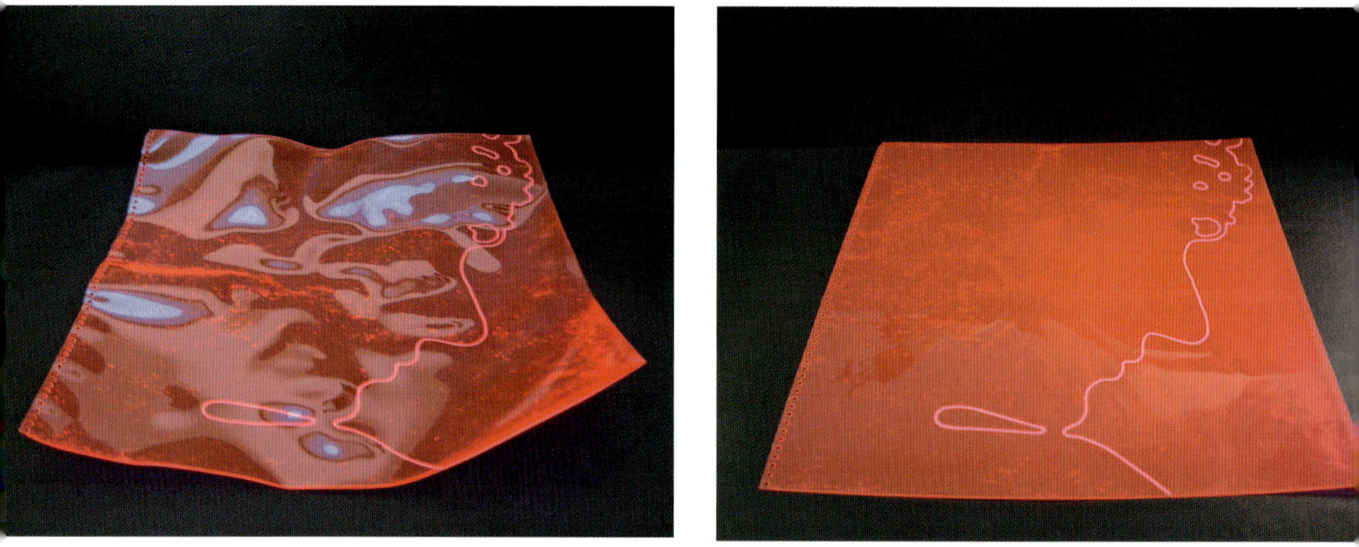

a. b.

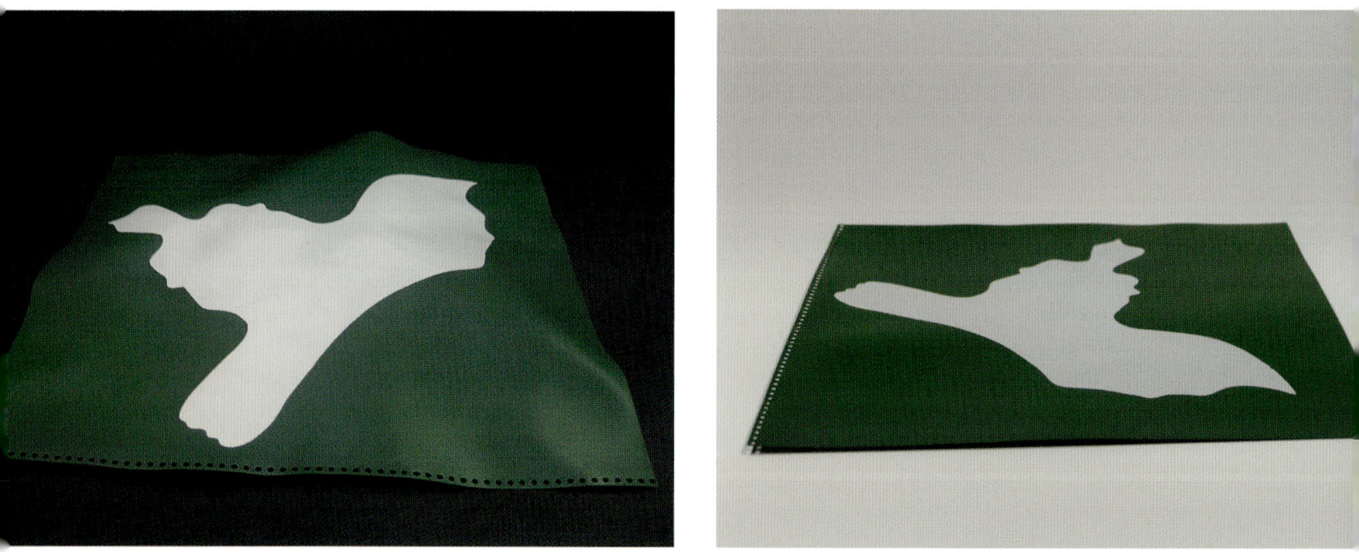

a. b.

FIGURE 8.6
Transparent cellulose acetate sheet of *Ombres Transparentes* before flattening (a) and after (b).

FIGURE 8.7
Opaque cellulose acetate sheet of *Ombres Transparentes* before flattening (a) and after (b).

Material: Cellulose Acetate (CA)

Hardness

The hardness of CA can vary between 35 and 125 (Rockwell R), which is comparable to the hardness of polymethyl methacrylate.

Glass transition temperature, softening point, melting point

Depending on the grade of CA (more or less acetylated):

- T_g can vary between 95°C and 108°C.
- Melting point can vary between 138°C and 180°C.
- Softening point can vary between 78°C and 113°C.

Compressive strength

Compressive strength for cellulose acetate can vary between 15 and 248 MPa, which means that CA can withstand pressure comparable to the thermoplastics cellulose nitrate and polyamide-6 and most of the thermosetting materials (see fig. 4.16).

Alkyd paints

Alkyd paints are oil-modified thermoplastic polyesters made by heating polyhydric alcohols with polybasic acids (or their anhydrides); they are used as protective coatings for synthetic paints due to good weathering properties.

The T_g of an alkyd paint depends on its composition (long-, short-, or medium-chain oils from unsaturated vegetable oils and saturated oils) on the pigment the paint contained, but it is approximately 65°C (Duce et al. 2014). When aged, alkyd paints show very good thermal stability due to cross-linking, which increases glass transition temperatures (Izzo et al. 2011; Mutar and Hassan 2017).

Data

Source: fact sheet 13.1, 1967 and 2016 data. Data show minimum and maximum values due to differences in processing (manufacturer).

Equipment used in treatment

Siser TS One, manual heat press:

- Maximum temperature: 220°C
- Maximum pressure: 300 g/cm^2
- Plate dimensions: 38 × 50 cm

Raytek MT series portable noncontact infrared thermometer.
Liquid crystal thermometers from Hallcrest.
https://www.hallcrest.com/color-change-basics/liquid-crystal-thermometers

Chapter 9 | Adhering and Loss Compensation of Hard Polyurethane Elastomer

Anna Laganà

Liet Heringa (b. 1966) and Maarten van Kalsbeek (b. 1962) have been working as an artist duo under the name Heringa/van Kalsbeek since 1998. Their sculptures are the results of experiment and coincidence. The works evolve out of dynamic, spontaneous, unpredictable, and organic processes of action and reaction: one continues where the other one stops (van Lieverloo 2009). This spontaneous approach draws inspiration from the unpredictability of nature and its uncontrollable processes, such as growth, flowering, decay, and coagulation, which are translated by the artist duo into accumulation, assemblage, and montage. The resulting sculptures, including the one that is the subject of this case study, are hybrids of fine and applied art. Typically, these works are built up around a core and consist of a diversity of media such as porcelain figurines, feathers, coral, artificial flowers, and wood and metal objects that are embedded in what can best be described as a three-dimensional abstract painting (Pratt 2009).

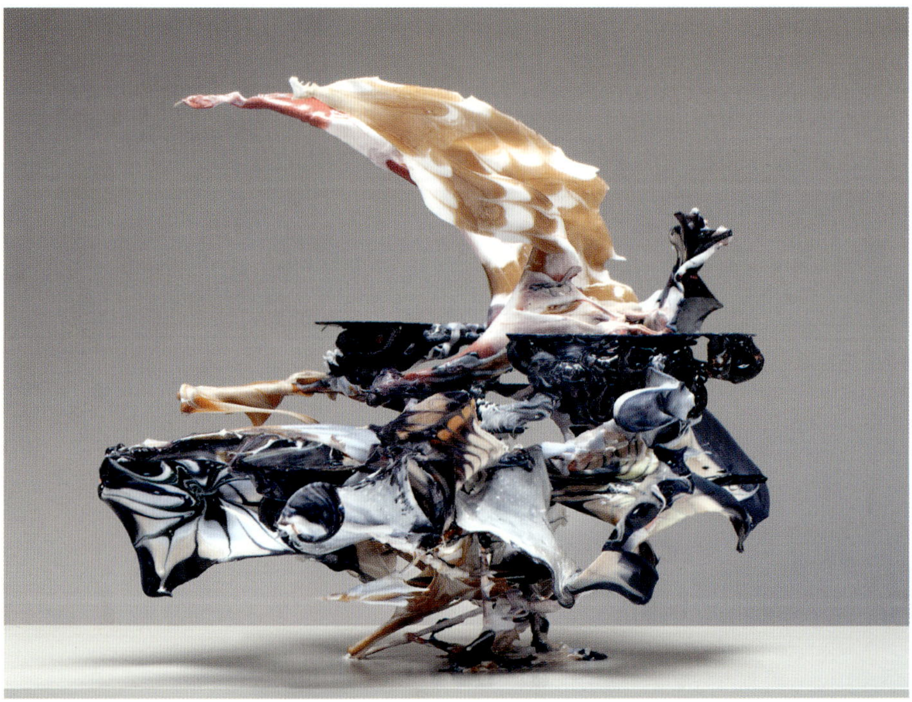

FIGURE 9.1
Heringa/van Kalsbeek (Dutch, b. 1966/b. 1962), *Untitled*, 2007. Porcelain, hard polyurethane elastomer, molded steel, wood, 60 × 77 × 55 cm. Private collection.

9.1 DESCRIPTION OF THE ARTWORK

Untitled (60 × 77 × 55 cm) is a sculpture created by the Dutch artist duo Heringa/van Kalsbeek in 2007 and owned by a private collector (fig. 9.1).

Material and Technique

Untitled is made of polyurethane (PUR), porcelain, molded steel, and wood. The object was created by mixing pigments with a clear hard-casting polyurethane elastomer (Crystal Clear 200, from Smooth-On) to produce a variety of colored mixtures—opaque and translucent—that were then used by the artists to create a dynamic overgrowth of color and form.

These colored mixtures of polyurethane resin were brushed and/or dripped onto a main structure composed of a molded steel support with porcelain and wood attachments, as well as separately cast in large, thin sections. Casting was performed by applying the colored mixtures in alternating stripes onto a tabletop using a spatula; a wooden skewer was then dragged across the stripes to mix the colors, creating a pattern. The table was covered with a release agent (carnauba wax) so that each cast section could be easily removed with a spatula. Sections were removed while still flexible (before completely cured) so that they could be applied to the main structure and then stretched, bent, and shaped into the desired form. Once the curing was complete, the colored polyurethane resin became hard and permanently adhered to the support.

Material Condition

In 2013, the artist duo was contacted by the owner of *Untitled* to repair the sculpture, which had been significantly damaged by a heavy impact force. Heringa/van Kalsbeek carefully examined the sculpture's condition and came to the conclusion that a conservator was needed. Consequently, they reexamined the sculpture together with a conservator (the author) to discuss and determine whether restoring the piece was possible, considering both its poor condition and the current lack of knowledge on materials and methods to repair hard polyurethane elastomer.

The condition assessment showed that the impact force resulted in several mechanical damages on one of the large, curved, translucent sections of colored polyurethane resin. This section was broken at its lower part (fig. 9.2), as well as in the middle along the edge and at its tip. The tip broke into three pieces, and unfortunately two of them were lost after the impact occurred (fig. 9.3).

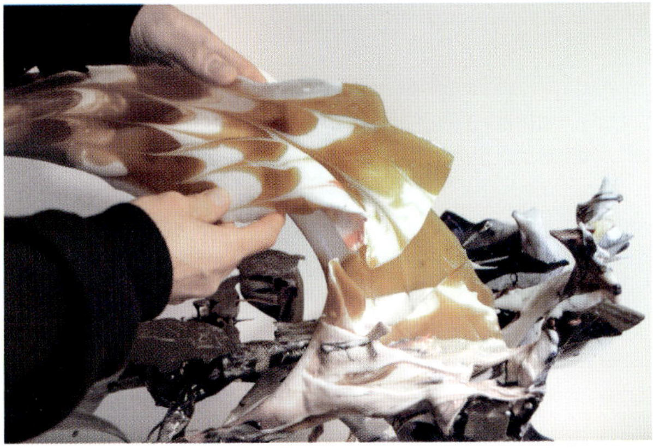

FIGURE 9.2
The artist duo documenting the damage to the large, thin, translucent cast-and-shaped section.

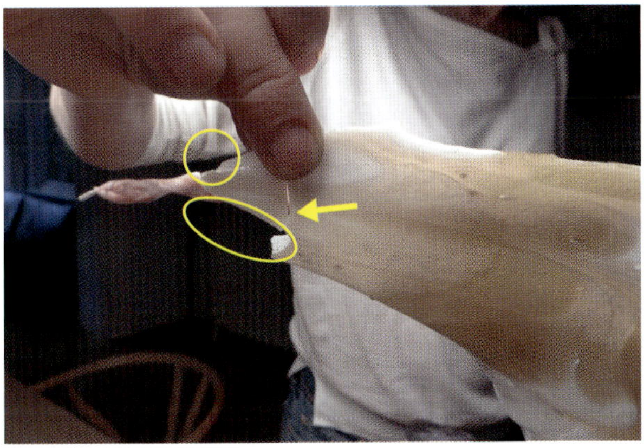

FIGURE 9.3
The artist duo documenting the damages to the broken tip. The yellow arrow points to the broken piece and the yellow circles highlight the two losses.

Conservation Proposal

The condition assessment and discussion with the artists and the owner about the possibility of repairing the severely damaged sculpture resulted in a conservation proposal based on the following considerations.

Selection of adhesive | The choice of a suitable adhesive involved the consideration of the type, shape, and position of the damaged areas; the properties of the plastic to be repaired (such as chemical resistance, hardness, transparency, refractive index); and the properties needed from an adhesive to perform safe, structurally stable, and aesthetically satisfactory repairs.

It was decided that to repair the broken polyurethane pieces, an adhesive would need to have the following properties:

- The ability to be applied to the polyurethane without dissolving or otherwise damaging the original material of the sculpture.
- Hardness comparable to that of the polyurethane.
- Optical clarity and colorlessness, with a refractive index (RI) closely matching the RI of the PUR in order to achieve invisible joints that would not interfere with the translucency created by the clear PUR mixed with pigments. To achieve this, the difference between the two refractive indices should not be greater than 0.02 (Laganà and van Oosten 2011; van Oosten and Laganà 2011).
- High strength in order to create a joint able to hold the broken parts together (see below).
- Low viscosity to be able to evenly cover all the tiny irregularities on the break surfaces.
- Low shrinkage.
- A curing process without exothermic reaction.
- Suitable working time to allow an optimal alignment of the broken parts.
- Nonyellowing over time.
- Chemically and physically stable over time.

The strength of the adhesive was a particularly important consideration for this piece. The strength required by the adhesive depends on several factors: the relationship between the bonding area (the surface area to be joined) and the masses of the two parts that the adhesive holds together, the position or location of the break, the weight of the broken pieces, and those pieces' geometry and shape. When joining two parts with considerable masses compared to the surface area to be joined, the adhesive must create a strong joint; the weight of the broken parts could lead to a failure of the joint, especially if it is compromised by any mechanical stress.

In this case, the break surfaces to be adhered were very small due to the extreme thinness of each cast section of polyurethane resin (just a couple of millimeters thick). Because one of the breaks was on the lower part (almost at the bottom) of the large and long curved section, a high-structural-strength bond was required to hold and permanently secure this large piece in place.

Based on all of the considerations listed, the crystal-clear epoxy adhesive HXTAL NYL-1 was selected for its compatibility with the hard polyurethane elastomer, having almost comparable hardness, high structural strength, low viscosity, appropriate type of curing, suitable curing and working time, transparency, and stability over time, as demonstrated in previous studies (Laganà et al. 2014, 2017, 2021).

Selection of filling technique | Once the adhesive was chosen, the very challenging task of reconstructing the two missing fragments was discussed. These losses were located in a very visible spot—one of the tips at the top of the piece—which was considered by both artists a very significant and expressive part of their dynamic sculpture. During the discussion of the importance of bringing back the sculpture's original appearance by reconstructing this part, the need for invisible repairs was brought to the fore. To achieve this, it would be of primary importance to use a filling material that perfectly matched the translucency of the cast, colored-polyurethane section.

Regarding the selection of the filling method, the direct filling technique was considered very difficult to carry out due to the curved and twisted shape of the losses, as well as their small size and their position, which would not allow one to easily make a backing or mold and perform a direct cast on the sculpture. Therefore, it was decided that this task could more easily be accomplished with a different technique: indirect filling or casting.

This method allows the separate reconstruction of the fragments, which can then be adhered in place. Fragments could in this way be reconstructed by casting using the same material and pigments originally employed by the artists—perfectly matching the original piece—and then joining them to the object using the same adhesive selected to repair the broken parts.

9.2 TREATMENT

In the treatment process, first the broken pieces were sorted and the order in which the pieces should be adhered was decided. Then the broken surfaces were cleaned; a small cotton swab moistened with demineralized water was enough to remove the grease from recent handling.

Adhering the Pieces

Before the epoxy could be applied, how to remove any excess or residue of adhesive from the original material had to be evaluated. Unfortunately, the organic solvents able to dissolve the epoxy resin HXTAL NYL-1 can also affect the original material. Therefore, to remove any excess safely and easily from the original surface, it was decided to protect the area around each joint by applying a very thin layer of a release agent with a cotton swab. The release agent would prevent any excess adhesive from sticking to the polyurethane, and would allow the excess to be easily peeled off after it set. The release agent could then be removed from the surface using a small cotton swab moistened with an aromatic-free mineral spirit—a solvent that, based on chemical resistance and solubility parameters, can be safely used on polyurethane elastomer. Carnauba wax, the same substance used by the artists to prevent the polyurethane resin sections from bonding to the tabletop during casting, was chosen as the release agent.

The adhesive was applied sparingly to the broken surfaces with a wooden skewer to minimize any excess epoxy, even though the surface was protected. Then the broken pieces were positioned and clamped, except for the small fragment at the top of the tip, which was difficult to clamp due to its curved shape. To hold this fragment in position while the epoxy was curing, tiny dots of soft, moldable beeswax were placed at one end of the break.

Excess adhesive was left to set overnight and then was easily removed the next day by gently peeling it from the surface with a plastic spatula. The epoxy adhesive was allowed

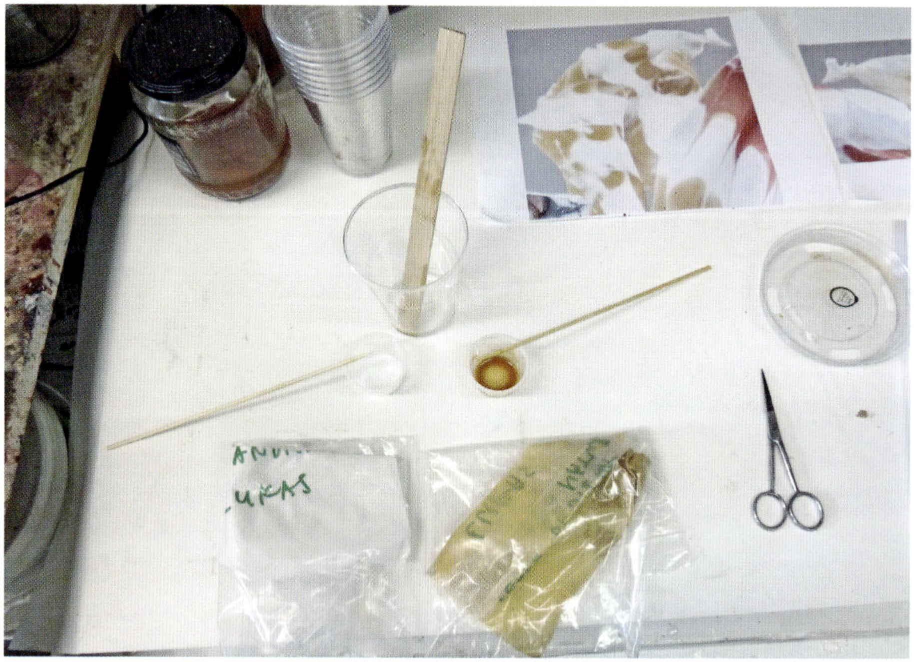

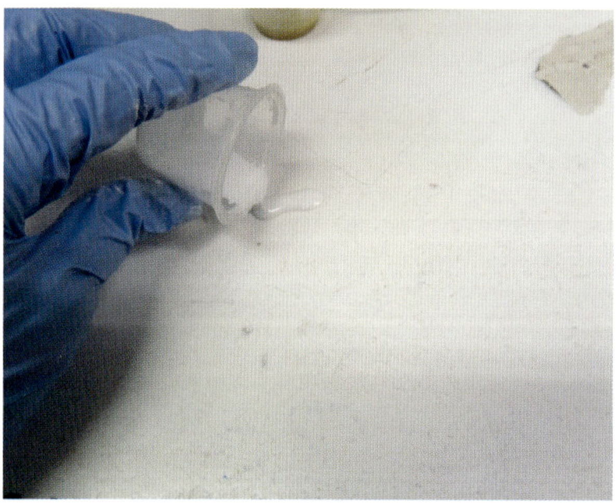

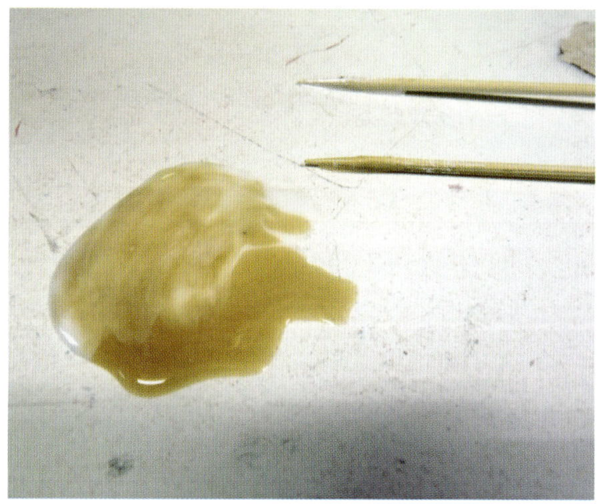

FIGURE 9.4
Making the missing fragments by mixing Crystal Clear 200 with white and ochre pigments following the same procedure used by the artists to make the sculpture.

FIGURE 9.5
Casting thin stripes of each colored mixture of Crystal Clear 200 on a working table previously covered with the mold release, carnauba wax.

FIGURE 9.6
One of the cast pieces after a wooden skewer was dragged across the white and ochre stripes to mix colors, creating the same pattern the original missing piece had.

to cure for a week before the sculpture was handled again to reconstruct the missing fragments. The adhered bonds appeared strong and secure and almost invisible to the naked eye. The thin layer of release agent was then removed from the surface using a small cotton swab moistened with aromatic-free mineral spirit.

Reconstructing the Missing Fragments

After evaluating different methods for loss compensation, it was decided to make the missing fragments following the same procedure the artists used to make the sculpture. The artist duo provided video showing their working process. The first step was to prepare the two components of the casting polyurethane resin (Crystal Clear 200), mixing the proper ratio of part A to part B (100A:90B) by weight, as indicated in the data sheet. The resin was mixed slowly for at least ninety seconds. Then the resin was mixed with pigments suggested by the artists; half of the resin was mixed with titanium white and the other half with yellow ochre (fig. 9.4). According to the data sheet, pigments should be added to part B before it is mixed with part A, but Heringa/van Kalsbeek's process was followed instead. Pigments were added to the clear resin until it was evaluated that both mixtures had the same appearance in color and translucency as the original part to be reconstructed.

For the casting step, thin stripes of each colored mixture of resin (white and yellow ochre) were poured adjacent to each other on a working table previously covered with the mold release, carnauba wax (fig. 9.5). The mixtures were poured with the same thickness as the cast in the sculpture. A wooden skewer was then dragged across the stripes to mix colors, creating the same pattern that the original had (fig. 9.6).

The next day, the two small cast pieces were removed from the table to be shaped as in the artists' making process. The resin takes sixteen hours to cure, and castings reach their ultimate physical properties at room temperature in five to seven days. This means that after seven hours the material is still flexible and can be easily cut and stretched into any shape.

Early in the process, the shape of the missing fragments was drawn freehand on paper using the damaged sculpture as a guide. The template thus created was used to cut the fills to the right shape. Then the shape was gently stretched and bent by hand and with tweezers to match the required curves. The shape was matched by looking at detailed photographs of the sculpture before it was damaged (fig. 9.7).

Once the fragments were made and the artists evaluated that the reconstructed pieces perfectly matched the originals, the fills were adhered to the tip with HXTAL NYL-1 following the same procedures used to adhere the original broken parts.

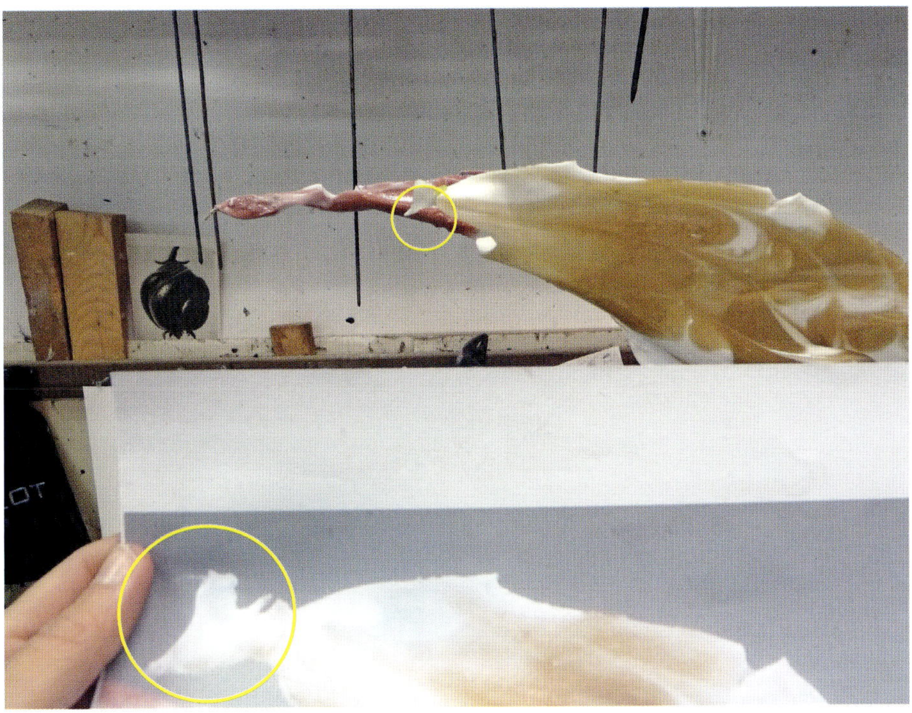

FIGURE 9.7
A photograph of the undamaged sculpture is used to evaluate the match between a missing piece from the tip and its reconstruction (highlighted by the circles). Prior to adhering, the reconstructed fragment was temporarily held in place with beeswax.

Material: Hard Polyurethane Elastomer (PUR)

Flexural strength

The flexural strength of PUR Crystal Clear 200 is 75.8 MPa, which means that it is rather difficult to bend, comparable to the average value of polyamide-6 (PA-6; see fig. 4.19).

Hardness

The hardness of Crystal Clear 200 PUR is 80 (Shore D), which is an extra-hard material, difficult to scratch, comparable to the average hardness value of polycarbonate (PC; see fig. 4.27).

Impact strength

The work of art broke due to a high-impact force. The piece did not collapse completely due to the high flexural strength of the hard PUR. The whole object can withstand a lot of stress; however, due to its construction, with rather long projections, some areas are subject to impact forces.

Crystal Clear 200 has a Charpy impact strength of 2.55 J/cm^2 (see fig. 4.12b).[1]

Refractive index

The refractive index is important when adhering transparent and translucent plastics. The refractive index of Crystal Clear 200 is 1.49962 and the refractive index of HXTAL NYL-1 is 1.515.

Tensile strength

The tensile strength of Crystal Clear 200 is 17.2 MPa, which is comparable to the average value of the strength of high-density polyethylene (PE-HD) (see fig. 4.7).

Data

Source: Values from the data sheet of Crystal Clear 200 from Smooth-On (https://www.smooth-on.com/tb/files/CRYSTAL_CLEAR_200_TB.pdf).

Materials used in treatment

Crystal Clear 200, Smooth-On.
https://www.smooth-on.com/products/crystal-clear-200/

HXTAL NYL-1.
https://www.talasonline.com/Hxtal-NYL-1

Kremer White Spirit (mineral spirit) 100/140, free of aromatics (below 0.1%), CAS-Nr: 64742-49-0.
https://www.kremer-pigmente.com/en/solvents-chemicals-und-additives/solvents/2229/white-spirit-100/140

1. While Smooth-On sells Crystal Clear 200 polyurethane resin, that company does not produce the material. Therefore, data given here for impact strength reflect that provided by the polymer manufacturer, Sika. No Izod impact strength was given (see fig. 4.12a).

Chapter 10 | Issues of a Natural Rubber Installation

Lydia Beerkens

The contemporary artist Kaari Upson (1970–2021) lived and worked in Los Angeles, where she made large installations, paintings, and sculptures. She found inspiration for her work in the large upholstered furniture, such as couches and especially mattresses, discarded all over the city of Los Angeles—silent witnesses of other people's lives, including the lives of the homeless. Upson's soft sculptures, made using latex (natural rubber), are exact casts, with every single detail, stitch, and trace of wear and tear visible. These massive, pliable couches and mattresses fill large architectural constructions, which drape and bend under their own weight. Occasionally adding paint and writing in charcoal, Upson introduced marks of their former use as a connection to the life of the other human being who has left the objects behind (Upson 2017).

In 2003, Upson began an ongoing series called The Larry Project, in which she reinvents the possible life of a man and the burned-out house in which he lived. This project entails Upson's psychological investigation and creation of a world between reality and illusion (Lucas 2017). The installation *Recollection Hysteria* (2012) is part of The Larry Project.

10.1 DESCRIPTION OF THE ARTWORK

Recollection Hysteria represents one room in "Larry's" house, reproduced in latex. The large elements of the installation were produced by casting off the complete interior of one room in a home in latex: walls, ceiling, and floor, including striking details such as tiles, battens, carpeting, and a light switch and light bulb. Extra elements, also in latex—bottles, boxes, clothes, rubbish on the floor, a mattress, and sneakers with socks—give the space a realistic appearance, as if the occupant has just left the room. Upson has manipulated the room by giving it a 90-degree turn, positioning the large latex sheets in a twisted way. The open door hangs down from the ceiling, the doorknob is a useless element, while the carpeted floor with all the waste on it is perpendicular to the door. The adjacent wall is the blackened ceiling, with its bare light bulb. One piece of latex, another wall of the room, is draped as a curtain, opening up to museum visitors (fig. 10.1). The colors Upson has used are limited; they include white, pink, beige, yellow, light and dark brown, and a localized black-stained area with a graffito in black charcoal: "I HATE LIFE." The strong, penetrating odor in the installation is recognized by many as the smell typical of rubber.

Material and Technique

After the arrival and installation of the artwork in the Kunstmuseum Den Haag, information on the materials and technique was retrieved from the artist and the artist's studio.

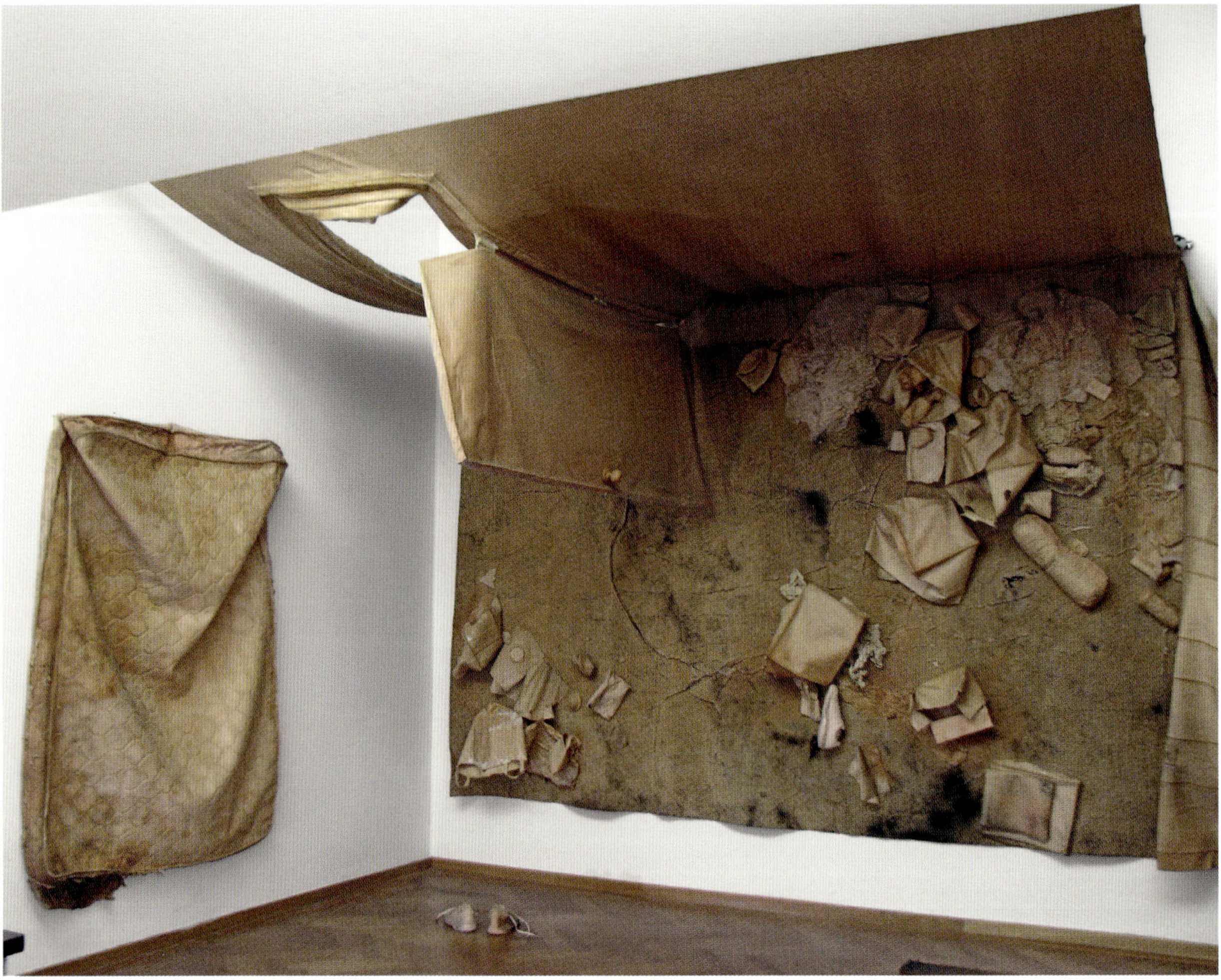

FIGURE 10.1
Kaari Upson (American, 1970–2021), *Recollection Hysteria*, 2012. Natural rubber installation. Kunstmuseum Den Haag.

Recollection Hysteria showcases the artist's preferred material, latex, which is flexible and easy to apply and peel off. Through the application of latex in thin layers, the artist built up the large elements of the installation. Kaari Upson created the work in her studio in Los Angeles, California, in the spring of 2012. She used RLM-460, a latex product purchased from Silpak, in North Hollywood.

According to the studio, the entirety of the piece was made out of two 5-gallon kits of RLM-460, and the manufacturer's instructions were followed. A small amount of acrylic paint (Golden Fluid Acrylics) was mixed into the latex. After this, it was directly applied to the mold in layers. Each layer was brushed on and left to dry. Once the first was dry, the next layer was applied. Each wall or part of the artwork was completed in this way over one day. The parts were connected after they were all complete. No additional materials were applied in between or on top of the latex. The entire surface of the artwork was covered with talcum powder and a small amount of charcoal powder, an aesthetic choice. New charcoal can be added, according to the studio, should it come off.[1]

1. Hans Janssen, personal communication, 2014.

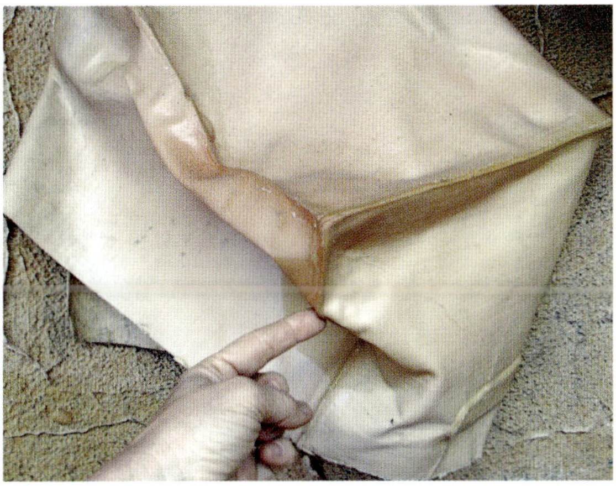

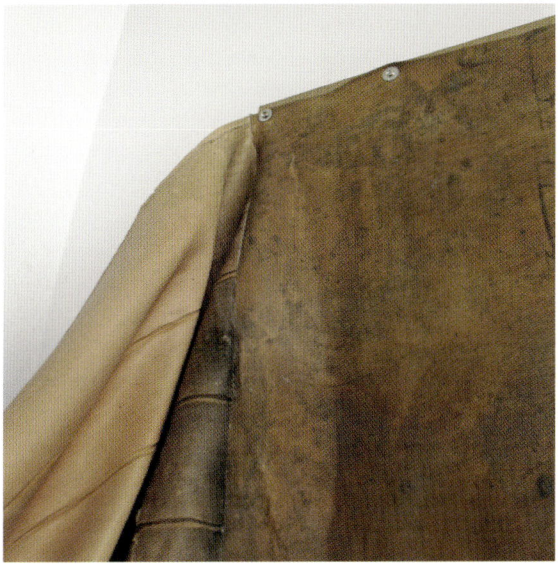

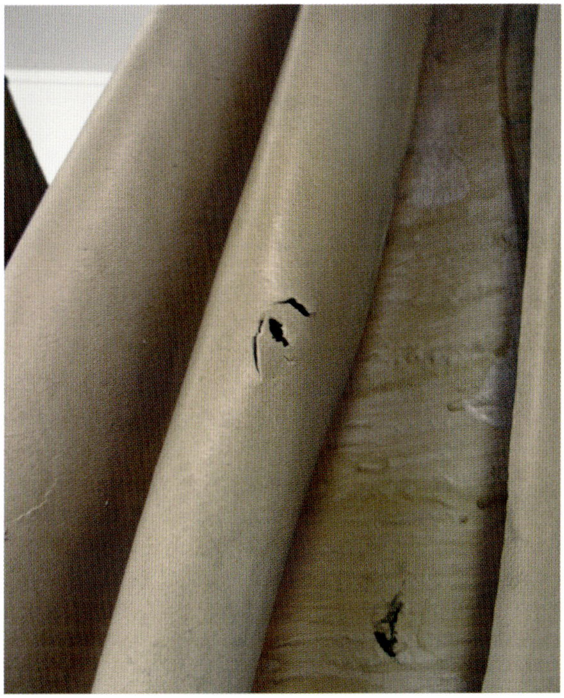

FIGURE 10.2
Detail of the artwork with variable thickness in different elements, rendering the rubber more stiff or more loose.

FIGURE 10.3
Hanging system with screw and protection ring through the rubber piece.

FIGURE 10.4
Rubber torn loose from its hanging due to its weight and limited tear resistance.

To strengthen the hanging construction, some edges of the larger elements have been lined with canvas, and metal rings have been used to enlarge the areas with the pinholes for mounting, supporting the areas subjected to strong forces while hanging (figs. 10.2, 10.3).

Material Condition

The artwork has been installed in the museum for several years now, and the installation and its separate elements are in good condition. The latex is, in general, strong and flexible and still has its natural elastic quality. The construction of the installation functions well. The canvas linings and the metal rings added to the separate latex elements as constructive parts are strong enough to hold the heavy weight of the pieces. The door and the mattress have a few edges that are slightly sticky. The colors in their various shades of beige and brown seem not to have changed since the work was produced. In the latex piece fixed as a curtain (the entrance to and exit from the installation), one mounting hole has torn. It now hangs from only one fixation point (fig. 10.4).

From the beginning, there were concerns about the latex, as it is not a long-lasting material and has a strong odor. The smell of the freshly produced material decreases over the years, but while it lingers it tends to upset museum visitors, guards, and conservation managers.

The artist's studio is aware of the deterioration intrinsic to latex and warns against direct sunlight and UV radiation. Studio staff are also familiar with early signs of decay, such as cracking, darkening, and stickiness. The studio discussed the possibility of cleaning with a solution of 50% isopropanol diluted in water, together with the option of applying a product to seal the latex surface. To conserve the installation as long as possible, it is key to estimate the expected lifetime of the latex, as well as the effects of aging on the optical appearance (changes) and on the construction of the installation (Beerkens 2014).

Analysis and Interpretation

In 2016, the installation was visually inspected and the flexibility was checked through careful touching and manipulation. The strong odor was still very present four years after the creation of the work. As time passes, the odor will diminish, but the material deterioration will progress and will affect the stability of the whole artwork. The latex will become less flexible; it will darken, stiffen, and ultimately crack and break. The large, heavy sheets of latex put a lot of weight on the few fixation points of the hanging construction, which is attached to the walls with screws and nails through the latex. These points are weak spots where tears may occur first, ultimately causing the sheets to come loose. Depending on the conditions, this process may take a few decades.

Optical criteria also play a role. The color and the amount of discoloration as a function of time can indicate the progress of deterioration. The amount of change may also depend on how much acrylic paint was mixed into the latex. A slight color change may be acceptable as a general sign of age without being disturbing to the viewer. However, when color change is accompanied by a cracking surface and loss of flexibility and smoothness, damage occurs.

10.2 CONSERVATION AND OPTIONS FOR TREATMENT

Recollection Hysteria is on permanent exhibition under stable museum climate conditions. Regular monitoring of the material will allow changes to be registered and addressed at an early stage. Cleaning, dusting, and checking the hanging points will contribute to good maintenance of the artwork. Optical changes may be dealt with through minor interventions using charcoal, as indicated by the artist. Structural damage will be more difficult to treat, as a tear in the latex is problematic to mend. Once a small tear has occurred, the latex lacks interior strength to prevent any further tearing. Few adhesives are available to mend these tears. Reversibility may well be considered of less importance, as the artwork inherently has a limited expected lifetime—decades rather than centuries.

The overall conclusion may be that the main material in this artwork—latex—has been used and applied according to the producer's instructions, that the slabs have good strength due to the application of several layers, and that the museum conditions prevent any extreme environmental impact. The permanent exhibition has several advantages over storage in a low-oxygen packing. With this type of packing, there is risk of handling, as the large, heavy parts of the artwork must be lifted and moved for installation or deinstallation. With permanent exhibition, regular monitoring is easy to perform, and the artwork can be viewed at any time during its limited (material) lifetime.

Material: Latex/Natural Rubber (NR)

Density (specific weight)

The density of NR is 0.95 g/cm^3, which means that a rubber piece of 2 × 1.5 m and a thickness of 1 cm has a weight of 28.5 kg. A piece like this can withstand 285 newtons (28.5 × 10). This amount of force has to be distributed over several pinholes. The tensile strength of natural rubber is 28 MPa, meaning that at least ten pinholes should be made at hanging, or a bar should be placed on the reverse side of the piece to equally divide the force over the whole piece.

Elongation

The outstanding elongation property (stretching until it breaks) of natural rubber is large (up to 800%) and lasts as long as the NR is in good condition; the rubber piece will return to its original shape after stretching.

Tensile strength

Tensile strength is important for a material that is going to be stretched or under tension. The tensile strength of 28 MPa makes NR a good material for the large-scale parts of this artwork. Combined with the good elongation property, it enables the separate parts to perform as soft, loose hanging curtains, folding around corners by their own weight and gravity.

Tear strength

Natural rubber is a strong elastomer (tear strength ranging from 151 to 159 kNm^{-1}); however, as NR is not reinforced by its own structure, there is chance of tearing once a small break in the latex occurs by force or by deterioration of the material itself. This can cause further tearing of the material due to its own weight or with handling.

Data

Source: fact sheet 15.10. Data show minimum and maximum values due to differences in processing (manufacturer).

Materials used by artist

RLM-460, Silpak, Inc.
http://www.silpak.com/pdfs/RLM460.pdf

Chapter 11 | Adhering Polyurethane Elastomer

Lydia Beerkens

Rachel Whiteread (b. 1963) is a sculptor and one of the leading contemporary artists in Britain today. Since winning the Turner Prize in 1993, she has built a large oeuvre of internationally acclaimed sculptures (Tate 2017b). Through her sculptures and monumental artworks, she investigates the relationship between volume, space, and "negative space"—the space included in or surrounding objects, under furniture, between architectural elements, and in interiors. Whiteread's solid sculptural casts of the space under a chair, in a stairway, against bookshelves, and around a bathtub represent new volumes, different shapes, that are of the same size but opposite to the original three-dimensional structures and constructions.

Her sculptures are made in one-material casts; however, the materials vary from polyester and concrete to plaster, glass, and rubber. With the addition of colors and through the casting technique, the sculptures gain a specific surface and skin, smooth or rough. This, in combination with their opacity, translucency, or transparency, alienates the sculpture further from the original objects' look and function (Mullins 2004).

Whiteread's work has been described as "minimalism with a heart" (Tate 2017a). The reference point of the objects she casts is always human. Familiar household objects such as mattresses, beds, sinks, and chairs retain their intimate, physical relationship to the human body despite their stark, sculpted forms. Larger objects—houses, windows, and staircases—have had a life prior to the artist's use of them, and they bring their own histories to her work. Whiteread addresses particular human stories, historical contexts, or sociopolitical issues by creating casts that memorialize loss and moments of life and death.

11.1 DESCRIPTION OF THE ARTWORK

The artwork *Untitled (Slab II)*, by Rachel Whiteread, is a large, elongated floor piece (13.5 × 75.5 × 197 cm) with one blunt end and one rounded end. As with many of Whiteread's sculptures, it is not immediately clear what kind of object it refers to. Here the shape and volume of the artwork represent the "negative space" of the top of a dissection table. The table had a rounded top, smooth round edges around the sides, and a blunt end at the foot. The rubbery material Whiteread used has specific material and aesthetic qualities. The surface is smooth and soft. The color of the material is deep brown in the center of the work, becoming more translucent and a lighter, more ochre color toward the edges and ridges. These visual properties give the artwork the impression of being made from an unknown material, almost resembling smooth amber (fig. 11.1).

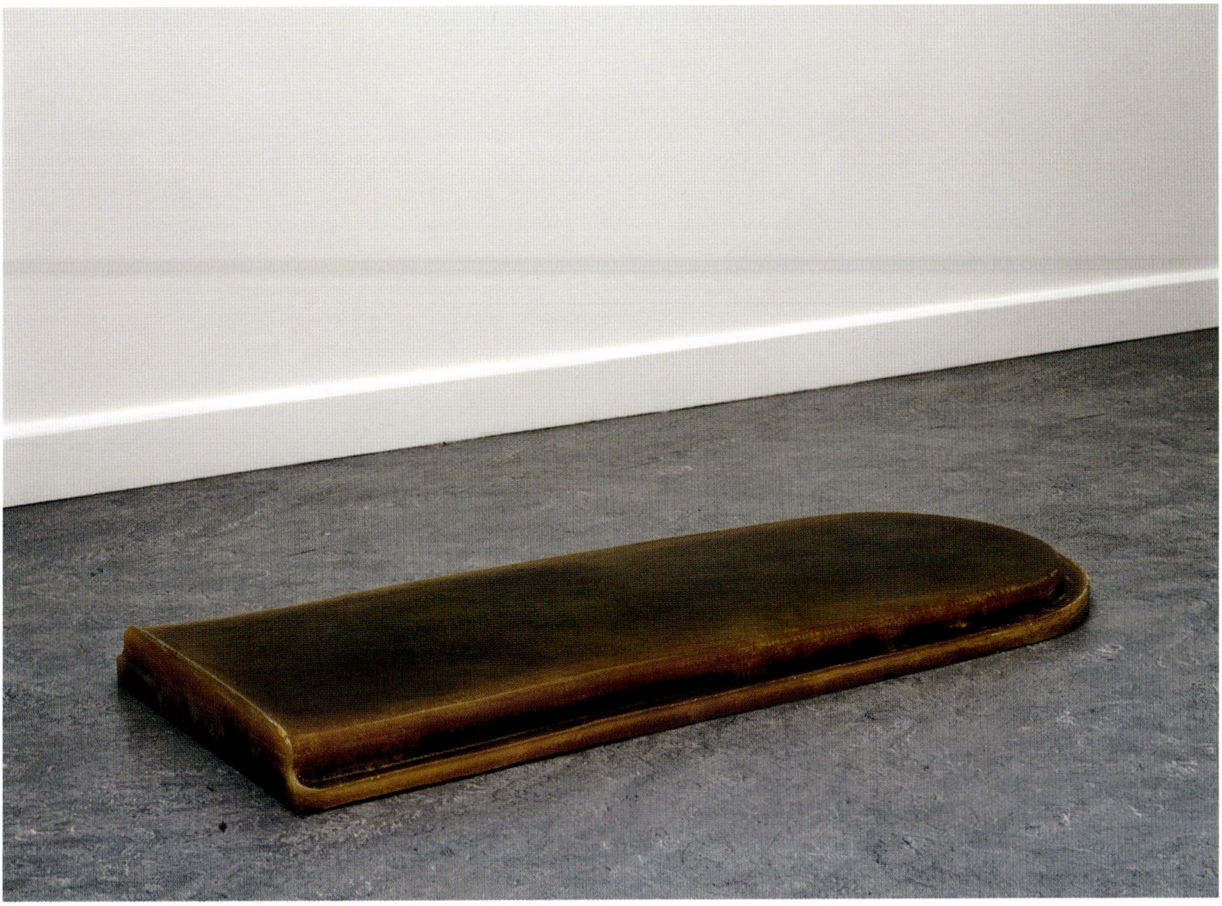

FIGURE 11.1
Rachel Whiteread (British, b. 1963), *Untitled (Slab II)*, 1991. Two-component polyether-polyurethane elastomer, 13.5 × 75.5 × 197 cm. Eindhoven, Van Abbemuseum.

Material and Technique

Whiteread produced *Untitled (Slab II)* by making a full cast of the top of a dissection table in a thermoplastic polyurethane—specifically, a two-component polyether-polyurethane elastomer. The amber-like, slightly translucent hue of the two-component material gives the object an extraordinary quality. The surface is semimatte and seems to have a vague imprint of waves (fig. 11.2). This may be a result of the casting process: the pouring of the fluid polyurethane mixture into the mold. Another characteristic of the material in this artwork is the fine structure of air bubbles enclosed in the set material in the rounded-off edge, probably from the mixing of the two polyurethane components (fig. 11.3). These parts are lighter in color than the rest of the edges in the sculpture.

Material Condition

The artwork was made in 1991 and was acquired by the Van Abbemuseum in Eindhoven, Netherlands, in 1992. The sculpture is now thirty years old and in generally good condition. According to the artist's instructions, *Untitled (Slab II)* is to be presented without plinth, flat on the museum floor. This method of presentation over the years caused some small losses on the thin rubber edges of the artwork, as visitors accidentally stumble over the object, sometimes even leaving an imprint of footsteps on the smooth, soft surface.

FIGURE 11.2
Top of *Untitled (Slab II)*.

FIGURE 11.3
Air bubbles trapped under the surface from mixing the two liquid elastomer components.

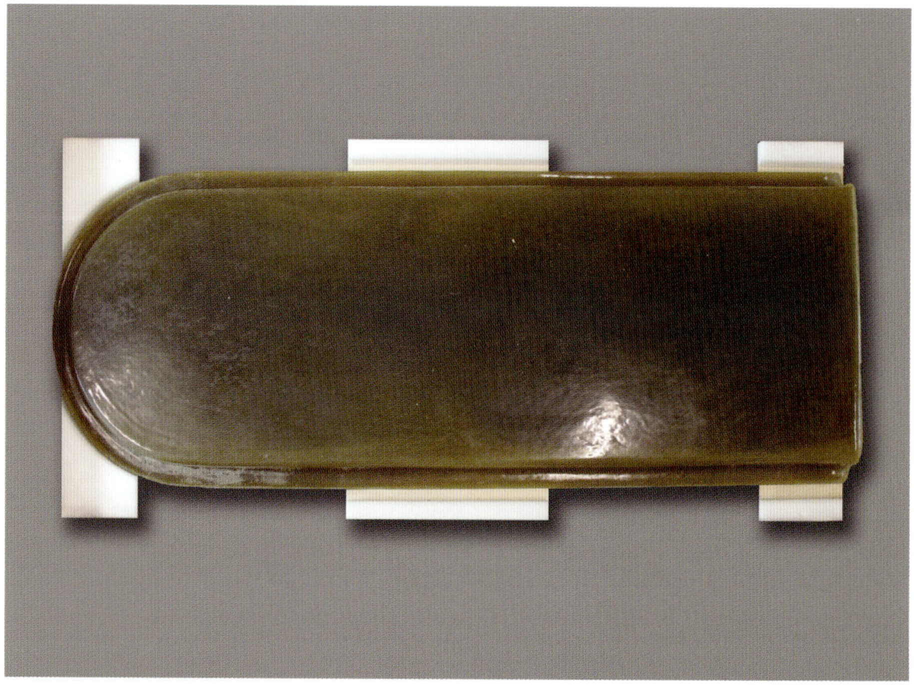

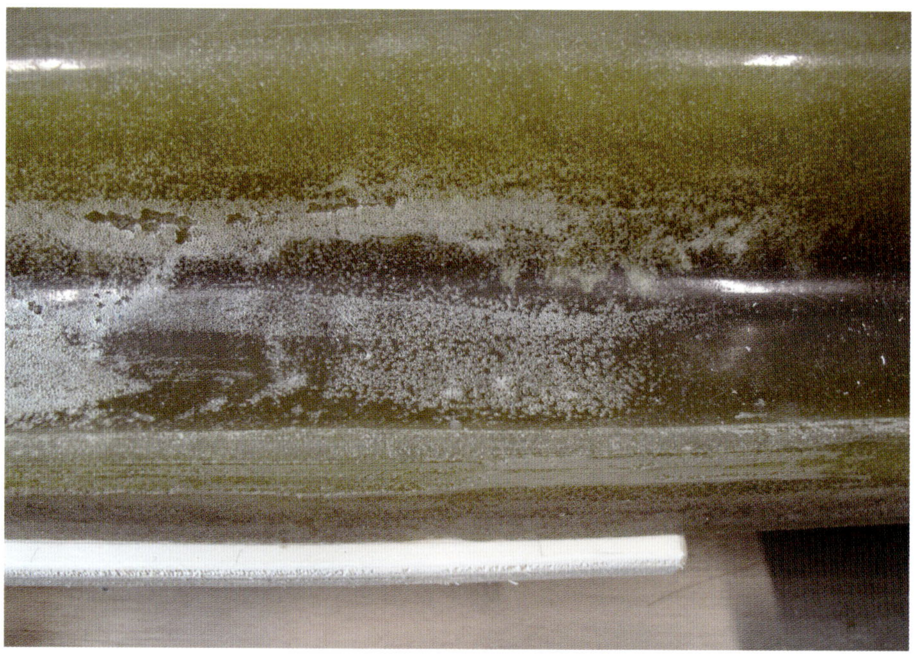

These small areas of damage and footprints on the surfaces required conservation treatment. Further, the unknown life span of the elastomer used posed questions about future deterioration, influence of climate, possible discoloration, and surface change: How long will this artwork last?

The work is not on permanent exhibition. Handling and packing require some preparation, as the rounded shape, vulnerable edges, and the combination of the piece's heavy weight, size, and elastic properties make it hard to handle for transport. In 1999, a conservation management plan for long-term preservation and proper storage was requested, together with a proposal for treatment of the small damages.

Analysis and Result

The type of elastomer used as the sole material in this artwork had to be identified in order to answer questions about its longevity and to formulate a good standard for exhibition, packing, storage, and long-term preservation of the piece. Fourier-transform infrared (FTIR) spectroscopy analysis of the material was performed in 1999 (van Oosten 1999) and revealed the use of the two-component polyether-polyurethane elastomer. The analysis result was translated into a proposal for the best preservation of this artwork and allowed for a specific conservation treatment.

The longevity of the polyurethane elastomer as a material depends on climate and storage conditions; however, in the case of *Untitled (Slab II)*, the thickness of 13.5 cm of solid material in the artwork is a crucial benefit. Deterioration caused by humidity and photooxidation occurs only on the material's surface, in this case only on the top of the solid cast artwork. It may take many decades before the deterioration enters into the elastomeric material of *Untitled (Slab II)*. The expected lifetime is therefore long, and optical changes may start to be distinguishable only after a long time. This differs greatly from the situation for polyurethane foam materials, in which deterioration begins immediately within the thin cell walls, endangering the structure of the foam rubber and the entire object. Taking this into account, the small ridges at the edges of *Untitled (Slab II)*, which have minute pockets caused by air bubbles introduced during the casting procedure, in this case are a more vulnerable part of the artwork, and where visible changes are expected to take place first (fig. 11.4).

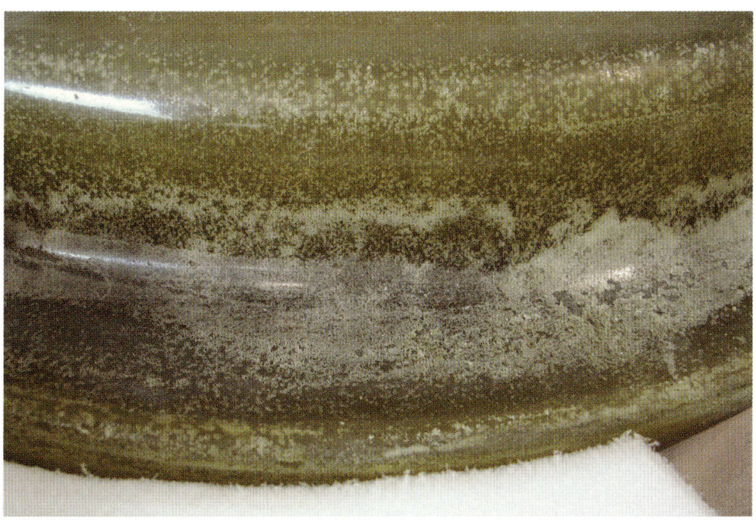

FIGURE 11.4
Edge with enclosed air pockets.

11.2 TREATMENT

Some very small, loose edges were consolidated back in position with Lascaux 360 HV, an acrylic dispersion adhesive, which stayed sticky and flexible after drying (fig. 11.5). The adherence bond was not very strong, but this was not required, as there was no stress or tension on the edges and manipulation could be avoided.

It is difficult to achieve for the best optical match and mimic the brown translucent material after drying while adhering a small loss in position. Locally, tiny tears in the material were treated with Aquapasto (a blending medium of gum arabic and silicate), with the main aim being to consolidate some loose material and match the translucency at the same time. All treated areas were small, and in the overall view do not stand out. The footstep marks were successfully removed with a soft, damp cloth, leaving only a few vague scratches of the imprint.

The general recommendation to strive for an oxygen-free packing for polyurethane during longer periods of storage would in this case entail the design of a large, sealed crate that has to be flushed regularly with nitrogen. Although the museum has good storage facilities, the practicalities would be too complex to maintain. An alternative is to wrap the sculpture in a large, airtight plastic bag. Although it may be expected with this particular material that there will be no off-gassing, there is no guarantee that the artist or production studio exactly followed the mixing and curing procedures, nor that the commercial product batches were flawless and without unknown additions. A vacuum procedure has not been tested on this material in, for example, accelerated aging. The artwork *Untitled (Slab II)* is currently packed in a standard transport crate, with a Tyvek wrapping and a lid that enables the exchange of air during long-term storage.

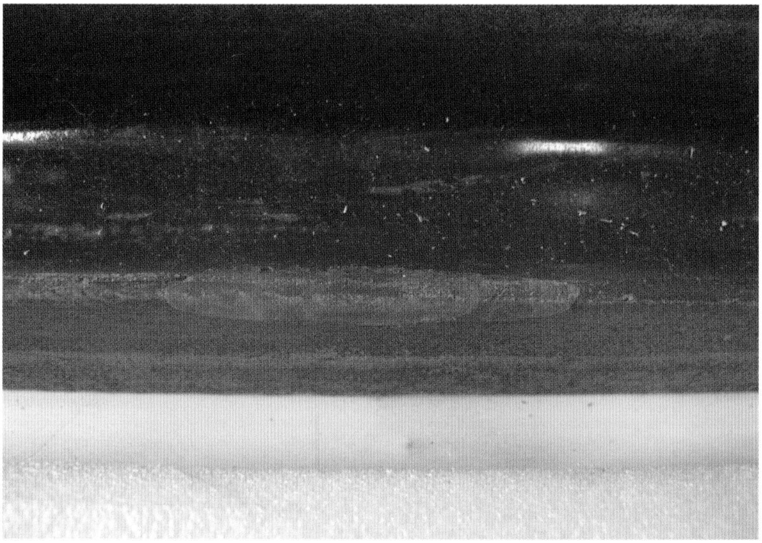

FIGURE 11.5
Adhering a loose part of the polyurethane elastomer.

Material: Thermoplastic Polyether Polyurethane Elastomer (TPU)

Density

The work of art *Untitled (Slab II)* measures 13.5 × 75.5 × 197 cm, and polyether urethane elastomer used can have a density between 1.04 and 1.27 g/cm^3, meaning that the work of art will have a weight between 209 and 255 kg. Transport of an object like *Untitled (Slab II)* needs special care to prevent falling and prevent losses from the edges of the artwork.

Hardness

The Shore D hardness of thermoplastic polyether urethanes (see fig. 4.24) can vary between 22 (medium hard) and 83 (extra hard), and Shore A hardness (see fig. 4.25) can vary between 25 (very soft) and 101 (extra hard), depending on the components used and the manufacturer. Thus, polyurethane elastomers can be easily scratched or not scratched at all, depending on the hardness.

Softening point

The softening point of a polyether urethane elastomer can vary between 110°C and 166°C; however, the softening point becomes lower when weight is applied. Eventually, for example, footprints will leave an imprint on the surface of *Untitled (Slab II).*

Tensile strength

The tensile strength of a polyether urethane elastomer can vary between 0.8 and 68.9 MPa, and TPU has the strongest (average) tensile strength of the elastomers (see fig. 4.7).

Elongation at break can vary between 77% and 1000%, meaning that polyether urethane elastomer can be stretched if the thickness of the piece allows it. In the case of this artwork, that is most likely to occur at the edges.

Data

Source: fact sheet 15.11. Data show minimum and maximum values due to differences in processing (manufacturer).

Materials used in treatment and storage

Aquapasto (blending medium of gum arabic and silicate), Winsor and Newton.
https://the-bank.azurewebsites.net/download/43818
Lascaux 360 HV, Talas.
Tyvek, DuPont.

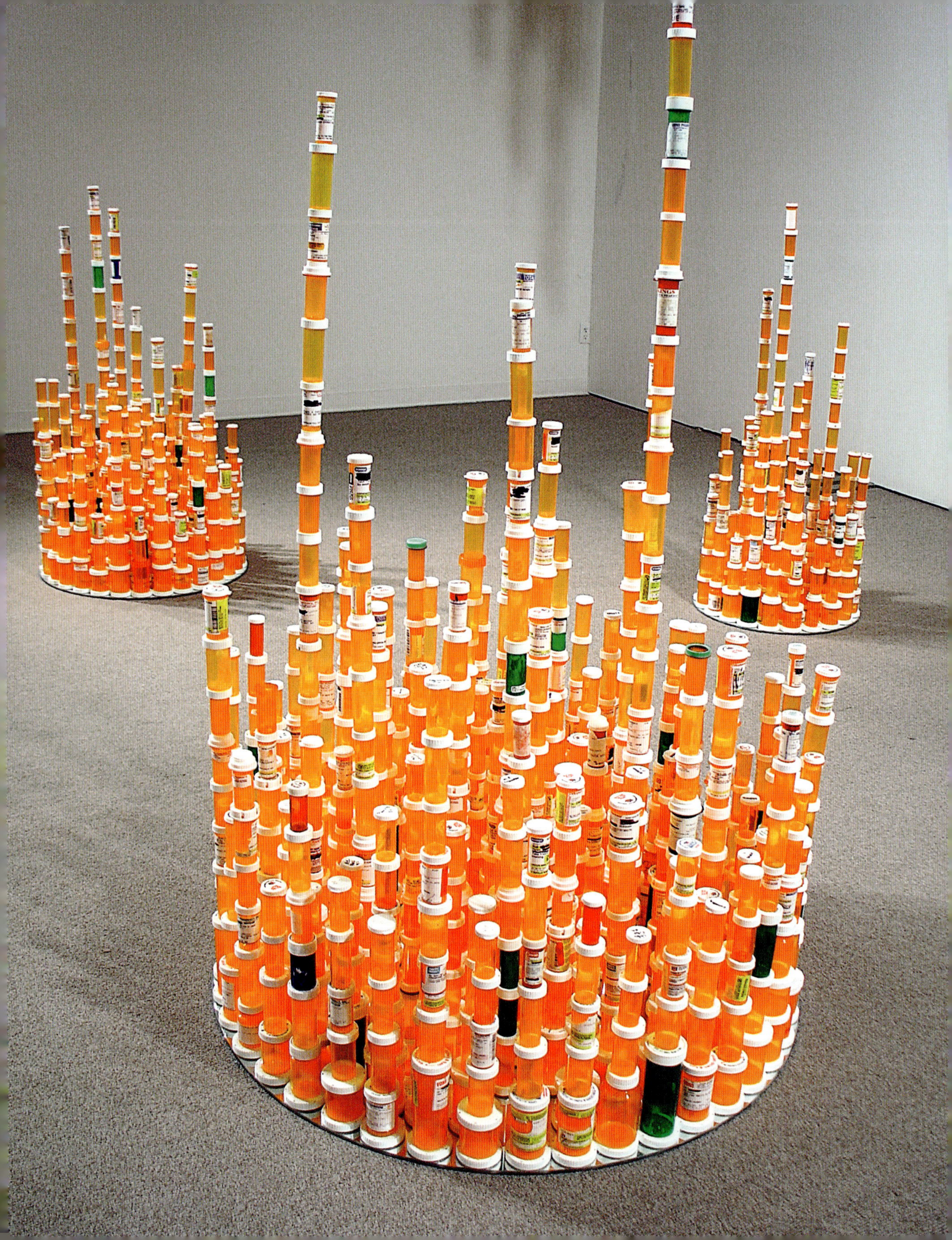

Part III

Fact Sheets of Plastics and Elastomers

The fact sheets that follow in chapters 12–15 compile data on the most important properties of plastics and elastomers (including natural and synthetic rubbers). They are designed to help readers better understand and recognize the types of plastics and elastomers.

Of the hundreds of types plastics in existence, the fifty-six most commonly found in cultural heritage objects are covered in the fact sheets: natural polymers,[1] thermoplastics, thermosets, and elastomers (see table III.1 for a complete table of contents for the fact sheets). Each fact sheet characterizes a specific plastic or elastomer and includes, among other information, its chemical abbreviation; Fourier-transform infrared (FTIR) spectrum and chemical structure; photos of sample objects; and data for the most essential properties: mechanical, physical, thermal, and optical. The information in the fact sheets provides a summary view of these materials, offering a better understanding of how they age and behave during conservation.

How the Data Were Compiled

This overview of data of physical, mechanical, thermal, and optical properties from the beginning of plastics use until today was collected from a variety of sources, primarily the *Modern Plastics Encyclopedia* (1947), a Shell Chemicals booklet (1967), and the *Handbook of Polymers* (Wypych 2016). Data from the latter book were compiled from open source literature (published articles, conference papers, and books), literature available from manufacturers, and patents. In a few cases, data from these sources were augmented with data from other publications listed in the fact sheets' source notes.

All data from the earlier sources (1947 and 1967) were expressed in units not commonly used today; therefore, those units have been converted into standard international (SI) units, with the exception of permeability measures (see below).

1. Natural polymers were used for the production of various "precious" objects, and some early plastics were developed as substitutes. Bone and ivory are not included because those materials are mostly carved into objects and are easily recognized by conservators. One important physical property in which they differ from plastics is their weight: bone and ivory have greater densities than imitations made of plastic.

How to Use the Fact Sheets

The first page of each fact sheet provides general information about the material, including the following:

- Date of first manufacture. Dates of invention/discovery, patent, and original manufacture often differ by country, so multiple dates are often listed. When a material has fallen out of use or there has been a revival of interest, that is also noted.
- Alternate/trade name examples. Today, more than 4,200 entries for trade names of plastics and elastomers exist,[1] so it is impossible to be exhaustive. Only some of the most well known are mentioned in the fact sheets, along with any other names that are or were in common use.
- Main degradation agents commonly affecting plastics and elastomers are given in order from most common to least.
- Some fact sheets feature photographs showing degradation. In addition, publications that catalog and illustrate the many types of damage to plastics and elastomers are available and worthy of consultation, including *Preservation of Plastic Artefacts in Museum Collections* (Lavéndrine et al. 2012), *Conservation of Plastics: Materials Science, Degradation, and Preservation* (Shashoua 2008), and the damage atlas compiled by the Preservation of Plastic Artefacts in Museum Collections (POPART) project at https://popart-highlights.mnhn.fr/collection-survey/damage-atlas/index.html.

The second page contains the compiled properties data in table form. Dashes (—) are used to indicate where data are not available because either the material was too new or it was no longer in production.

1. http://www.ets-corp.com/tradenames/pa.htm.

There are a few items to keep in mind when using the data provided:

- Values of properties can vary due to measurement conditions and differences in processing; therefore, the values compiled in the fact sheets are typically shown as ranges.
- The theory of the various properties and their units, as used in the fact sheets, is explained in chapter 4. The bar graphs in chapter 4 of specific properties can be used to compare the value of a property for a specific plastic or elastomer from the fact sheet with that of other materials.
- Permeability measurements in the past were performed under different measurement conditions (e.g., pressure, thickness of sample) and reported using different units than those currently in use, and conversion to SI units was not feasible; therefore, the unit as well as the data for permeability are shown as in the original source. While this makes it more difficult to compare the values, it is still possible to use these data to understand a plastic's or elastomer's degree of permeability to water vapor, oxygen, or carbon dioxide.
- The distinct mechanical properties of elastomers, in contrast to those of thermoplastics and thermosets, require different terminology on some elastomer fact sheets (15.1–15.15):
    - *Compressive strength* (ASTM D695), a measure of the amount of force needed to compress plastic until it breaks, has been replaced by *compression set* (ASTM D395), a measure of the amount of resilience after force has been applied.
    - *Flexural modulus* (ASTM D790) often replaces *tensile modulus*.
    - *Tear strength* (ASTM D790), a measure of the amount of force required to initiate a tear, has been added.

Finally, please note that all data must be used with care. These are not absolute values but ranges of *possible* values. Knowing a range of possible values for a specific property of a plastic or elastomer can help in choosing proper conservation materials with the same range of values for preventive and curative treatments.

Table III.1. Fact sheets by chapter

12.1 | Horn

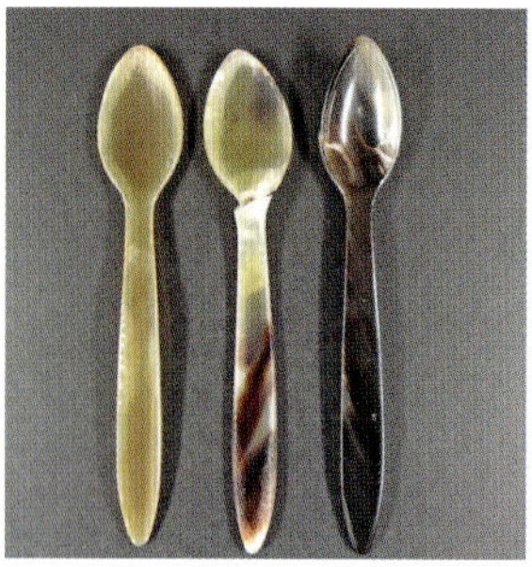

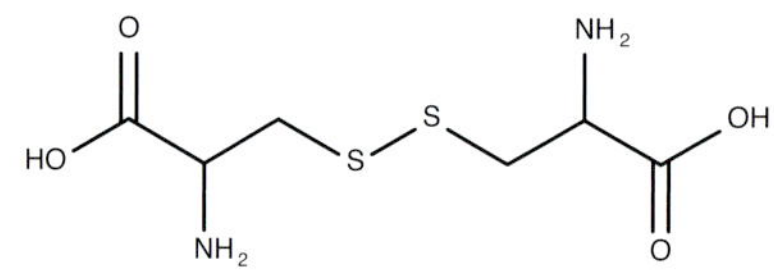

CHEMICAL STRUCTURE

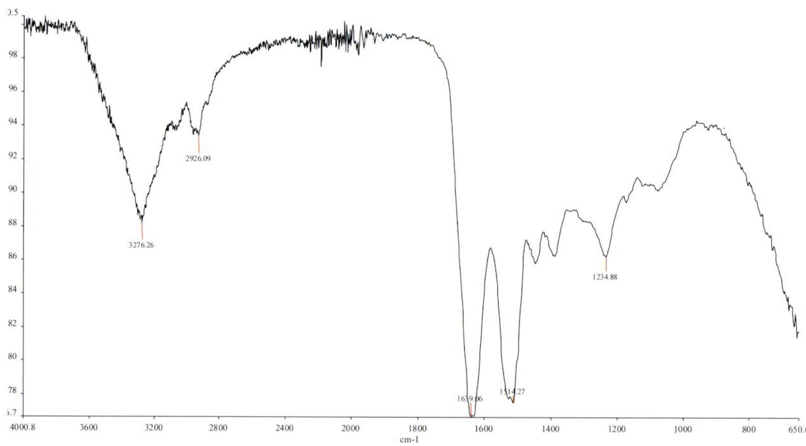

FTIR SPECTRUM

Date of first manufacture
Since ancient times; molding from the early seventeenth century.

Alternate/trade name examples
N/A

Polymerization
Animal (cattle, sheep, antelope) horn consists of the protein keratin. The amino acids of keratin are polymerized by condensation reaction, and the amino acid cysteine, which contains sulfur, provides some disulfide bridges (cross-links) between the polymer chains.

Processing
Horn, a thermoplastic material, was used to imitate tortoiseshell. To do so, horn was made into plates using heat and pressure (compression molding). Horn can be stamped, engraved, painted, and carved.

Selected applications
Buttons, combs, drinking vessels, furniture decoration, grips, handles, translucent panels for lanterns.

Feel/look/sound/smell
Translucent or opaque, soft polish/glossy to matte surface. Rigid but flexible if sufficiently thin.

Main degradation agents
Light, UV radiation, moisture (changes in relative humidity), oxygen, bacteria, fungi, stress.

Main signs of degradation and damage
Swelling, shrinkage, splitting, cracking, dull surfaces.

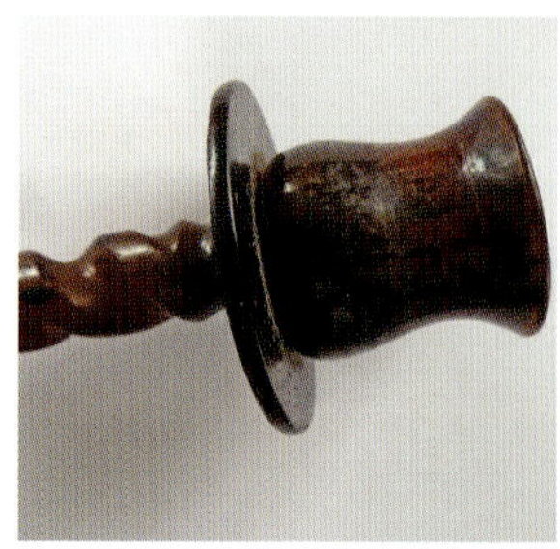

Dull surface of a candle holder.

12.1 | Horn

	PROPERTY	UNITS (SI)	DATA 1947	DATA 1967	DATA 2016
MECHANICAL	**Compressive strength** (ASTM D695)	MPa	—	—	—
	Elongation (ASTM D638)	%	—	—	—
	Flexural strength (ASTM D790)	MPa	—	—	—
	Hardness (ASTM D785)	Rockwell	—	—	—
	Impact strength (Izod notched) (ASTM D256)	J/cm	—	—	—
	Tensile modulus (rigidity) (ASTM D638)	GPa	—	—	—
	Tensile strength (ASTM D638)	MPa	—	—	—
PHYSICAL	**Density** (ASTM D792)	g/cm^3	—	—	—
	Hildebrand solubility parameter (δ)	$(MPa)^{1/2}$	—	—	—
	Permeability: H_2O vapor, O_2, CO_2 (ASTM D1434)	$cm^3 \cdot cm \cdot cm^{-2} \cdot s^{-1} \cdot Pa^{-1}$	—	—	—
	Resistance to solvents p = poor, f = fair, g = good	Not commonly mentioned			
	Surface tension	mNm^{-1}	—	—	—
	Water absorption (equilibrium 24 hours) ⅛-inch (3.175 mm) thickness (ASTM D570)	%	—	—	—
THERMAL	**Coefficient of linear thermal expansion** (ASTM D696)	$°C^{-1}$ ($\times 10^{-5}$)	—	—	—
	Glass transition temperature (Tg)	°C	—	—	—
	Low-temperature performance / Brittleness temperature (ASTM D746)	°C	—	—	—
	Melting point (T_m) (ASTM D1238)	°C	—	—	—
	Softening point (Vicat) (ASTM D1525)	°C	—	—	—
OPTICAL	**Clarity/transparency**	—	—	—	—
	Transmittance (ASTM D1003)	%	—	—	—
	Refractive index (23°C) (ASTM D542)	—	—	—	1.535–1.570

Data sources:
1947: *Modern Plastics Encyclopedia*, vol. 3, *Charts* (New York: Plastics Catalogue Corporation).
1967: Shell Chemicals, *Properties of Plastics*, rev. metric ed. (London: Shell International Chemical Company and Shell Chemicals UK).
2016: George Wypych, *Handbook of Polymers*, 2nd ed. (Toronto: ChemTec).

12.2 | Shellac

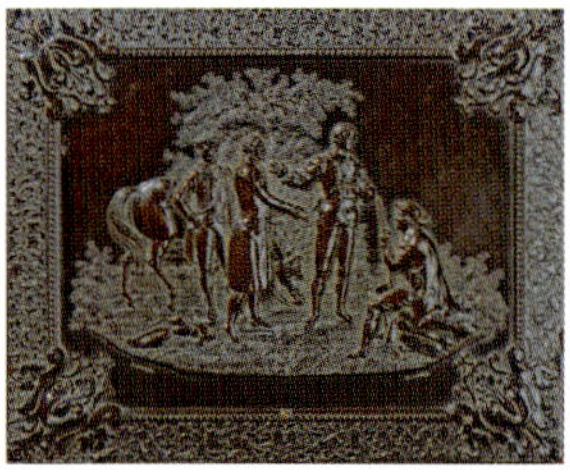

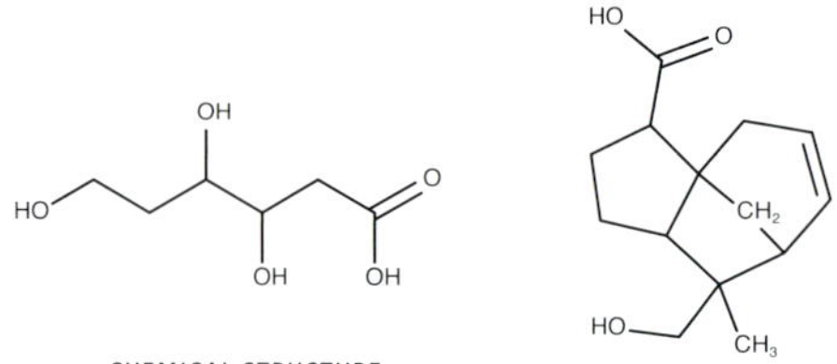

CHEMICAL STRUCTURE

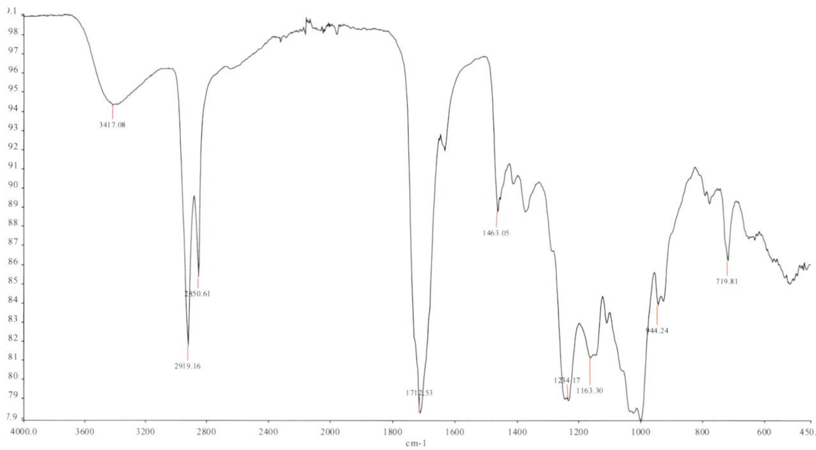

FTIR SPECTRUM

Date of first manufacture

Since ancient times; used commercially from the 1860s to the 1940s. Used today for lacquers and inlay material.

Alternate/trade name examples

Diatite, Florence Compound, Complac, Compo-Site.

Processing

Shellac is a resinous product of zoological origin, specifically the exudation of the females of certain insects (lac bugs) that live on host trees, principally in India. It consists of resin, wax, and colorants. The lac incrustations are collected and purified by a hot filtration process to give shellac. Shellac is a thermoplastic resin, but on prolonged heating it thermosets. Shellac can be hot molded.

Selected applications

Gramophone records, lacquer and inlay material for furniture, Union cases for daguerreotypes.

Feel/look/sound/smell

Shellac is a rigid, hard material and comes in various red colors; becomes pale yellow when bleached.

Main degradation agents

Light, UV radiation, oxygen, ozone, moisture (changes in relative humidity), temperature, stress, chemicals.

Main signs of degradation and damage

Darkening, opacity, yellowing, brittleness.

12.2 | Shellac

	PROPERTY	UNITS (SI)	DATA 1947	DATA 1961	DATA 1967	DATA 2016
MECHANICAL	**Compressive strength** (ASTM D695)	MPa	68.9–117.2	—	—	—
	Elongation (ASTM D638)	%	—	—	—	—
	Flexural strength (ASTM D790)	MPa	—	—	—	—
	Hardness (ASTM D785)	Rockwell	—	—	—	—
	Impact strength (Izod notched) (ASTM D256)	J/cm	1.4–1.55	—	—	—
	Tensile modulus (rigidity) (ASTM D638)	GPa	3.4–4.14	—	—	—
	Tensile strength (ASTM D638)	MPa	6.2–13.8	—	—	—
PHYSICAL	**Density** (ASTM D792)	g/cm^3	—	1.207 (not bleached) 1.196 (bleached) 1.217 (dry refined)	—	—
	Hildebrand solubility parameter (δ)	$(MPa)^{1/2}$	—	—	—	—
	Permeability: H_2O vapor O_2 CO_2 (ASTM D1434)	$cm^3 \cdot cm \cdot cm^{-2} \cdot s^{-1} \cdot Pa^{-1}$	— — —	— — —	— — —	— — —
	Resistance to solvents p = poor, f = fair, g = good	Alcohol (p), ketones (p), mineral spirit (g)				
	Surface tension	mNm^{-1}	—	—	—	—
	Water absorption (equilibrium 24 hours) ⅛-inch (3.175 mm) thickness (ASTM D570)	%	—	2 (not bleached) 6 (bleached)	—	—
THERMAL	**Coefficient of linear thermal expansion** (ASTM D696)	$°C^{-1}$ ($\times 10^{-5}$)	—	—	—	—
	Glass transition temperature (T_g)	°C	—	—	—	—
	Low-temperature performance / Brittleness temperature (ASTM D746)	°C	—	—	—	—
	Melting point (T_m) (ASTM D1238)	°C	—	—	—	—
	Softening point (Vicat) (ASTM D1525)	°C	—	70–80	—	—
OPTICAL	**Clarity/transparency**	—	Opaque	—	—	—
	Transmittance (ASTM D1003)	%	—	—	—	—
	Refractive index (23°C) (ASTM D542)	—	—	1.516 (not bleached) 1.534 (bleached)	—	—

Data sources:
1947: *Modern Plastics Encyclopedia*, vol. 3, *Charts* (New York: Plastics Catalogue Corporation).
1961: Edward Hicks, *Shellac: Its Origin and Applications* (New York: Chemical Publishing Company).
1967: Shell Chemicals, *Properties of Plastics*, rev. metric ed. (London: Shell International Chemical Company and Shell Chemicals UK).
2016: George Wypych, *Handbook of Polymers*, 2nd ed. (Toronto: ChemTec).

12.3 | Tortoiseshell

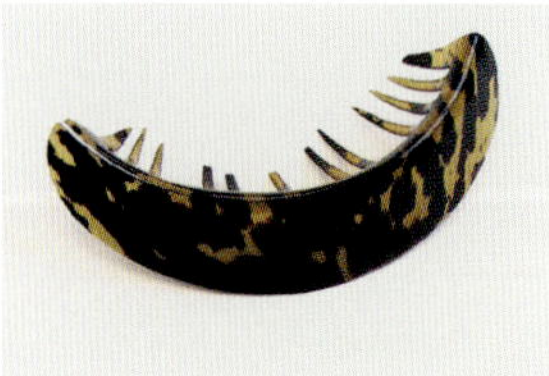

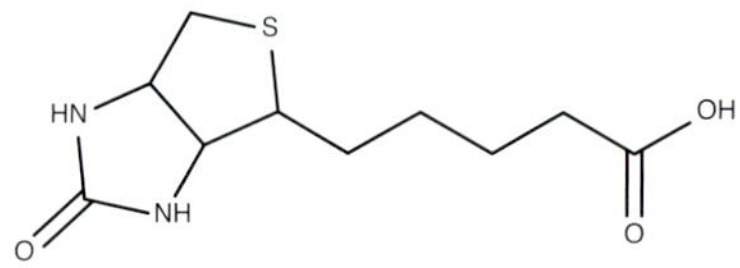

CHEMICAL STRUCTURE

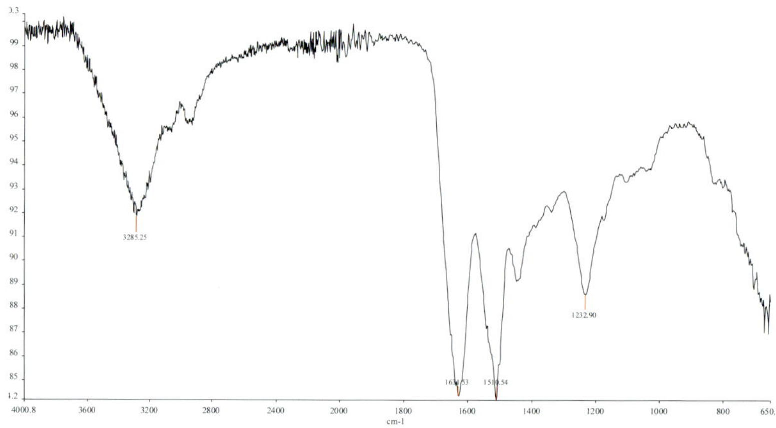

FTIR SPECTRUM

Date of first manufacture
From ancient times. Trade was banned worldwide in the 1970s.

Alternate/trade name examples
N/A

Natural polymer
Tortoiseshell is material from the shells of turtles and tortoises; it consists of the proteins keratin and chitin.

Processing
Tortoiseshell is a natural thermoplastic; it can be softened by the application of heat and molded into various forms. On cooling, the material retains the form into which it has been shaped. Tortoiseshell was cut in pieces or slices and then softened by boiling in salt water. The softened tortoiseshell pieces were then flattened in a press.

Applications
Combs, guitar picks/plectra, eyeglass frames, inlaid veneers for furniture and small boxes, knitting needles, picture frames.

Feel/look/sound/smell
Usually mottled, but also uniform dark brown, pale yellow, amber-like, and reddish brown. High polish. True tortoiseshell is heavier than its imitations (e.g., cellulose nitrate, cellulose acetate).

Main degradation agent(s)
Light, UV radiation, oxygen, ozone, moisture (changes in relative humidity), temperature, stress, chemicals.

Main signs of degradation and damage
Discoloration, dull/matte surface.

12.3 | Tortoiseshell

	PROPERTY	UNITS (SI)	DATA 1947	DATA 1967	DATA 1994
MECHANICAL	**Compressive strength** (ASTM D695)	MPa	—	—	—
	Elongation (ASTM D638)	%	—	—	—
	Flexural strength (ASTM D790)	MPa	—	—	—
	Hardness (ASTM D785)	Rockwell	—	—	—
	Impact strength (Izod notched) (ASTM D256)	J/cm	—	—	—
	Tensile modulus (rigidity) (ASTM D638)	GPa	—	—	—
	Tensile strength (ASTM D638)	MPa	—	—	—
PHYSICAL	**Density** (ASTM D792)	g/cm^3	—	—	1.26–1.35
	Hildebrand solubility parameter (δ)	(MPa)$^{1/2}$	—	—	—
	Permeability: H_2O vapor O_2 CO_2 (ASTM D1434)	cm^3·cm·cm^{-2}·s^{-1}·Pa^{-1}	— — —	— — —	— — —
	Resistance to solvents p = poor, f = fair, g = good	Not commonly mentioned			
	Surface tension	mNm^{-1}	—	—	—
	Water absorption (equilibrium 24 hours) ⅛-inch (3.175 mm) thickness (ASTM D570)	%	—	—	—
THERMAL	**Coefficient of linear thermal expansion** (ASTM D696)	°C^{-1} (× 10^{-5})	—	—	—
	Glass transition temperature (T_g)	°C	—	—	—
	Low-temperature performance / Brittleness temperature (ASTM D746)	°C	—	—	—
	Melting point (T_m) (ASTM D1238)	°C	—	—	—
	Softening point (Vicat) (ASTM D1525)	°C	—	—	—
OPTICAL	**Clarity/transparency**	—	—	—	Semitransparent, semitranslucent
	Transmittance (ASTM D1003)	%	—	—	—
	Refractive index (23°C) (ASTM D542)	—	—	—	1.54

Data sources:
1947: *Modern Plastics Encyclopedia*, vol. 3, *Charts* (New York: Plastics Catalogue Corporation).
1967: Shell Chemicals, *Properties of Plastics*, rev. metric ed. (London: Shell International Chemical Company and Shell Chemicals UK).
1994: Robert Webster, *Gems: Their Sources, Descriptions and Identification*, 5th ed., rev. by Peter George Read (Oxford: Butterworth-Heinemann).

12.4 | Vulcanized fiber

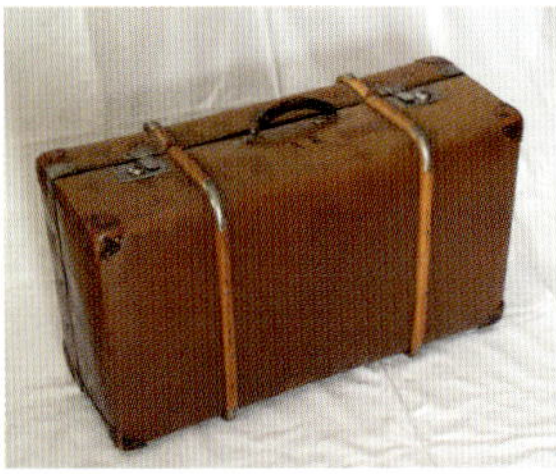

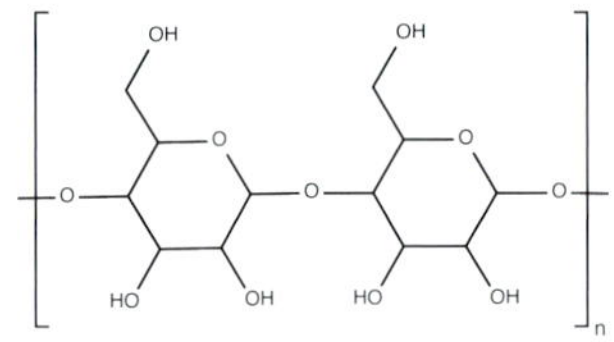

CHEMICAL STRUCTURE

Cellulose

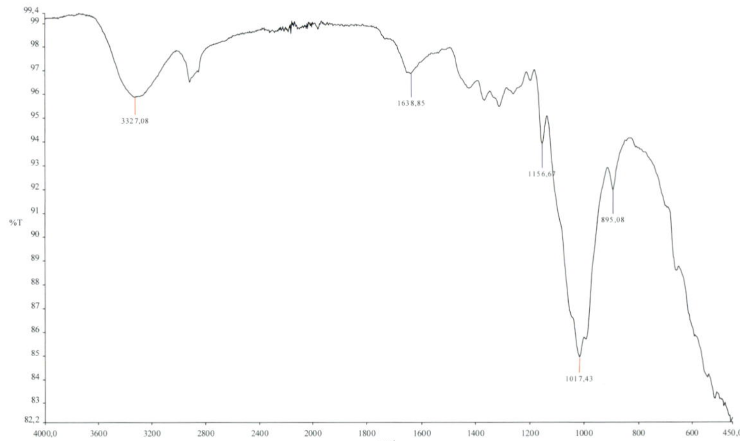

FTIR SPECTRUM

Date of first manufacture
1859 (England), 1871 (USA). Remained in use until the advent of thermoplastics after World War II. Still in use for technical and electrical applications.

Alternate/trade name examples
Vulcan fiber, vulcanized fiber, DYNOS, DYNAL, VULCOID.

Processing
Vulcanized fiber consists of thin sheets of high-quality paper that are first gelatinized by immersion in a strong aqueous zinc chloride ($ZnCl_2$) solution. Layers of the still-gelatinous paper are then passed through rollers to merge them. At the next stage, the sheets are soaked in water to remove the gelatinizing agent. After leaching, the material is dried on steam-heated drums, calendered, and taken off in rolls or sheets.

Selected applications
Electrical insulation, flexible abrasive discs, lightweight attaché cases and suitcases.

Feel/look/sound/smell
Tough, somewhat leathery.

Main degradation agents
Moisture (changes in relative humidity), temperature, light, UV radiation, oxygen, ozone, stress, chemicals.

Main signs of degradation and damage
Distortion, discoloration, breaks, tearing.

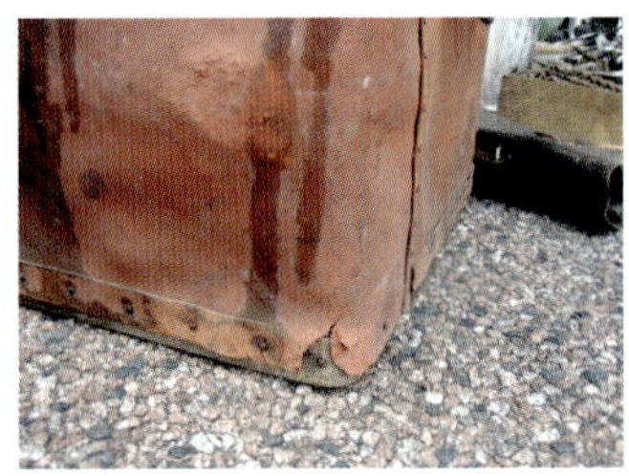

Discolored, distorted, and torn suitcase.

12.4 | Vulcanized fiber

	PROPERTY	UNITS (SI)	DATA 1947	DATA 1988	DATA 2016
MECHANICAL	**Compressive strength** (ASTM D695)	MPa	138-222	—	—
	Elongation (ASTM D638)	%	—	—	—
	Flexural strength (ASTM D790)	MPa	76-152	—	—
	Hardness (ASTM D785)	Rockwell R	R60-R100	R80-R140	—
	Impact strength (Izod notched) (ASTM D256)	J/cm	0.42-4.3	—	—
	Tensile modulus (rigidity) (ASTM D638)	GPa	—	—	—
	Tensile strength (ASTM D638)	MPa	34-83	85-100	—
PHYSICAL	**Density** (ASTM D792)	g/cm^3	1.0-1.5	1.1-1.45	—
	Hildebrand solubility parameter (δ)	$(MPa)^{1/2}$	—	—	—
	Permeability: H_2O vapor O_2 CO_2 (ASTM D1434)	$cm^3 \cdot cm \cdot cm^{-2} \cdot s^{-1} \cdot Pa^{-1}$	— — —	— — —	— — —
	Resistance to solvents p = poor, f = fair, g = good	Oils (g), fats (g), organic solvents (g), diluted acids and alkalis (g)			
	Surface tension	mNm^{-1}	—	—	—
	Water absorption (equilibrium 24 hours) ⅛-inch (3.175 mm) thickness (ASTM D570)	%	—	7-9	—
THERMAL	**Coefficient of linear thermal expansion** (ASTM D696)	$°C^{-1}$ ($\times 10^{-5}$)	—	—	—
	Glass transition temperature (T_g)	°C	—	—	—
	Low-temperature performance / Brittleness temperature (ASTM D746)	°C	—	—	—
	Melting point (T_m) (ASTM D1238)	°C	—	—	—
	Softening point (Vicat) (ASTM D1525)	°C	—	—	—
OPTICAL	**Clarity/transparency**	—	Opaque	Opaque	—
	Transmittance (ASTM D1003)	%	—	—	—
	Refractive index (23°C) (ASTM D542)	—	—	—	—

Data sources:
1947: *Modern Plastics Encyclopedia*, vol. 3, *Charts* (New York: Plastics Catalogue Corporation).
1988: Hans Domininghaus, Guide chart of physical properties of polymers, in *Die Kunststoffe und ihre Eigenschaften* (Berlin, Heidelberg: Springer-Verlag).
2016: George Wypych, *Handbook of Polymers*, 2nd ed. (Toronto: ChemTec).

13.1 | Cellulose acetate (CA)

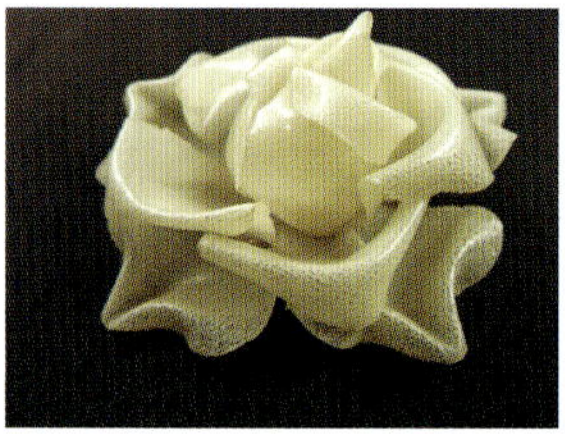

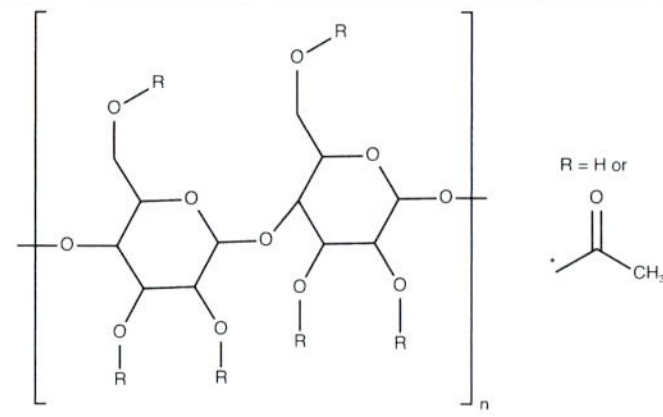

CHEMICAL STRUCTURE

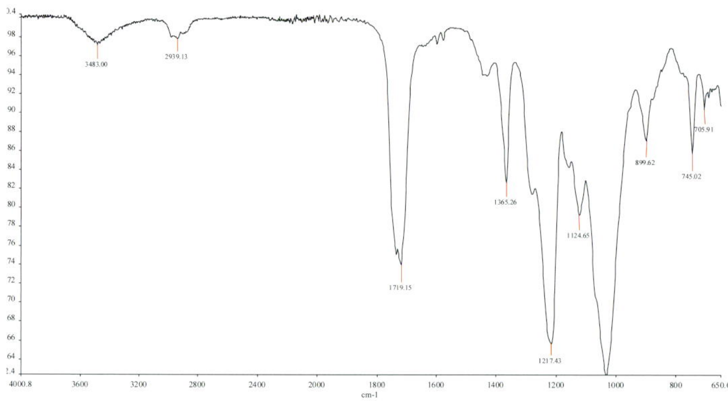

FTIR SPECTRUM

Date of first manufacture
Fibers: 1865 (France), 1892 (UK), 1904 (Switzerland, Germany, USA). Injection molding: 1927 (UK). Since the 1970s new interest in CA and other biobased plastics has been shown due to sustainability concerns.

Alternate/trade name examples
Lumarith, Celanese, Cellit, Estron, Plastacele, Bexoid, Tenite, Clarifoil.

Polymerization
Cellulose acetate is made from wood pulp (cellulose) and acetic acid by condensation polymerization (catalyst: sulfuric acid).

Processing
Extrusion, cast from solution, injection molding, blow molding.
Plasticizers—di(2-ethylhexyl) phthalate, diethyl phthalate, dimethyl phthalate, dimethyl sebacate, dioctyl sebacate, tri(2-ethylhexyl) phosphate, triphenyl phosphate, etc.—are added before processing.

Selected applications
Artworks, automotive parts, combs, cutlery handles, electrical parts, eyeglass frames, handbags, jewelry, knobs, packaging sheet, pens and pencils, photographic film, pressure-sensitive tape, shoe heels, steering wheels, toothbrushes, toys.

Feel/look/sound/smell
Resembles tortoiseshell, ivory, amber, or horn; without colorants it is transparent. When aged, it has an acetic acid smell indicative of vinegar syndrome.

Main degradation agents
Moisture (changes in relative humidity), light, UV radiation, oxygen, ozone, temperature, stress, chemicals.

Main signs of degradation and damage
Cracking, warping, shrinkage, delamination, plasticizer exudation, acetic acid droplets, white bloom.

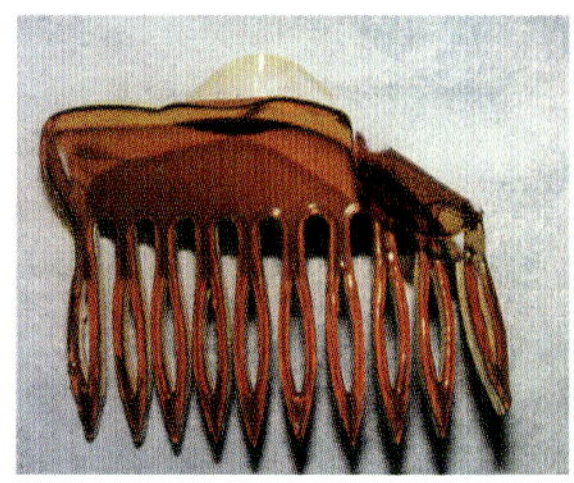

Warped, delaminated haircomb.

13.1 | Cellulose acetate (CA)

	PROPERTY	UNITS (SI)	DATA 1947	DATA 1967	DATA 2016
MECHANICAL	**Compressive strength** (ASTM D695)	MPa	34.5-248.2	15.2-248	—
	Elongation (ASTM D638)	%	4-45	6-70	15-70
	Flexural strength (ASTM D790)	MPa	13.7-89.6	13.8-110.5	27-72
	Hardness (ASTM D785)	Rockwell R	R80-R120	R35-R125	R31-R117
	Impact strength (Izod notched) (ASTM D256)	J/cm	0.37-3.20	0.21-2.78	0.48-3.52
	Tensile modulus (rigidity) (ASTM D638)	GPa	0.69-2.41	4.49-27.6	1.6-2.46
	Tensile strength (ASTM D638)	MPa	20.7-55.2	13.1-58.8	24-53
PHYSICAL	**Density** (ASTM D792)	g/cm^3	1.27	1.23-1.34	1.27-1.34
	Hildebrand solubility parameter (δ)	$(MPa)^{1/2}$	—	—	23.5-27.83
	Permeability: H_2O vapor O_2 CO_2 (ASTM D1434) N.T.P. = normal temperature and pressure	1967: cc(N.T.P.)$\cdot cm^{-2}\cdot sec^{-1}\cdot mil^{-1}\cdot cm\ Hg^{-1}$ at 25°C × 10^{-8} 2016: $cm^3\cdot cm\cdot cm^{-2}\cdot s^{-1}\cdot Pa^{-1}$ × 10^{-13}	— — —	29000-47000 3-4 27-28	4.000-5.000 0.6 —
	Resistance to solvents p = poor, f = fair, g = good	Ketones (p), esters (p), alcohols (f), hydrocarbons (g)			
	Surface tension	mNm^{-1}	—	—	45.9
	Water absorption (equilibrium 24 hours) ⅛-inch (3.175 mm) thickness (ASTM D570)	%	1.5-2.9	1.9-6.5	2-6.5
THERMAL	**Coefficient of linear thermal expansion** (ASTM D696)	$°C^{-1}$ (× 10^{-5})	10-15	8-16	8-18
	Glass transition temperature (T_g)	°C	—	—	95-108
	Low-temperature performance / Brittleness temperature (ASTM D746)	°C	—	Good	−40
	Melting point (T_m) (ASTM D1238)	°C	—	—	138-180
	Softening point (Vicat) (ASTM D1525)	°C	—	70	78-113
OPTICAL	**Clarity/transparency**	—	Transparent or translucent	Transparent	—
	Transmittance (ASTM D1003)	%	—	—	>90%
	Refractive index (23°C) (ASTM D542)	—	1.47	1.46-1.50	1.46-1.49

Data sources:
1947: *Modern Plastics Encyclopedia*, vol. 3, *Charts* (New York: Plastics Catalogue Corporation).
1967: Shell Chemicals, *Properties of Plastics*, rev. metric ed. (London: Shell International Chemical Company and Shell Chemicals UK).
2016: George Wypych, *Handbook of Polymers*, 2nd ed. (Toronto: ChemTec).

13.2 | Cellulose acetate-butyrate (CAB)

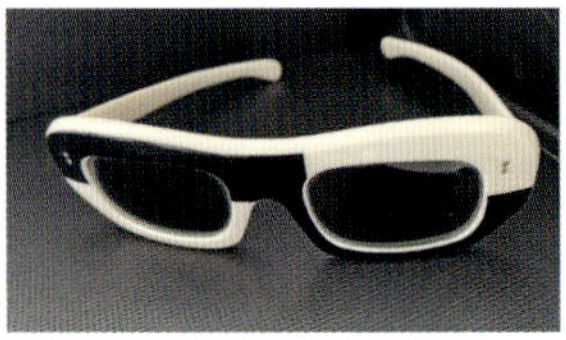

7
OTHER

PLASTICS RECYCLING CODE

CHEMICAL STRUCTURE

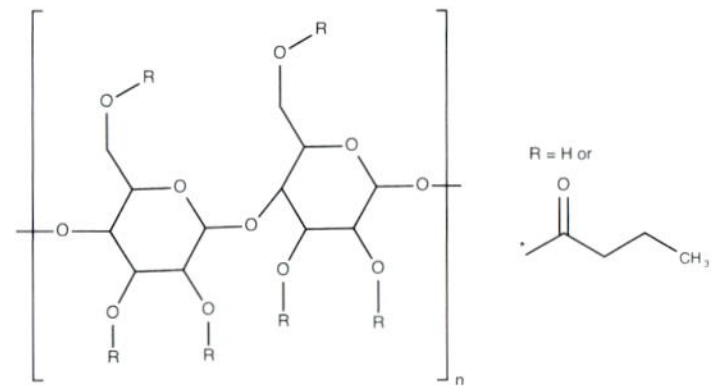

FTIR SPECTRUM

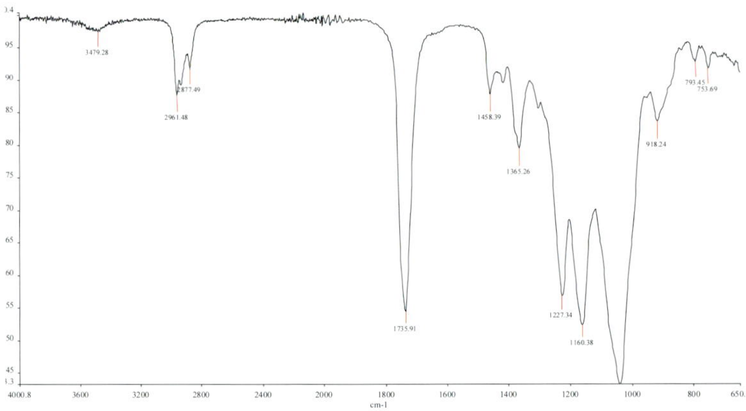

Date of first manufacture
1929 (invented USA), 1932 (production USA).

Alternate/trade name examples
Brandobutyrar, Cellidor, Cellidor B, Cabulite, Tenite Butyrate, Hercose C, Tenite II, Tenite, Uvex, Excelon, Spartech.

Polymerization
Condensation polymerization from cellulose and mixture of butyric acid and acetic anhydride with sulfuric acid as catalyst. Usually 10%–30% of plasticizer is added to decrease the molding temperature. Hydroxy ethyl cellulose for antistatic properties and release agents such as microcrystalline wax, polyethylene wax, or silicone can be added.

Processing
Injection and compression molding, extrusion.

Selected applications
Blister packaging (laminated with aluminum foil), gas masks, lacquers, laminated safety glass and eyeglass frames, mugs, outdoor signs, parts for reflectors, piping, recording tape, tool handles, toothbrushes.

Feel/look/sound/smell
Transparent, glossy.

Main degradation agents
Moisture (changes in relative humidity), light, UV radiation, oxygen, ozone, temperature, stress, chemicals.

Main signs of degradation and damage
Smell of rancid butter, yellowing; more stable (no warping, no delamination) than cellulose acetate.

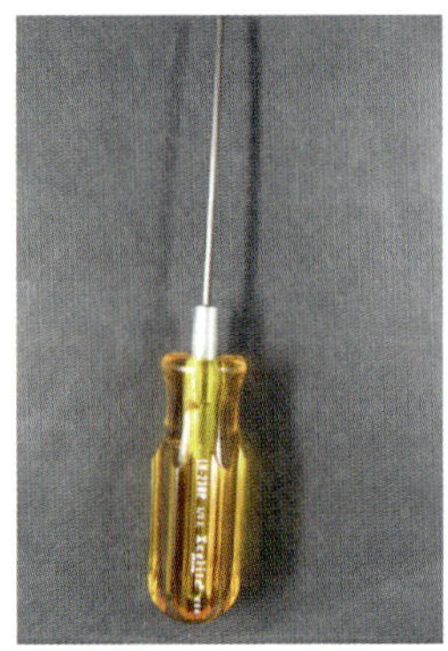

A degraded screwdriver handle may smell of rancid butter.

13.2 | Cellulose acetate-butyrate (CAB)

	PROPERTY	UNITS (SI)	DATA 1947	DATA 1967	DATA 2016
MECHANICAL	**Compressive strength** (ASTM D695)	MPa	51.7-151.7	14.5-152	14-52
	Elongation (ASTM D638)	%	35-95	40-90	19-90
	Flexural strength (ASTM D790)	MPa	11.0-82.7	12.4-64.3	21-70
	Hardness (ASTM D785)	Rockwell R	R35-R120	R31-R116	R26-R116
	Impact strength (Izod notched) (ASTM D256)	J/cm	0.21-5.02	0.43-3.36	0.8-5.30
	Tensile modulus (rigidity) (ASTM D638)	GPa	0.34-1.38	0.35-1.38	0.3-2.0
	Tensile strength (ASTM D638)	MPa	9.7-45.5	17.9-47.7	16-51
PHYSICAL	**Density** (ASTM D792)	g/cm^3	1.15-1.21	1.15-1.22	1.16-1.26
	Hildebrand solubility parameter (δ)	(MPa)$^{1/2}$	—	—	22.2
	Permeability: H_2O vapor O_2 CO_2 (ASTM D1434)	cm^3·cm·cm^{-2}·s^{-1}·Pa^{-1} × 10^{12}	— — —	— — —	— 0.356 —
	Resistance to solvents p = poor, f = fair, g = good	Ketones (p), esters (p), alcohols (p), little affected by hydrocarbons			
	Surface tension	mNm^{-1}	—	—	34.0
	Water absorption (equilibrium 24 hours) ⅛-inch (3.175 mm) thickness (ASTM D570)	%	1.0-2.4	0.9-2.2	1.3-2.2
THERMAL	**Coefficient of linear thermal expansion** (ASTM D696)	°C^{-1} (× 10^{-5})	11-17	11-17	12-17
	Glass transition temperature (T_g)	°C	—	—	85-161
	Low-temperature performance / Brittleness temperature (ASTM D746)	°C	—	Good	−40
	Melting point (T_m) (ASTM D1238)	°C	—	—	150-240
	Softening point (Vicat) (ASTM D1525)	°C	—	70	65-102
OPTICAL	**Clarity/transparency**	—	Transparent, translucent, or opaque	Transparent	—
	Transmittance (ASTM D1003)	%	—	—	>90%
	Refractive index (23°C) (ASTM D542)	—	1.47-1.49	1.46-1.49	1.47-1.48

Data sources:
1947: *Modern Plastics Encyclopedia*, vol. 3, *Charts* (New York: Plastics Catalogue Corporation).
1967: Shell Chemicals, *Properties of Plastics*, rev. metric ed. (London: Shell International Chemical Company and Shell Chemicals UK).
2016: George Wypych, *Handbook of Polymers*, 2nd ed. (Toronto: ChemTec).

13.3 | Cellulose acetate-propionate (CAP)

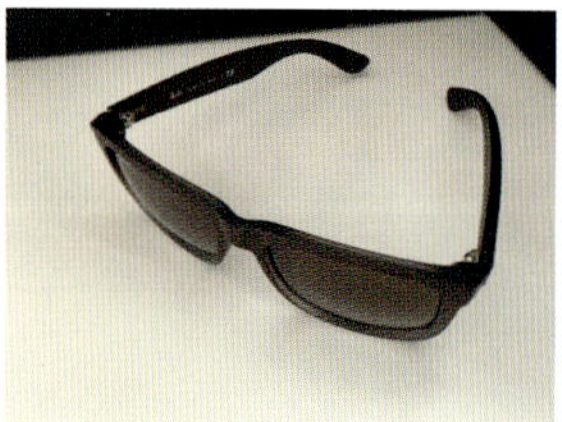

7
OTHER

PLASTICS RECYCLING CODE

CHEMICAL STRUCTURE

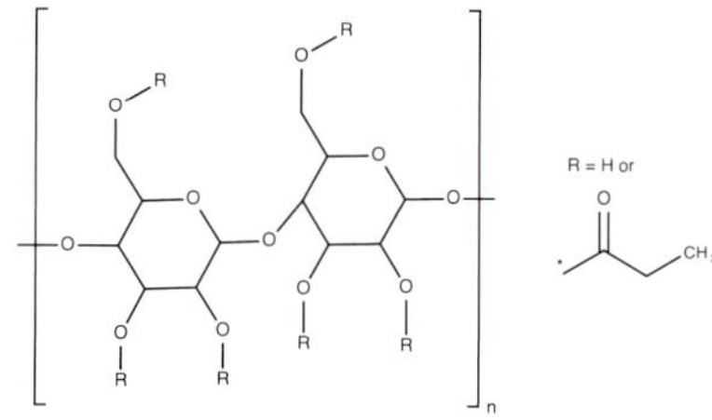

FTIR SPECTRUM

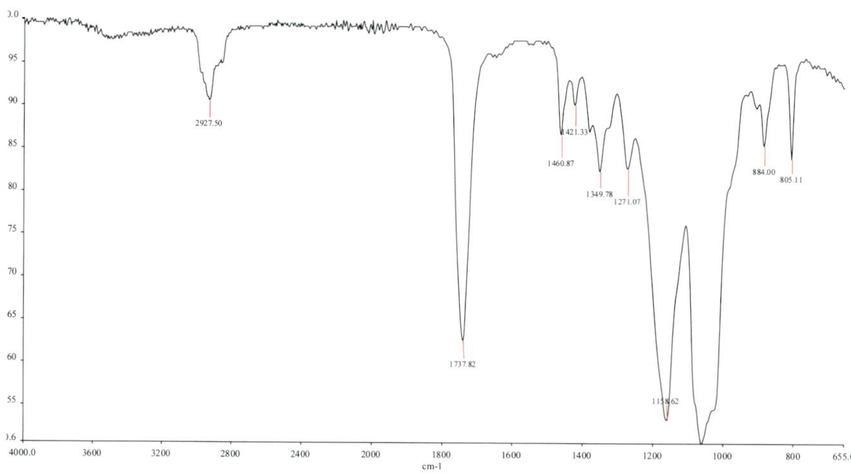

Date of first manufacture
1924 (invented USA), 1937 (production USA).

Alternate/trade name examples
Cellidor, Tenite Propionate, Forticel.

Polymerization
Condensation polymerization from cellulose, propionic acid, and acetic acid.

Processing
Injection molding.

Selected applications
Coatings, eyeglass frames, films, medical applications, nail polish, pharmaceuticals, blister packs for dispensing tablets.

Feel/look/sound/smell
Tough, glossy, transparent.

Main degradation agents
Light, UV radiation, oxygen, ozone, moisture, temperature, stress, chemicals.

Main signs of degradation and damage
None; more stable than cellulose acetate and cellulose acetate-butyrate.

13.3 | Cellulose acetate-propionate (CAP)

	PROPERTY	UNITS (SI)	DATA 1947	DATA 1967	DATA 2016
MECHANICAL	**Compressive strength** (ASTM D695)	MPa	—	21.4-152	58
	Elongation (ASTM D638)	%	13-53	30-100	3-45
	Flexural strength (ASTM D790)	MPa	27.6-67.6	26.9-76	29-58
	Hardness (ASTM D785)	Rockwell R	R63-104	R20-120	R40-96
	Impact strength (Izod notched) (ASTM D256)	J/cm	0.63-8.98	0.34-5.33	1.3-5.2
	Tensile modulus (rigidity) (ASTM D638)	GPa	0.10-1.93	0.41-1.48	1-2.2
	Tensile strength (ASTM D638)	MPa	19.3-41.4	13.8-50.5	22-66
PHYSICAL	**Density** (ASTM D792)	g/cm^3	1.17-1.22	1.18-1.24	1.17-1.25
	Hildebrand solubility parameter (δ)	$(MPa)^{1/2}$	—	—	—
	Permeability: H_2O vapor O_2 CO_2 (ASTM D1434)	$cm^3 \cdot cm \cdot cm^{-2} \cdot s^{-1} \cdot Pa^{-1}$	— — —	— — —	— — —
	Resistance to solvents p = poor, f = fair, g = good	Ketones (p), esters (p), alcohols (f), hydrocarbons (f)			
	Surface tension	mNm^{-1}	—	—	—
	Water absorption (equilibrium 24 hours) ⅛-inch (3.175 mm) thickness (ASTM D570)	%	1.0-1.7	1.2-2.8	2.2-5.0
THERMAL	**Coefficient of linear thermal expansion** (ASTM D696)	$°C^{-1}$ ($\times 10^{-5}$)	12-19	11-17	12-18
	Glass transition temperature (T_g)	°C	—	—	128-145
	Low-temperature performance / Brittleness temperature (ASTM D746)	°C	—	Good	−40
	Melting point (T_m) (ASTM D1238)	°C	—	—	184-210
	Softening point (Vicat) (ASTM D1525)	°C	—	—	87-96
OPTICAL	**Clarity/transparency**	—	Transparent, translucent, or opaque	Transparent	—
	Transmittance (ASTM D1003)	%	—	—	90%
	Refractive index (23°C) (ASTM D542)	—	1.475-1.490	1.46-1.49	1.46-1.475

Data sources:
1947: *Modern Plastics Encyclopedia*, vol. 3, *Charts* (New York: Plastics Catalogue Corporation).
1967: Shell Chemicals, *Properties of Plastics*, rev. metric ed. (London: Shell International Chemical Company and Shell Chemicals UK).
2016: George Wypych, *Handbook of Polymers*, 2nd ed. (Toronto: ChemTec).

13.4 | Cellulose nitrate (CN)

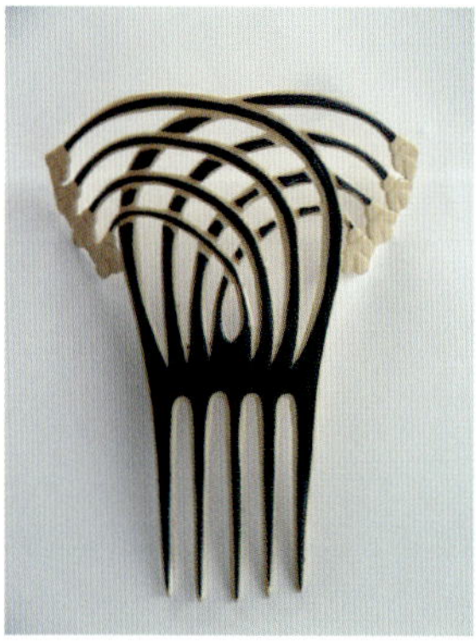

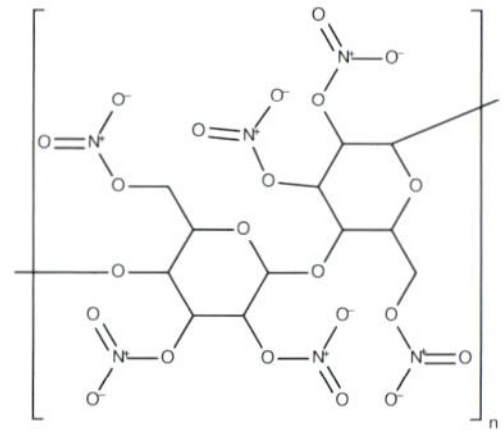

CHEMICAL STRUCTURE

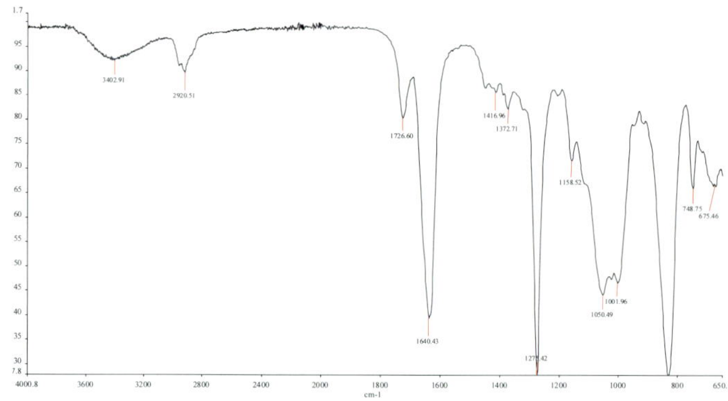

FTIR SPECTRUM

Date of first manufacture
1862 (England), 1870 (USA). Since 1940 replaced by cellulose acetate due to its flammability. Some cellulose nitrate products are still being made, such as lacquers and Ping-Pong balls.

Alternate/trade name examples
Celluloid, Parkesine, French Ivory, Xylonite, Pyralin.

Polymerization
Condensation polymerization of nitric acid and cellulose (cotton cellulose fibers), with sulfuric acid as catalyst and camphor and other plasticizers added at processing.

Processing
Blow molding, casting, compounding, compression molding, machining; made into blocks that are sliced into thin sheets.

Selected applications
Bathroom accessories, belts, combs, dolls, eyeglass frames, film material, handbags, knife handles, mirror frames, powder boxes, toothbrushes, toys. Has also been used for coatings on textile to imitate leather for upholstery in cars and furniture.

Feel/look/sound/smell
Transparent to opaque; imitation tortoiseshell, ivory, or amber; any color, including mottled and pearlescent; smell of camphor.

Main degradation agents
Moisture (changes in relative humidity), light, UV radiation, oxygen, ozone, temperature, stress, chemicals.

Main signs of degradation and damage
Characteristic cubic degradation pattern (like sugar crystals), pungent nitric acid smell, crazing, cracking, haziness (loss of transparency), brittleness, blooming, camphor crystals, plasticizer efflorescence.

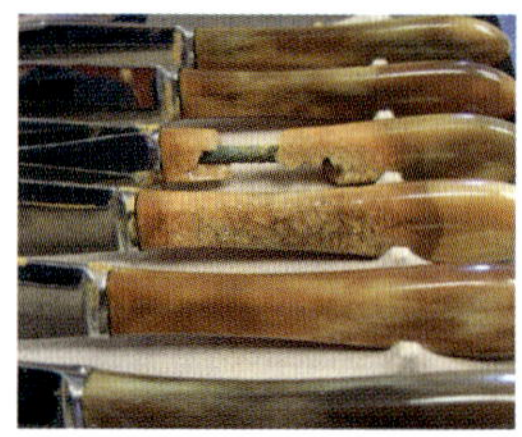

Knife handles with cracking, crazing, and crystal formations.

13.4 | Cellulose nitrate (CN)

	PROPERTY	UNITS (SI)	DATA 1947	DATA 1967	DATA 2016
MECHANICAL	**Compressive strength** (ASTM D695)	MPa	137.9–206.8	—	14–55
	Elongation (ASTM D638)	%	25–50	—	10–40
	Flexural strength (ASTM D790)	MPa	41.4–103.4	—	62–76
	Hardness (ASTM D785)	Rockwell R	R90–R120	—	R95–R115
	Impact strength (Izod notched) (ASTM D256)	J/cm	1.07–4.27	—	2.70–3.70
	Tensile modulus (rigidity) (ASTM D638)	GPa	1.38–2.76	—	—
	Tensile strength (ASTM D638)	MPa	20.7–69.0	—	35–70
PHYSICAL	**Density** (ASTM D792)	g/cm^3	1.35–1.40	—	1.35–1.40
	Hildebrand solubility parameter (δ)	$(MPa)^{1/2}$	—	—	21.4–23.5
	Permeability: H_2O vapor O_2 CO_2 (ASTM D1434)	$cm^3 \cdot cm \cdot cm^{-2} \cdot s^{-1} \cdot Pa^{-1} \times 10^{12}$	— — —	— — —	0.146 472 —
	Resistance to solvents p = poor, f = fair, g = good	Ketones (p), esters (p), alcohols (f), hydrocarbons (f)			
	Surface tension	mNm^{-1}	—	—	38.0
	Water absorption (equilibrium 24 hours) ⅛-inch (3.175 mm) thickness (ASTM D570)	%	0.7–4.0	—	0.6–2.0
THERMAL	**Coefficient of linear thermal expansion** (ASTM D696)	$°C^{-1}$ ($\times 10^{-5}$)	9–16	—	6–7
	Glass transition temperature (T_g)	°C	—	—	53–66
	Low-temperature performance / Brittleness temperature (ASTM D746)	°C	—	—	−20
	Melting point (T_m) (ASTM D1238)	°C	—	—	142–170 (ignites)
	Softening point (Vicat) (ASTM D1525)	°C	—	—	—
OPTICAL	**Clarity/transparency**	—	Transparent, translucent, or opaque	—	—
	Transmittance (ASTM D1003)	%	—	—	—
	Refractive index (23°C) (ASTM D542)	—	1.50	—	1.49–1.51

Data sources:
1947: *Modern Plastics Encyclopedia*, vol. 3, *Charts* (New York: Plastics Catalogue Corporation).
1967: Shell Chemicals, *Properties of Plastics*, rev. metric ed. (London: Shell International Chemical Company and Shell Chemicals UK).
2016: George Wypych, *Handbook of Polymers*, 2nd ed. (Toronto: ChemTec).

13.5 | Ethyl cellulose (EC)

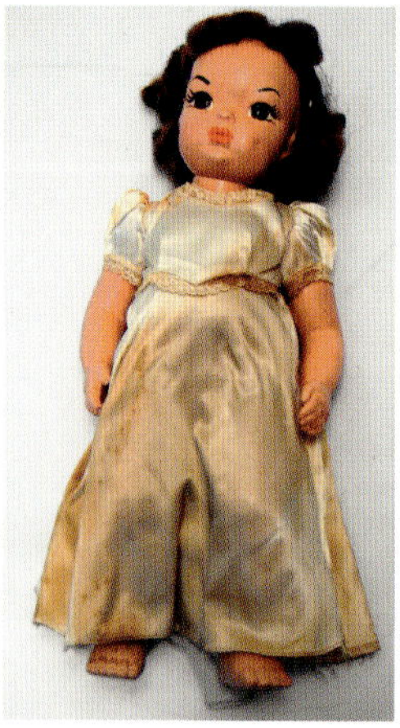

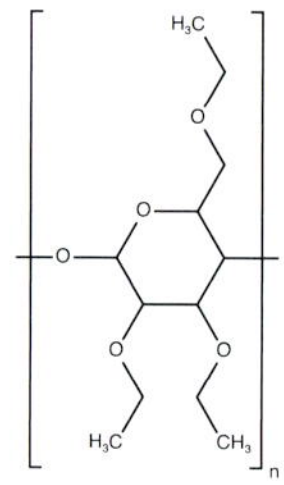

CHEMICAL STRUCTURE

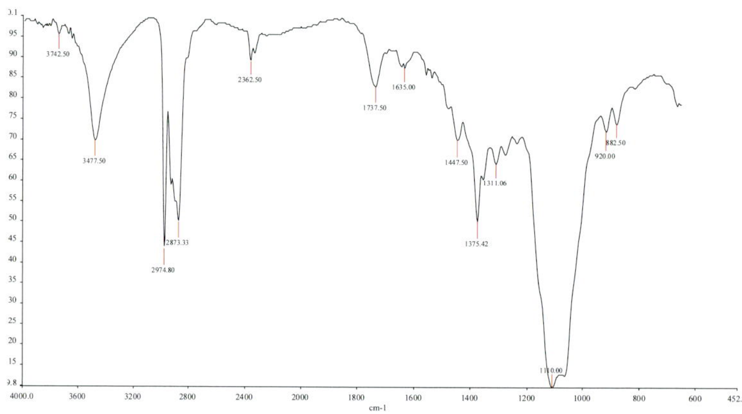

FTIR SPECTRUM

Date of first manufacture
1912 (USA, discovery), 1935 (USA, production).

Alternate/trade name examples
ETHOCEL, Nixon E/C.

Polymerization
Ethyl cellulose, a cellulose ether, is polymerized from cellulose, ethyl chloride, and sodium hydroxide.

Processing
Injection and compression molding, sheet extrusion, solvent casting.

Selected applications
Ethyl cellulose was used to manufacture plastic dolls from the early days until the switch to cellulose nitrate was made in the 1950s; early EC dolls were painted with a colored varnish. Ethyl cellulose was otherwise mostly used as flexible coatings for paper, textiles, and leather. Today, it is used for pharmaceutical applications such as coatings on tablets. EC is the toughest thermoplastic of the cellulose derivatives and has outstanding properties as an adhesive and emulsifier.

Feel/look/sound/smell
N/A

Main degradation agents
Light, UV radiation, oxygen, ozone, moisture, temperature, stress, chemicals.

Main signs of degradation and damage
Slow rate of deterioration, damage not commonly detected.

13.5 | Ethyl cellulose (EC)

	PROPERTY	UNITS (SI)	DATA 1947	DATA 1967	DATA 2016
MECHANICAL	**Compressive strength** (ASTM D695)	MPa	20.7-137.9	—	—
	Elongation (ASTM D638)	%	5-100	—	5-40
	Flexural strength (ASTM D790)	MPa	20.7-82.7	—	28-83
	Hardness (ASTM D785)	Rockwell R	R50-R110	—	R50-R115
	Impact strength (Izod notched) (ASTM D256)	J/cm	1.07-6.14	—	0.2
	Tensile modulus (rigidity) (ASTM D638)	GPa	0.41-3.44	—	1.4-1.8
	Tensile strength (ASTM D638)	MPa	13.8-68.9	—	14-62
PHYSICAL	**Density** (ASTM D792)	g/cm^3	1.07-1.18	—	1.07-1.18
	Hildebrand solubility parameter (δ)	$(MPa)^{1/2}$	—	—	20.4-21.1
	Permeability: H_2O vapor O_2 CO_2 (ASTM D1434)	$cm^3 \cdot cm \cdot cm^{-2} \cdot s^{-1} \cdot Pa^{-1}$ $\times 10^{12}$	— — —	— — —	670 1.1 —
	Resistance to solvents p = poor, f = fair, g = good	Ketones (p), esters (p), alcohols (f), hydrocarbons (g)			
	Surface tension	mNm^{-1}	—	—	31.8-34.1
	Water absorption (equilibrium 24 hours) ⅛-inch (3.175 mm) thickness (ASTM D570)	%	1.0-2.5	—	0.8-1.8
THERMAL	**Coefficient of linear thermal expansion** (ASTM D696)	$°C^{-1}$ ($\times 10^{-5}$)	10-20	—	10-20
	Glass transition temperature (T_g)	°C	—	—	113-142
	Low-temperature performance / Brittleness temperature (ASTM D746)	°C	—	—	—
	Melting point (T_m) (ASTM D1238)	°C	—	—	240-255
	Softening point (Vicat) (ASTM D1525)	°C	—	—	152-180
OPTICAL	**Clarity/transparency**	—	Transparent, translucent, or opaque	—	—
	Transmittance (ASTM D1003)	%	—	—	—
	Refractive index (23°C) (ASTM D542)	—	1.470	—	1.479

Data sources:
1947: *Modern Plastics Encyclopedia*, vol. 3, *Charts* (New York: Plastics Catalogue Corporation).
1967: Shell Chemicals, *Properties of Plastics*, rev. metric ed. (London: Shell International Chemical Company and Shell Chemicals UK).
2016: George Wypych, *Handbook of Polymers*, 2nd ed. (Toronto: ChemTec).

13.6 | Polyamide-6 / Nylon-6 (PA-6)

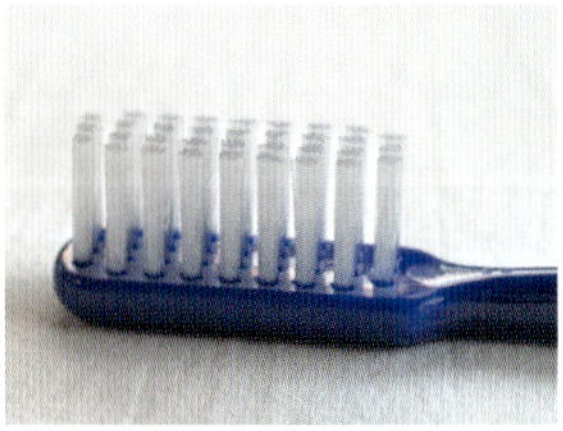

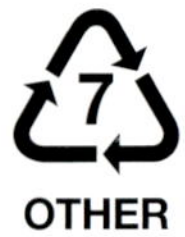

PLASTICS RECYCLING CODE

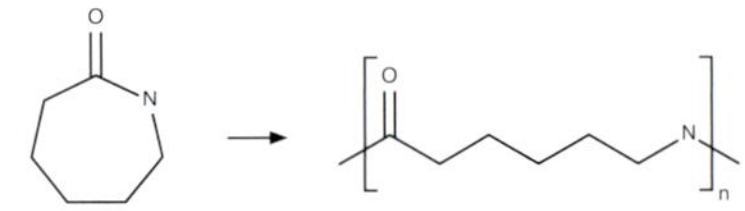

CHEMICAL STRUCTURE

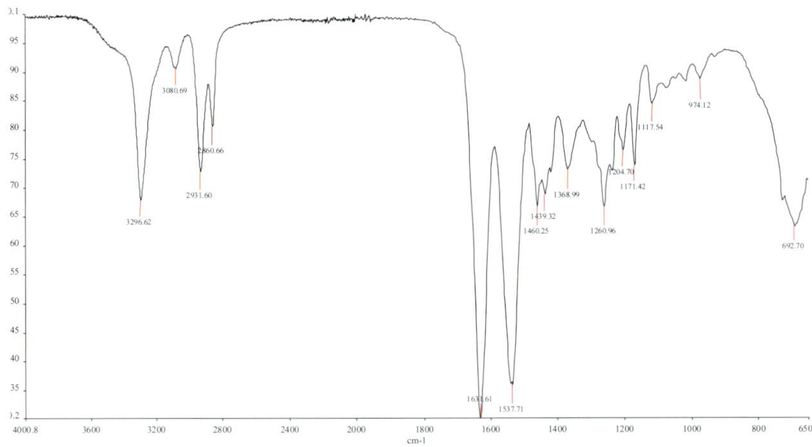

FTIR SPECTRUM

Date of first manufacture
1938 (Germany).

Alternate/trade name examples
Zytel, Akulon, Durethan, Capran, Ultramid B.

Polymerization
Nylon-6 is polymerized by ring-opening polymerization of ε-caprolactam in the presence of a catalyst.

Processing
Injection molding, extrusion, additive manufacturing (rapid prototyping, 3D printing).

Selected applications
Fasteners, film for food packaging, fishing gear, flexible tubing, textiles (carpets, parachutes, sportswear), toothbrushes, zip-tie fasteners.

Feel/look/sound/smell
Translucent to opaque with a waxy feel.

Main degradation agents
Moisture (changes in relative humidity), light, UV radiation, oxygen, ozone temperature, stress, chemicals.

Main signs of degradation and damage
Yellowing, brittleness, cracking. In storage (without light) nylon discolors (darkens).

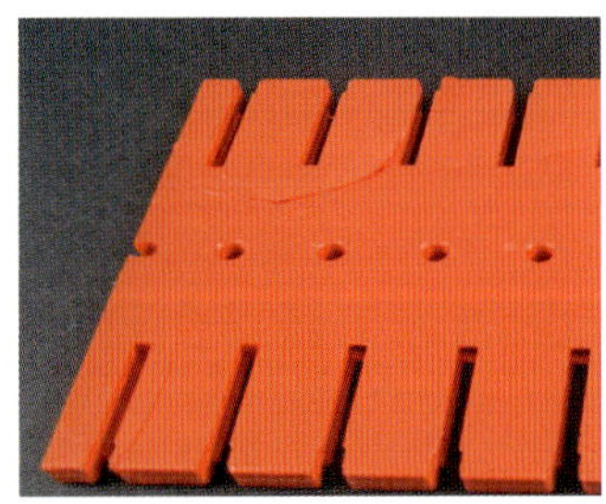

A cracked piece of building toy.

13.6 | Polyamide-6 / Nylon-6 (PA-6)

	PROPERTY	UNITS (SI)	DATA 1947	DATA 1967	DATA 2016
MECHANICAL	**Compressive strength** (ASTM D695)	MPa	96.5-124.1	48.4-96.6	55
	Elongation (ASTM D638)	%	45-55	90-320	10-160 (dry) 327 (conditioned)
	Flexural strength (ASTM D790)	MPa	75.8-89.6	55.3-110.5	100 (dry)
	Hardness (ASTM D785)	Rockwell M, R	M80-M90	R103-R118	—
	Impact strength (Izod notched) (ASTM D256)	J/cm	0.43-0.53	0.54-1.93	17-20 (unnotched)
	Tensile modulus (rigidity) (ASTM D638)	GPa	2.1-2.24	1.04-2.48	2.35-3.37
	Tensile strength (ASTM D638)	MPa	62.1-72.4	70.5-83.9	74-106
PHYSICAL	**Density** (ASTM D792)	g/cm^3	1.14	1.13	—
	Hildebrand solubility parameter (δ)	$(MPa)^{1/2}$	—	—	20.3
	Permeability: H_2O vapor O_2 CO_2 (ASTM D1434) N.T.P. = normal temperature and pressure	$cc(N.T.P.)\cdot cm^{-2}\cdot sec^{-1}\cdot mil^{-1}\cdot cm\ Hg^{-1}$ at 25°C × 10^{-8}	— — —	2,800 0.12 0.5	— — —
	Resistance to solvents p = poor, f = fair, g = good	Resistant to common solvents and oils			
	Surface tension	mNm^{-1}	—	—	40-47
	Water absorption (equilibrium 24 hours) ⅛-inch (3.175 mm) thickness (ASTM D570)	%	1.5	1.9-3.3	7.1-10
THERMAL	**Coefficient of linear thermal expansion** (ASTM D696)	$°C^{-1}$ ($\times 10^{-5}$)	10	8-13	8.3-9.3
	Glass transition temperature (T_g)	°C	—	—	50-75 (dry) 3-20 (50% RH)
	Low-temperature performance / Brittleness temperature (ASTM D746)	°C	—	Good	-30
	Melting point (T_m) (ASTM D1238)	°C	—	200-220	220-260
	Softening point (Vicat) (ASTM D1525)	°C	—	200-220	195-204
OPTICAL	**Clarity/transparency**	—	Translucent or opaque	Translucent or opaque	—
	Transmittance (ASTM D1003)	%	—	—	—
	Refractive index (23°C) (ASTM D542)	—	1.53	—	—

Data sources:
1947: *Modern Plastics Encyclopedia*, vol. 3, *Charts* (New York: Plastics Catalogue Corporation).
1967: Shell Chemicals, *Properties of Plastics*, rev. metric ed. (London: Shell International Chemical Company and Shell Chemicals UK).
2016: George Wypych, *Handbook of Polymers*, 2nd ed. (Toronto: ChemTec).

13.7 | Polyamide-6,6 / Nylon-6,6 (PA-6,6)

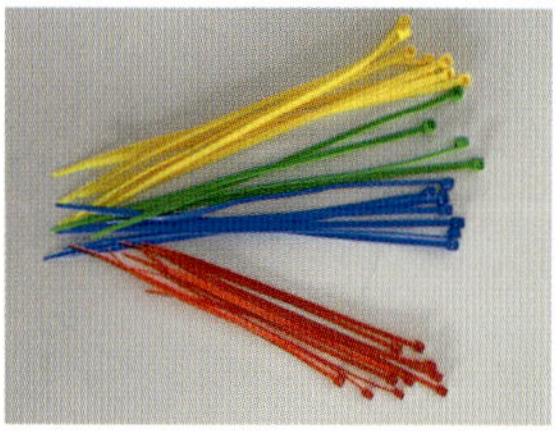

7 OTHER

PLASTICS RECYCLING CODE

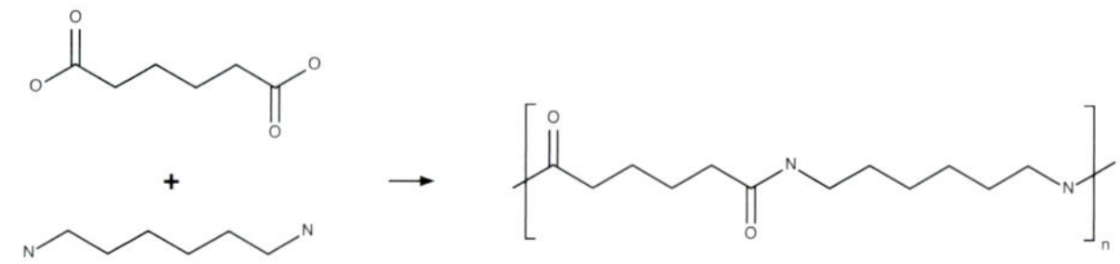

CHEMICAL STRUCTURE

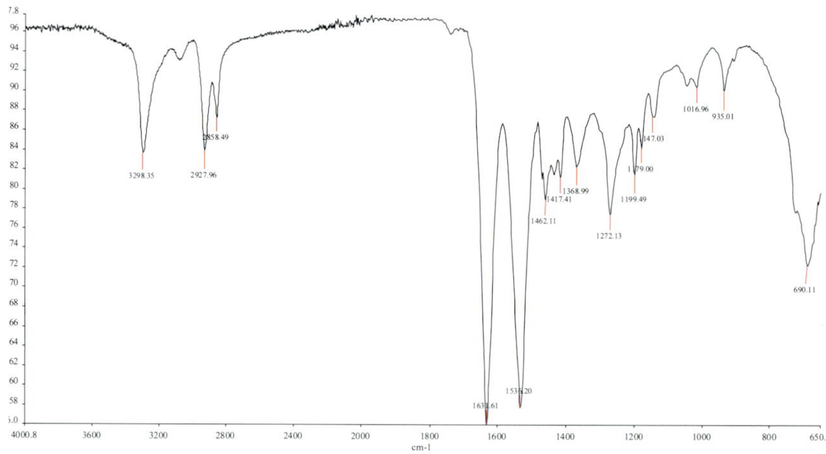

FTIR SPECTRUM

Date of first manufacture
1935 (patented USA), 1938 (production USA).

Alternate/trade name examples
Tecamid 66, Ultramid A, Dartek.

Polymerization
Nylon-6,6 is polymerized from the condensation reaction of hexamethylene diamine and adipic acid.

Processing
Injection molding, cast-film extrusion, extrusion, extrusion spun into fibers, additive manufacturing (rapid prototyping, 3D printing).

Selected applications
Automotive structures; bearings and gears; boats; film to wrap fine art; fishing equipment; musical instruments; stockings, underwear, and clothing; tennis racket strings; zip tie fasteners.

Feel/look/sound/smell
Translucent to opaque with waxy feel.

Main degradation agents
Moisture (changes in relative humidity), light, UV radiation, oxygen, ozone, temperature, stress, chemicals.

Main signs of degradation and damage
Yellowing, brittleness. In storage (without light) nylon discolors (darkens).

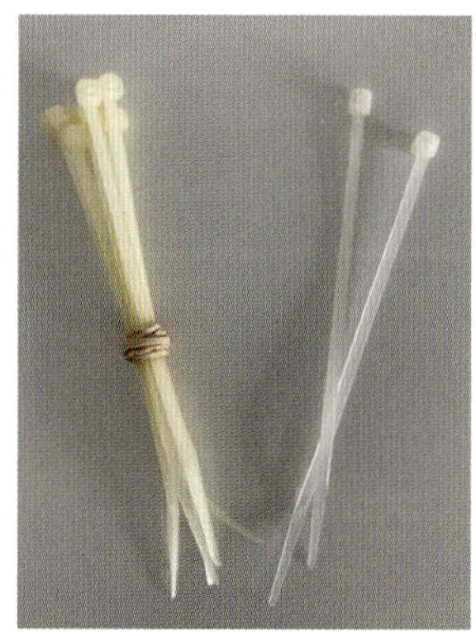

Yellowed zip-ties.

13.7 | Polyamide-6,6 / Nylon-6,6 (PA-6,6)

	PROPERTY	UNITS (SI)	DATA 1947*	DATA 1967	DATA 2016
MECHANICAL	**Compressive strength** (ASTM D695)	MPa	96.5-124.1	48.4-110.5	—
	Elongation (ASTM D638)	%	45-55	60-300	50-100
	Flexural strength (ASTM D790)	MPa	75.8-89.6	55.3-95.3	28-276†
	Hardness (ASTM D785)	Rockwell M, R	M80-M90	R108-R118	M105
	Impact strength (Izod notched) (ASTM D256)	J/cm	0.43-0.53	0.54-1.07	—
	Tensile modulus (rigidity) (ASTM D638)	GPa	2.1-2.24	1.79-2.76	1.07-5.08
	Tensile strength (ASTM D638)	MPa	62.1-72.4	48.0-83.9	70-88
PHYSICAL	**Density** (ASTM D792)	g/cm^3	1.14	1.14	1.05-1.14
	Hildebrand solubility parameter (δ)	$(MPa)^{1/2}$	—	—	22.87-25.8
	Permeability: H_2O vapor O_2 CO_2 (ASTM D1434)	$cm^3 \cdot cm \cdot cm^{-2} \cdot s^{-1} \cdot Pa^{-1}$	— — —	— — —	— — —
	Resistance to solvents p = poor, f = fair, g = good	Resistant to common solvents and oils			
	Surface tension	mNm^{-1}	—	—	46.5
	Water absorption (equilibrium 24 hours) ⅛-inch (3.175 mm) thickness (ASTM D570)	%	1.5	0.1-1.5	8.5-9.0
THERMAL	**Coefficient of linear thermal expansion** (ASTM D696)	$°C^{-1}$ (× 10^{-5})	10	10-15	10
	Glass transition temperature (T_g)	°C	—	—	60-70
	Low-temperature performance / Brittleness temperature (ASTM D746)	°C	—	Good	-30
	Melting point (T_m) (ASTM D1238)	°C	—	250-265	257-270
	Softening point (Vicat) (ASTM D1525)	°C	—	250-265	255
OPTICAL	**Clarity/transparency**	—	Translucent or opaque	Translucent or opaque	—
	Transmittance (ASTM D1003)	%	—	—	—
	Refractive index (23°C) (ASTM D542)	—	1.53	1.53	1.565-1.568

Data sources:
1947: *Modern Plastics Encyclopedia*, vol. 3, *Charts* (New York: Plastics Catalogue Corporation).
1967: Shell Chemicals, *Properties of Plastics*, rev. metric ed. (London: Shell International Chemical Company and Shell Chemicals UK).
2016: George Wypych, *Handbook of Polymers*, 2nd ed. (Toronto: ChemTec).
†2021: MatWeb, www.matweb.com.

*Type of nylon not specified.

13.8 | Polyamide-12 / Nylon-12 (PA-12)

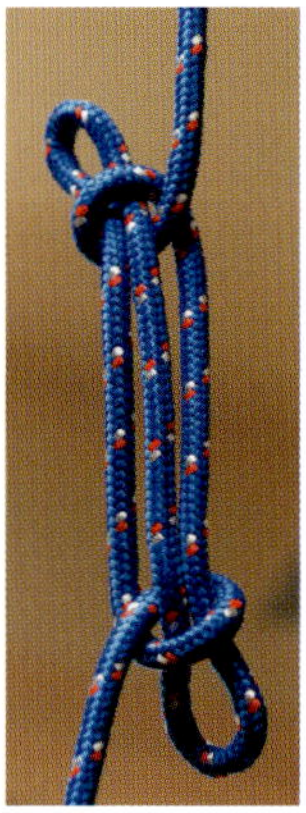

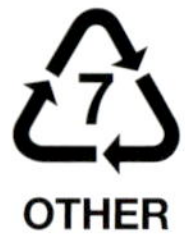

PLASTICS RECYCLING CODE

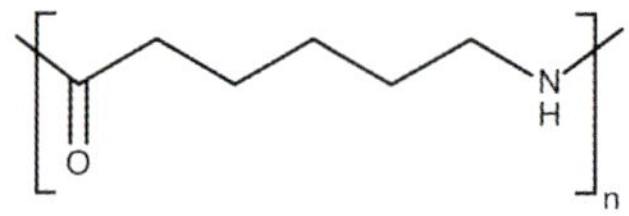

CHEMICAL STRUCTURE

FTIR SPECTRUM

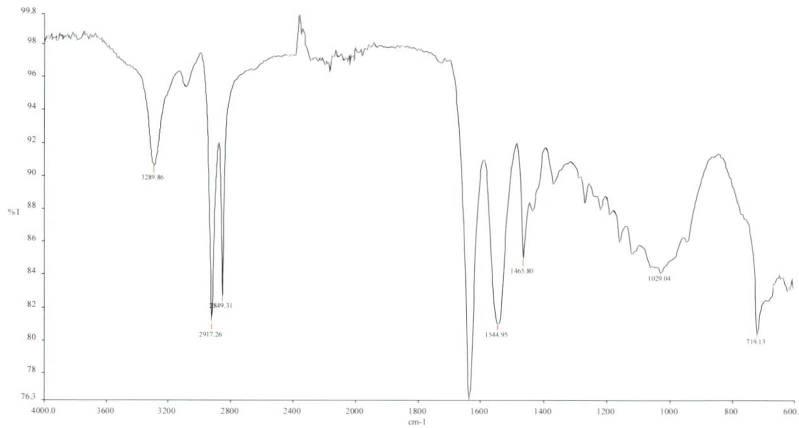

Date of first manufacture
1960s.

Alternate/trade name examples
VESTAMID, Rilsamid, Latamid Nylon 12.

Polymerization
Nylon-12 can be polymerized by polycondensation of ω-aminolauric acid, releasing water, as well as by ring-opening polymerization of laurolactam, without releasing water.

Processing
Extrusion, injection molding, spinning, additive manufacturing (rapid prototyping, 3D printing).

Selected applications
Filament for additive manufacturing; films, bags, and other packaging; sporting goods; textiles.

Feel/look/sound/smell
Translucent to opaque with waxy feel.

Main degradation agents
Light, UV radiation, oxygen, ozone, moisture (changes in relative humidity), temperature, stress, chemicals.

Main signs of degradation and damage
Yellowing. In storage (without light) nylon discolors (darkens).

13.8 | Polyamide-12 / Nylon-12 (PA-12)

	PROPERTY	UNITS (SI)	DATA 1947	DATA 1967	DATA 2016
MECHANICAL	**Compressive strength** (ASTM D695)	MPa	—	—	—
	Elongation (ASTM D638)	%	—	—	>50
	Flexural strength (ASTM D790)	MPa	—	—	—
	Hardness (ASTM D785)	Shore D	—	—	D61–D79
	Impact strength (Izod notched) (ASTM D256)	J/cm	—	—	1.60 (unnotched)
	Tensile modulus (rigidity) (ASTM D638)	GPa	—	—	0.46–1.9
	Tensile strength (ASTM D638)	MPa	—	—	50–70 (dry) 45–50 (conditioned)
PHYSICAL	**Density** (ASTM D792)	g/cm^3	—	—	1.01–1.03
	Hildebrand solubility parameter (δ)	$(MPa)^{1/2}$	—	—	22.2
	Permeability: H_2O vapor O_2 CO_2 (ASTM D1434)	$cm^3 \cdot cm \cdot cm^{-2} \cdot s^{-1} \cdot Pa^{-1}$	— — —	— — —	— — —
	Resistance to solvents p = poor, f = fair, g = good	Alcohols (p), aliphatic hydrocarbons (g), esters (f), ketones (g)			
	Surface tension	mNm^{-1}	—	—	35.8
	Water absorption (equilibrium 24 hours) ⅛-inch (3.175 mm) thickness (ASTM D570)	%	—	—	1.3–1.5
THERMAL	**Coefficient of linear thermal expansion** (ASTM D696)	$°C^{-1}$ ($\times 10^{-5}$)	—	—	0.11–0.15
	Glass transition temperature (T_g)	°C	—	—	55 (dry) 45 (at 50% RH)
	Low-temperature performance / Brittleness temperature (ASTM D746)	°C	—	—	−70
	Melting point (T_m) (ASTM D1238)	°C	—	—	174–185
	Softening point (Vicat) (ASTM D1525)	°C	—	—	170
OPTICAL	**Clarity/transparency**	—	—	—	—
	Transmittance (ASTM D1003)	%	—	—	—
	Refractive index (23°C) (ASTM D542)	—	—	—	1.52–1.53

Data sources:
1947: *Modern Plastics Encyclopedia*, vol. 3, *Charts* (New York: Plastics Catalogue Corporation).
1967: Shell Chemicals, *Properties of Plastics*, rev. metric ed. (London: Shell International Chemical Company and Shell Chemicals UK).
2016: George Wypych, *Handbook of Polymers*, 2nd ed. (Toronto: ChemTec).

13.9 | Polycarbonate (PC)

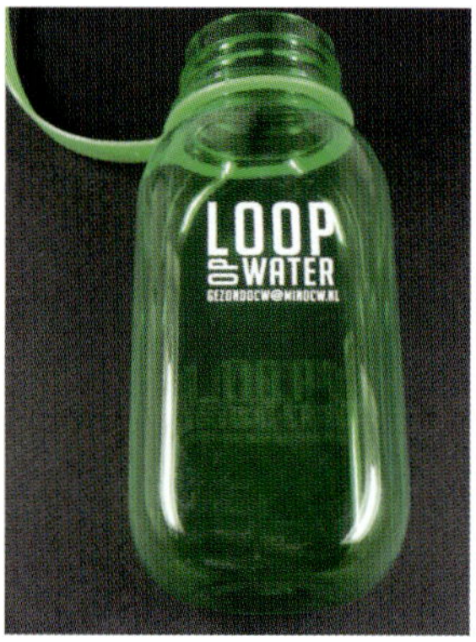

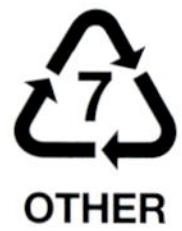

PLASTICS RECYCLING CODE

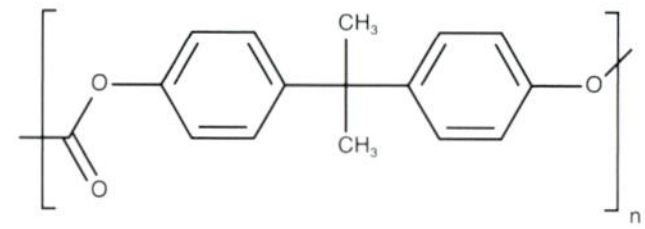

CHEMICAL STRUCTURE

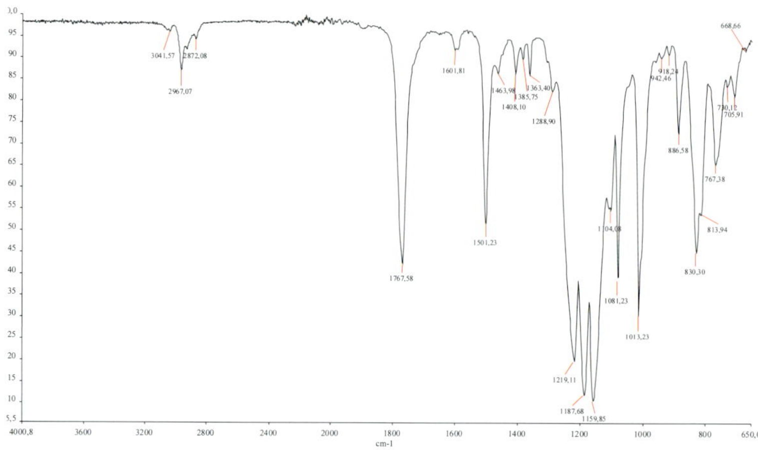

FTIR SPECTRUM

Date of first manufacture
1956 (Germany).

Alternate/trade name examples
Lexan, Makrolon.

Polymerization
Polycarbonate is the condensation polymerization product of phosgene and bisphenol A.

Processing
Extrusion, injection molding, blow molding, calendaring, casting, thermoforming (vacuum forming), additive manufacturing (rapid prototyping, 3D printing).

Selected applications
Corrugated plastics, dental applications, drinking glasses, DVDs, baby bottles, lamp covers, syringes, optical lenses, packaging film, safety glass, safety helmets, skylights, toys, water bottles, windows, windscreens.

Feel/look/sound/smell
Transparent.

Main degradation agents
Light, UV radiation, oxygen, ozone, moisture, temperature, stress, chemicals.

Main signs of degradation and damage
Cracking and crazing can be caused by some solvents.

13.9 | Polycarbonate (PC)

	PROPERTY	UNITS (SI)	DATA 1947	DATA 1967	DATA 2016
MECHANICAL	**Compressive strength** (ASTM D695)	MPa	—	76	70
	Elongation (ASTM D638)	%	—	60–100	66–140
	Flexural strength (ASTM D790)	MPa	—	76–89.7	94–120
	Hardness (ASTM D785)	Rockwell M Rockwell R	—	M80 R125	M85 R124
	Impact strength (Izod notched) (ASTM D256)	J/cm	—	3.8–8.43	1.2–7.36
	Tensile modulus (rigidity) (ASTM D638)	GPa	—	2.20	2.39–2.60
	Tensile strength (ASTM D638)	MPa	—	58.8–65.7	55–88
PHYSICAL	**Density** (ASTM D792)	g/cm^3	—	1.2	1.19–1.22
	Hildebrand solubility parameter (δ)	$(MPa)^{1/2}$	—	Not measured	22.2
	Permeability: H_2O vapor O_2 CO_2 (ASTM D1434)	$gm^{-2}\,24\,h^{-1}\cdot bar^{-1}$ $cm^3\,m^{-2}\,24\,h^{-1}\,bar^{-1}$	— — —	— — —	15 700 —
	Resistance to solvents p = poor, f = fair, g = good	Paraffins (f), aromatic hydrocarbons (p), ketones (p), alcohols (g)			
	Surface tension	mNm^{-1}	—	34.2	28.4–42.9
	Water absorption (equilibrium 24 hours) ⅛-inch (3.175 mm) thickness (ASTM D570)	%	—	0.3	0.12–0.40
THERMAL	**Coefficient of linear thermal expansion** (ASTM D696)	$°C^{-1}$ ($\times 10^{-5}$)	—	7.0	6.0–7.5
	Glass transition temperature (T_g)	°C	—	Not measured	137–154
	Low-temperature performance / Brittleness temperature (ASTM D746)	°C	—	Excellent	−100
	Melting point (T_m) (ASTM D1238)	°C	—	Not measured	255–267
	Softening point (Vicat) (ASTM D1525)	°C	—	165	138–150
OPTICAL	**Clarity/transparency**	—	—	Transparent	—
	Transmittance (ASTM D1003)	%	—	—	—
	Refractive index (23°C) (ASTM D542)	—	—	1.584	1.586–1.587

Data sources:
1947: *Modern Plastics Encyclopedia*, vol. 3, *Charts* (New York: Plastics Catalogue Corporation).
1967: Shell Chemicals, *Properties of Plastics*, rev. metric ed. (London: Shell International Chemical Company and Shell Chemicals UK).
2016: George Wypych, *Handbook of Polymers*, 2nd ed. (Toronto: ChemTec).

13.10 | High-density polyethylene (PE-HD)

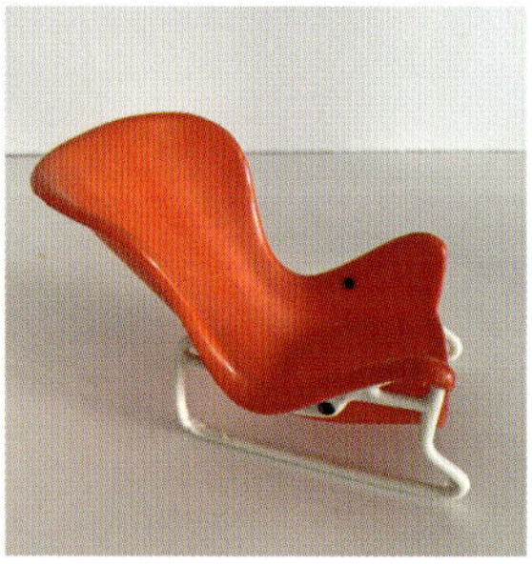

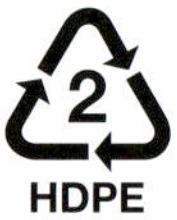

PLASTICS RECYCLING CODE

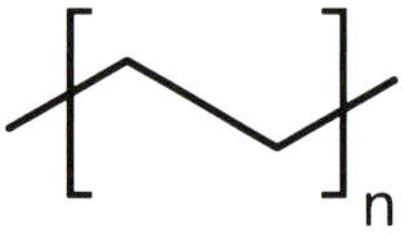

CHEMICAL STRUCTURE

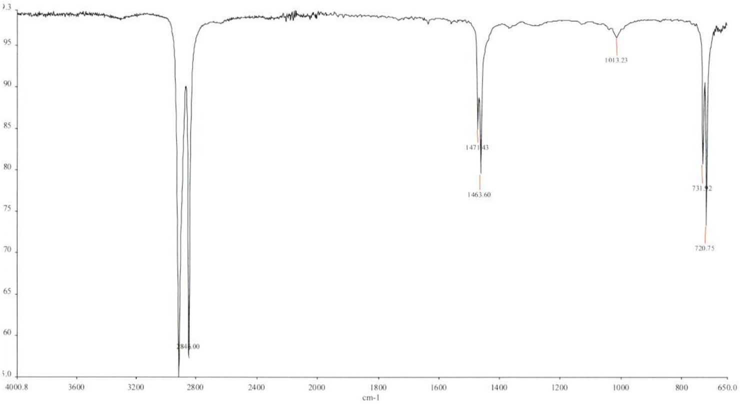

FTIR SPECTRUM

Date of first manufacture
1953 (Germany), 1954 (UK).

Alternate/trade name examples
Paxon, Resinol Type F, Tyvek.

Polymerization
High-density polyethylene is polymerized by polyaddition reaction from the monomer ethylene at low pressure and low temperature using Ziegler-Natta catalysts. PE-HD has greater rigidity than low-density polyethylene due to less branching and higher crystallinity.

Processing
Extrusion, blown-film extrusion, cast film extrusion, injection molding, blow molding, rotational molding, thermoforming (vacuum forming, foaming).

Selected applications
Appliance parts, bottles, carrying cases, closures, containers, crates, dustbins, freezer bags, housings, packaging film, piping, sheet, toys.

Feel/look/sound/smell
Translucent or opaque. Feels waxy.

Main degradation agents
Light, UV radiation, oxygen, ozone, moisture, temperature, stress, chemicals.

Main signs of degradation and damage
Brittleness, chalkiness. Smell of (old) plastic. Waxy, white exudate on surface (additive); fewer scratches than PE-LD.

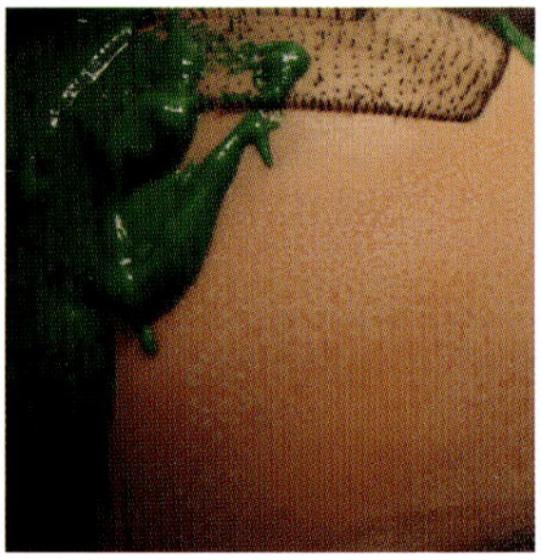

Waxy, white exudate (bloom) on the surface of the head of a mannequin.

13.10 | High-density polyethylene (PE-HD)

	PROPERTY	UNITS (SI)	DATA 1947	DATA 1967	DATA 2016
MECHANICAL	**Compressive strength** (ASTM D695)	MPa	—	16.5	20
	Elongation (ASTM D638)	%	—	50-800	250-1200
	Flexural strength (ASTM D790)	MPa	—	13.8-20.3	—
	Hardness (ASTM D785)	Rockwell D	—	D60-D70	D57-D69
	Impact strength (Izod notched) (ASTM D256)	J/cm	—	0.80-6.4	20-220
	Tensile modulus (rigidity) (ASTM D638)	GPa	—	0.55-1.04	0.8
	Tensile strength (ASTM D638)	MPa	—	21.4-38.0	13.0-51.0
PHYSICAL	**Density** (ASTM D792)	g/cm^3	—	0.941-0.965	0.94-0.965
	Hildebrand solubility parameter (δ)	$(MPa)^{1/2}$	—	—	16.0-16.8
	Permeability: H_2O vapor O_2 CO_2 (ASTM D1434) N.T.P. = normal temperature and pressure	1967: $cc(N.T.P.){\cdot}cm^{-2}{\cdot}sec^{-1}{\cdot}mil^{-1}{\cdot}cm\ Hg^{-1}$ at 25°C × 10^{-8} 2016: $cm^3{\cdot}cm{\cdot}cm^{-2}{\cdot}s^{-1}{\cdot}Pa^{-1} \times 10^{12}$	— — —	60 3 13	0.9 0.03 —
	Resistance to solvents p = poor, f = fair, g = good	Acetone (p), ethanol (g), isopropanol (g), mineral spirit (p), xylene (p)			
	Surface tension	mNm^{-1}	—	—	—
	Water absorption (equilibrium 24 hours) ⅛-inch (3.175 mm) thickness (ASTM D570)	%	—	<0.01	0.005-0.015
THERMAL	**Coefficient of linear thermal expansion** (ASTM D696)	°C^{-1} (× 10^{-5})	—	—	−20 to −76.1
	Glass transition temperature (T_g)	°C	—	11-13	—
	Low-temperature performance / Brittleness temperature (ASTM D746)	°C	—	—	−118 to −133
	Melting point (T_m) (ASTM D1238)	°C	—	Good	−20 to −76.1
	Softening point (Vicat) (ASTM D1525)	°C	—	—	125-135
OPTICAL	**Clarity/transparency**	—	—	Translucent to opaque	—
	Transmittance (ASTM D1003)	%	—	—	—
	Refractive index (23°C) (ASTM D542)	—	—	1.54	1.54

Data sources:
1947: *Modern Plastics Encyclopedia*, vol. 3, *Charts* (New York: Plastics Catalogue Corporation).
1967: Shell Chemicals, *Properties of Plastics*, rev. metric ed. (London: Shell International Chemical Company and Shell Chemicals UK).
2016: George Wypych, *Handbook of Polymers*, 2nd ed. (Toronto: ChemTec).

13.11 | Low-density polyethylene (PE-LD)

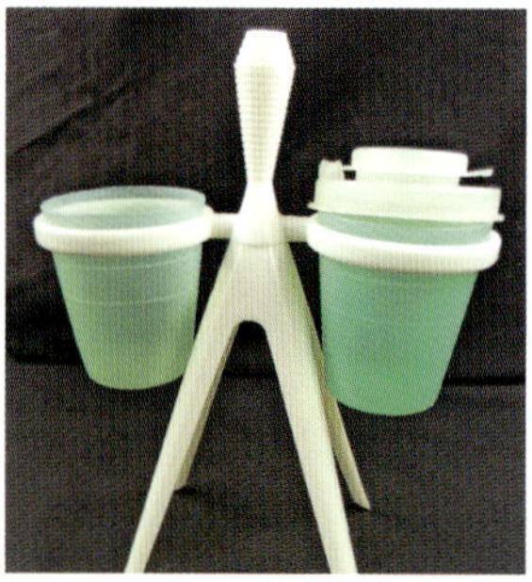

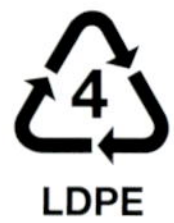

PLASTICS RECYCLING CODE

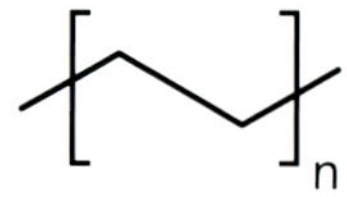

CHEMICAL STRUCTURE

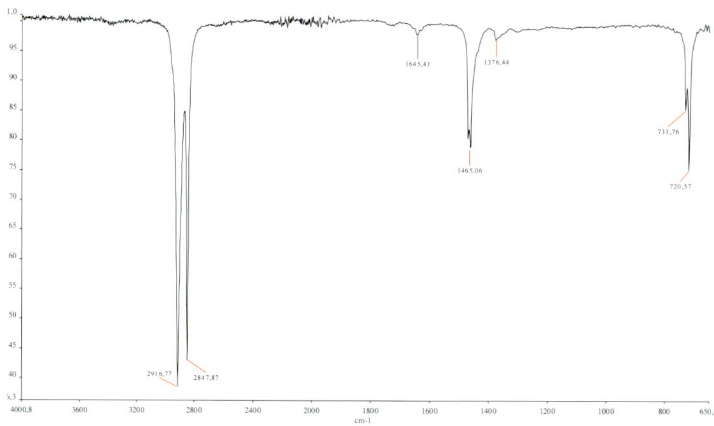

FTIR SPECTRUM

Date of first manufacture
1933 (UK).

Alternate/trade name examples
Polythene, Alkathene, Resinol Type A, Lupolene.

Polymerization
Low-density polyethylene is polymerized by polyaddition reaction from the monomer ethylene at high pressure and temperature. PE-LD has many branched polymer chains, and is therefore less crystalline and rigid than PE-HD.

Processing
Extrusion, blown-film extrusion, coextrusion, cast film extrusion, coating, injection molding, molding, lamination, rotational molding, foaming.

Selected applications
Bottles, bowls, buckets, cling film, closures for squeeze tubes, coating for paper milk cartons, cold water storage tanks, containers (e.g., Tupperware), garment bags, industrial and agricultural uses, monofilament, mugs, packaging film, plumbing, sandwich bags, sheet, toys.

Feel/look/sound/smell
Waxy surface feel. Translucent or opaque; transparent when very thin. Smell of wax when aged.

Main degradation agents
Light, UV radiation, oxygen, ozone, moisture, temperature, stress, chemicals.

Main signs of degradation and damage
Yellowing, stiffening, brittleness, crumbling, chalky surface, scratches.

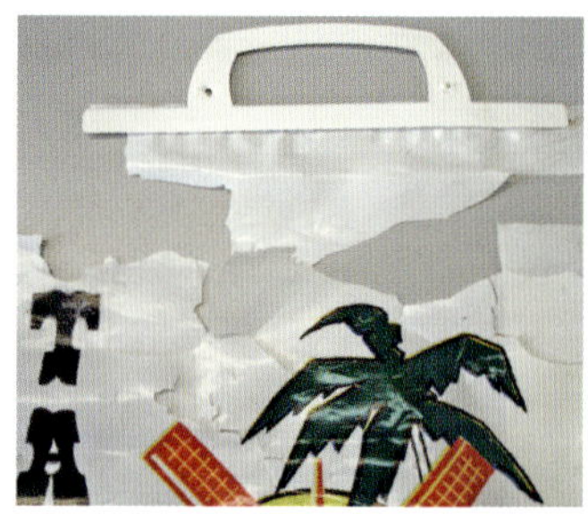

Brittle grocery bag fragments.

13.11 | Low-density polyethylene (PE-LD)

	PROPERTY	UNITS (SI)	DATA 1947	DATA 1967	DATA 2016
MECHANICAL	**Compressive strength** (ASTM D695)	MPa	—	Excessive cold flow	—
	Elongation (ASTM D638)	%	50-600	90-650	130-270
	Flexural strength (ASTM D790)	MPa	10.3-11.7	No fracture	—
	Hardness (ASTM D785)	Rockwell R, Shore D	R25-R35	D41-D46	D41-D50
	Impact strength (Izod notched) (ASTM D256)	J/cm	>1.60	No fracture	No breakage
	Tensile modulus (rigidity) (ASTM D638)	GPa	0.1	0.12-0.24	0.13-0.3
	Tensile strength (ASTM D638)	MPa	12.4-20.7	6.9-15.9	10-20
PHYSICAL	**Density** (ASTM D792)	g/cm^3	0.92-0.93	0.910-0.930	0.915-0.929
	Hildebrand solubility parameter (δ)	(MPa)$^{1/2}$	—	—	7.9-8.1
	Permeability: H_2O vapor O_2 CO_2 (ASTM D1434) N.T.P. = normal temperature and pressure	1967: cc(N.T.P.)·cm^{-2}·sec^{-1}·mil^{-1}·cm Hg^{-1}at 25°C × 10^{-8} 2016: cm^3·cm·cm^{-2}·s^{-1}·Pa^{-1} × 10^{-12}	— — —	420 15 55	6.8 0.22 —
	Resistance to solvents p = poor, f = fair, g = good	Acetone (p), ethanol (g), isopropanol (g), mineral spirit (p), xylene (p)			
	Surface tension	mNm^{-1}	—	—	35.3
	Water absorption (equilibrium 24 hours) ⅛-inch (3.175 mm) thickness (ASTM D570)	%	<0.01	<0.015	0.005-0.015
THERMAL	**Coefficient of linear thermal expansion** (ASTM D696)	°C^{-1} (× 10^{-5})	18	16-18	—
	Glass transition temperature (T_g)	°C	—	—	−103 to −133
	Low-temperature performance / Brittleness temperature (ASTM D746)	°C	—	Good	−34 to −60
	Melting point (T_m) (ASTM D1238)	°C	—	—	105-115
	Softening point (Vicat) (ASTM D1525)	°C	—	85-87	76-109
OPTICAL	**Clarity/transparency**	—	Translucent to opaque	Nearly transparent to opaque	—
	Transmittance (ASTM D1003)	%	—	—	90-91
	Refractive index (23°C) (ASTM D542)	—	1.52	1.51	1.517-1.526

Data sources:
1947: *Modern Plastics Encyclopedia*, vol. 3, *Charts* (New York: Plastics Catalogue Corporation).
1967: Shell Chemicals, *Properties of Plastics*, rev. metric ed. (London: Shell International Chemical Company and Shell Chemicals UK).
2016: George Wypych, *Handbook of Polymers*, 2nd ed. (Toronto: ChemTec).

13.12 | Ultra-high molecular weight polyethylene (PE-UHMW)

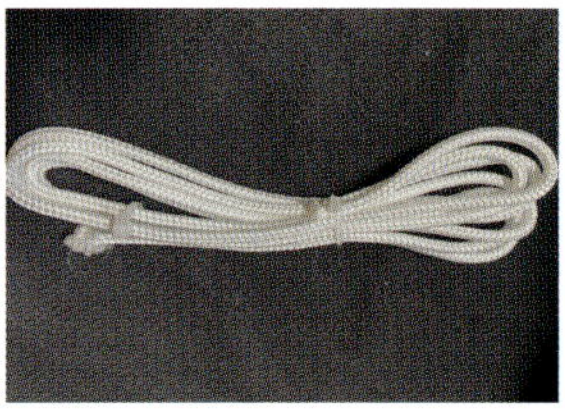

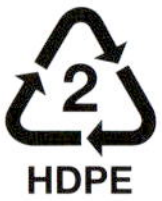

PLASTICS RECYCLING CODE

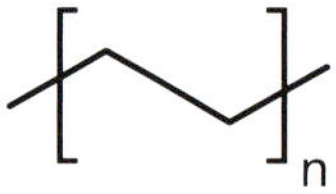

CHEMICAL STRUCTURE

FTIR SPECTRUM

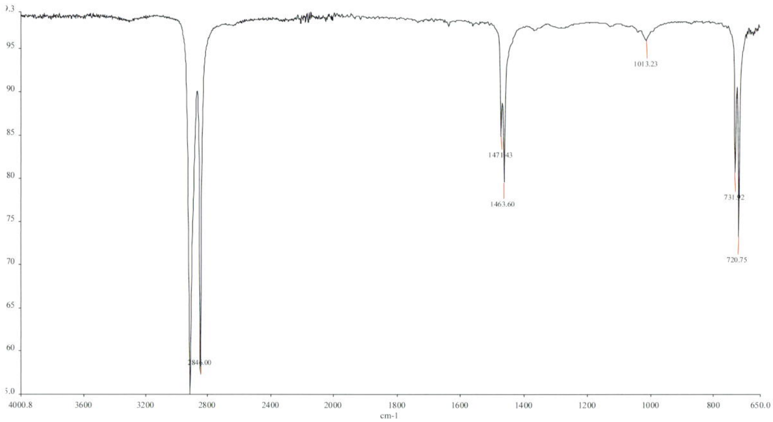

Date of first manufacture
1963 (Germany). First used for joint replacements in the 1960s. Became commercially available in the 1990s.

Alternate/trade name examples
Dyneema, TIVAR.

Polymerization
PE-UHMW is created through addition polymerization of the monomer ethylene using metallocene catalysts; molecular weight is between 1 and 10 million.

Processing
Extrusion, compression molding, spinning fibers.

Selected applications
Biomedical applications in total joint prosthesis, bulletproof vests, skis and snowboards. Unique material properties such as low-friction behavior and high resistance to wear and to impact strengths. High stability in body applications.

Feel/look/sound/smell
N/A

Main degradation agents
Light, UV radiation, oxygen, ozone, moisture, temperature, stress, chemicals.

Main signs of degradation and damage
Slow rate of deterioration, damage not commonly detected.

13.12 | Ultra-high molecular weight polyethylene (PE-UHMW)

	PROPERTY	UNITS (SI)	DATA 1947	DATA 1967	DATA 2016
MECHANICAL	**Compressive strength** (ASTM D695)	MPa	—	—	14–23
	Elongation (ASTM D638)	%	—	—	250–600
	Flexural strength (ASTM D790)	MPa	—	—	20–26.5
	Hardness (ASTM D785)	Shore D	—	—	D57–D68
	Impact strength (Izod notched) (ASTM D256)	J/cm	—	—	—
	Tensile modulus (rigidity) (ASTM D638)	GPa	—	—	1.8–3.3
	Tensile strength (ASTM D638)	MPa	—	—	21–50.2
PHYSICAL	**Density** (ASTM D792)	g/cm^3	—	—	0.93–0.94
	Hildebrand solubility parameter (δ)	(MPa)$^{1/2}$	—	—	—
	Permeability: H_2O vapor O_2 CO_2 (ASTM D1434)	cm^3·cm·cm^{-2}·s^{-1}·Pa^{-1}	— — —	— — —	— — —
	Resistance to solvents p = poor, f = fair, g = good	Alcohol (g), aliphatic hydrocarbons (p), esters (p), ketones (p)			
	Surface tension	mNm^{-1}	—	—	—
	Water absorption (equilibrium 24 hours) ⅛-inch (3.175 mm) thickness (ASTM D570)	%	—	—	0.01–0.02
THERMAL	**Coefficient of linear thermal expansion** (ASTM D696)	°C^{-1} (× 10^{-5})	—	—	11–20
	Glass transition temperature (T_g)	°C	—	—	−110
	Low-temperature performance / Brittleness temperature (ASTM D746)	°C	—	—	—
	Melting point (T_m) (ASTM D1238)	°C	—	—	133–140
	Softening point (Vicat) (ASTM D1525)	°C	—	—	126
OPTICAL	**Clarity/transparency**	—	—	—	—
	Transmittance (ASTM D1003)	%	—	—	—
	Refractive index (23°C) (ASTM D542)	—	—	—	—

Data sources:
1947: *Modern Plastics Encyclopedia*, vol. 3, *Charts* (New York: Plastics Catalogue Corporation).
1967: Shell Chemicals, *Properties of Plastics*, rev. metric ed.(London: Shell International Chemical Company and Shell Chemicals UK).
2016: George Wypych, *Handbook of Polymers*, 2nd ed. (Toronto: ChemTec).

13.13 | Polyethylene terephthalate / Saturated polyester (PET, PETE, PETP)

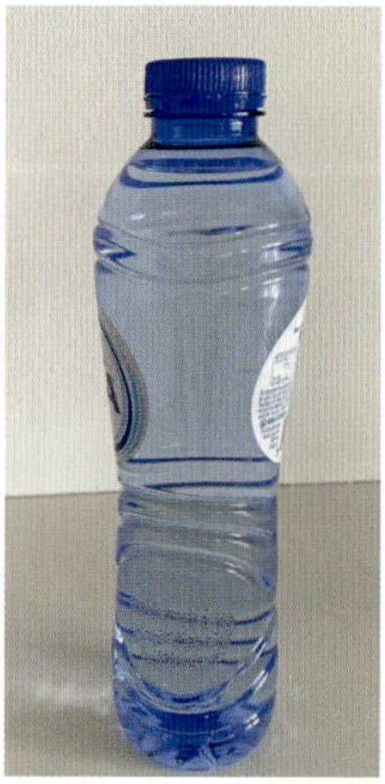

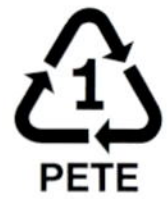

PLASTICS RECYCLING CODE

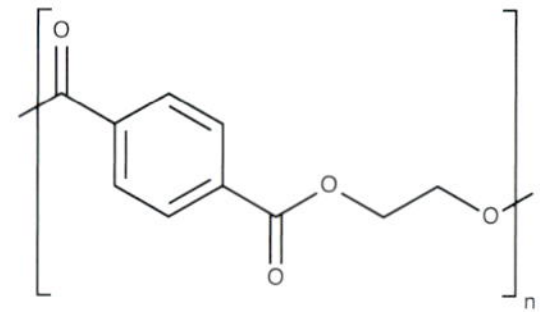

CHEMICAL STRUCTURE

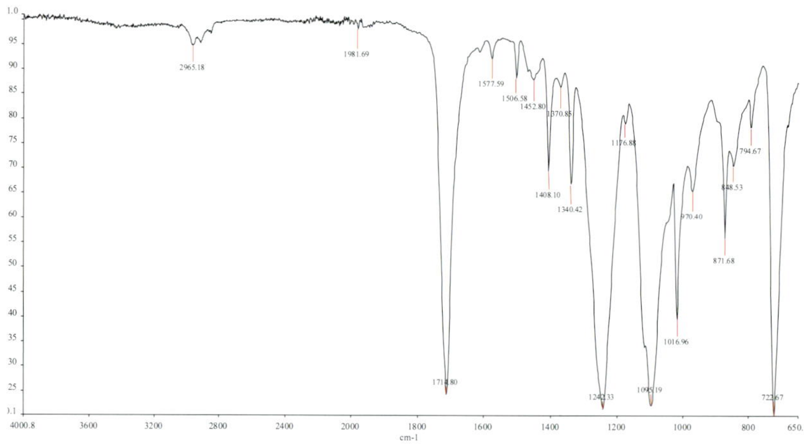

FTIR SPECTRUM

Date of first manufacture
1941 (as fiber, USA), 1947 (as film material, USA), 1980 (blow-molded bottles).

Alternate/trade name examples
Rynite, Melinex, Mylar.

Polymerization
Polyethylene terephthalate is a thermoplastic (saturated) polyester made by polycondensation reaction from dibenzoic acid and glycerol.

Processing
Extrusion, fiber extrusion, injection molding, blow molding, injection blow molding, thermoforming (vacuum forming).

Selected applications
Polyethylene terephthalate has a low gas permeability for CO_2, which makes it suitable for bottles for carbonated beverages. Also used for athletic wear and other textiles, bristles for brushes, housewares, housings, lighting, medical applications, packaging, video- and audiotape.

Feel/look/sound/smell
Rigid, transparent.

Main degradation agents
Light, UV radiation, oxygen, ozone, moisture, temperature, stress, chemicals.

Main signs of degradation and damage
Slight yellowing when recycled PET has been used for production.

13.13 | Polyethylene terephthalate / Saturated polyester (PET, PETE, PETP)

	PROPERTY	UNITS (SI)	DATA 1947	DATA 1967	DATA 2016
MECHANICAL	**Compressive strength** (ASTM D695)	MPa	—	—	80
	Elongation (ASTM D638)	%	—	—	100–250
	Flexural strength (ASTM D790)	MPa	—	—	68.9–78
	Hardness (ASTM D785)	Rockwell M, R	—	—	R110 M95–100
	Impact strength (Izod notched) (ASTM D256)	J/cm	—	—	6.4
	Tensile modulus (rigidity) (ASTM D638)	GPa	—	—	2.0–2.7
	Tensile strength (ASTM D638)	MPa	—	—	24–41.4
PHYSICAL	**Density** (ASTM D792)	g/cm^3	—	—	1.3–1.4
	Hildebrand solubility parameter (δ)	$(MPa)^{1/2}$	—	—	17.1–20.8
	Permeability: H_2O vapor O_2 CO_2 (ASTM D1434)	$cm^3 \cdot cm \cdot cm^{-2} \cdot s^{-1} \cdot Pa^{-1}$ $\times 10^{12}$ $m^3 \cdot m \cdot m^{-2} \cdot s^{-1} \cdot Pa^{-1}$ $\times 10^{18}$	— — —	— — —	11.3 0.4 —
	Resistance to solvents p = poor, f = fair, g = good	Alcohols (g), aliphatic hydrocarbons (g), ketones (g)			
	Surface tension	mNm^{-1}	—	—	39.5–42.1
	Water absorption (equilibrium 24 hours) ⅛-inch (3.175 mm) thickness (ASTM D570)	%	—	—	0.04–0.14
THERMAL	**Coefficient of linear thermal expansion** (ASTM D696)	$°C^{-1}$ ($\times 10^{-5}$)	—	—	1.7
	Glass transition temperature (T_g)	°C	—	—	60–76
	Low-temperature performance / Brittleness temperature (ASTM D746)	°C	—	—	−20
	Melting point (T_m) (ASTM D1238)	°C	—	—	245–265
	Softening point (Vicat) (ASTM D1525)	°C	—	—	77
OPTICAL	**Clarity/transparency**	—	—	—	—
	Transmittance (ASTM D1003)	%	—	—	—
	Refractive index (23°C) (ASTM D542)	—	—	—	1.575

Data sources:
1947: *Modern Plastics Encyclopedia*, vol. 3, *Charts* (New York: Plastics Catalogue Corporation).
1967: Shell Chemicals, *Properties of Plastics*, rev. metric ed. (London: Shell International Chemical Company and Shell Chemicals UK).
2016: George Wypych, *Handbook of Polymers*, 2nd ed. (Toronto: ChemTec).

13.14 | Polymethyl methacrylate (PMMA)

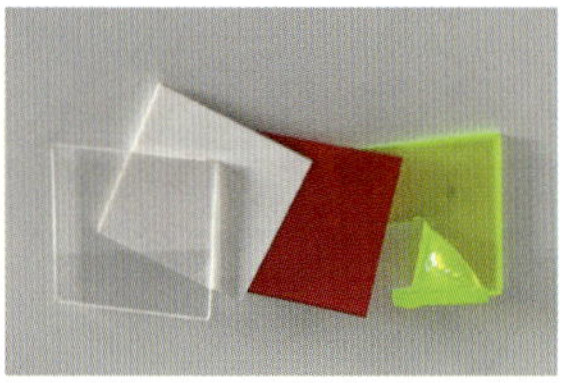

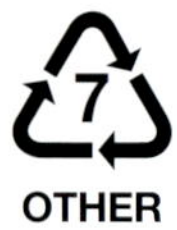

PLASTICS RECYCLING CODE

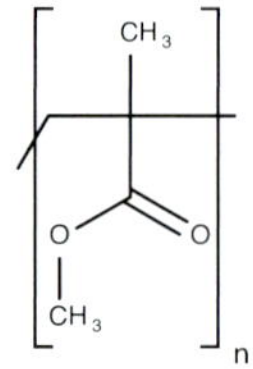

CHEMICAL STRUCTURE

FTIR SPECTRUM

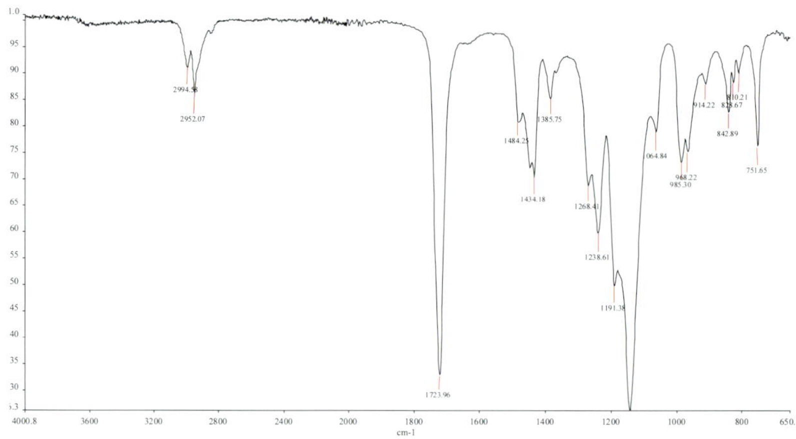

Date of first manufacture
1933 (Germany and England).

Alternate/trade name examples
Perspex, Plexiglas, Lucite, Lucite Diakon, Altuglas, Acrylite, Oroglas, Corian.

Polymerization
PMMA is made by addition polymerization of methyl methacrylate into polymethyl methacrylate.

Processing
Extrusion, casting, thermoforming (vacuum forming), injection molding, compression molding. PMMA comes in sheets, rods, and tubes. Acrylics are also used in additive manufacturing (rapid prototyping, 3D printing).

Selected applications
Aircraft windows, bathtubs, exhibition materials, household items, jewelry, light guides, modern art, optical fibers, plastic optics, sanitary and dental applications. As Corian, it is used as solid surfaces in kitchens and bathrooms. Also found in design objects.

Feel/look/sound/smell
Hard, rigid. High gloss, known for its transparency and optical clarity. Available in bright and fluorescent colors.

Main degradation agents
Moisture, light, UV radiation, oxygen, ozone, temperature, stress, chemicals.

Main signs of degradation and damage
Stress cracking (crazing), breakage, scratching, chipping off.

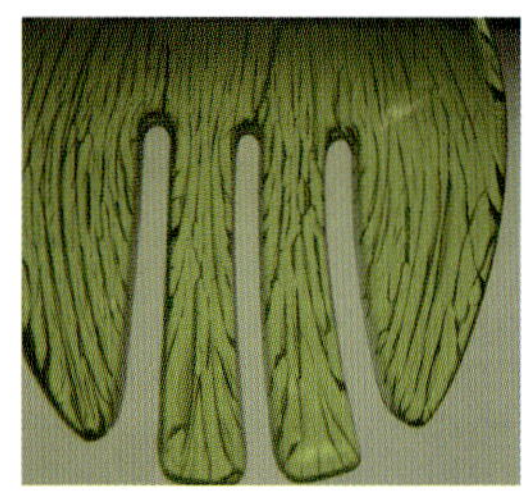

A salad fork with crazing.

13.14 | Polymethyl methacrylate (PMMA)

	PROPERTY	UNITS (SI)	DATA 1947	DATA 1967	DATA 2016
MECHANICAL	**Compressive strength** (ASTM D695)	MPa	75.8-89.6	83.9-138	—
	Elongation (ASTM D638)	%	1-15	3-10	2-6
	Flexural strength (ASTM D790)	MPa	68.9-103.4	89.7-117.5	107-117
	Hardness (ASTM D785)	Rockwell M	M60-M100	M85-M105	M89-M95
	Impact strength (Izod notched) (ASTM D256)	J/cm	0.11-0.32	0.16-0.34	0.19
	Tensile modulus (rigidity) (ASTM D638)	GPa	2.1-3.4	3.1	1.3
	Tensile strength (ASTM D638)	MPa	41.4-68.9	48-76	63-78
PHYSICAL	**Density** (ASTM D792)	g/cm^3	1.18-1.20	1.17-1.20	1.17-1.20
	Hildebrand solubility parameter (δ)	$(MPa)^{1/2}$	—	—	21.3
	Permeability: H_2O vapor O_2 CO_2 (ASTM D1434)	$cm^3 \cdot cm \cdot cm^{-2} \cdot s^{-1} \cdot Pa^{-1} \times 10^{12}$	— — —	— — —	48 0.0115 —
	Resistance to solvents p = poor, f = fair, g = good	Ketones (p), esters (p), chlorinated hydrocarbons (p), alcohols (f)			
	Surface tension	mNm^{-1}	—	—	22.69
	Water absorption (equilibrium 24 hours) ⅛-inch (3.175 mm) thickness (ASTM D570)	%	0.4	0.3-0.4	0.1-0.3
THERMAL	**Coefficient of linear thermal expansion** (ASTM D696)	$°C^{-1}$ ($\times 10^{-5}$)	8	5-9	7-60
	Glass transition temperature (T_g)	°C	—	—	104-105
	Low-temperature performance / Brittleness temperature (ASTM D746)	°C	—	Good	-40
	Melting point (T_m) (ASTM D1238)	°C	—	—	105-160
	Softening point (Vicat) (ASTM D1525)	°C	—	80-98	105-117
OPTICAL	**Clarity/transparency**	—	Transparent	Transparent	Optical clarity
	Transmittance (ASTM D1003)	%	90-92	92	92
	Refractive index (23°C) (ASTM D542)	—	1.49-1.50	1.49	1.448-1.50

Data sources:
1947: *Modern Plastics Encyclopedia*, vol. 3, *Charts* (New York: Plastics Catalogue Corporation).
1967: Shell Chemicals, *Properties of Plastics*, rev. metric ed. (London: Shell International Chemical Company and Shell Chemicals UK).
2016: George Wypych, *Handbook of Polymers*, 2nd ed. (Toronto: ChemTec).

13.15 | Polyoxymethylene (POM)/ Polyacetal / Acetal

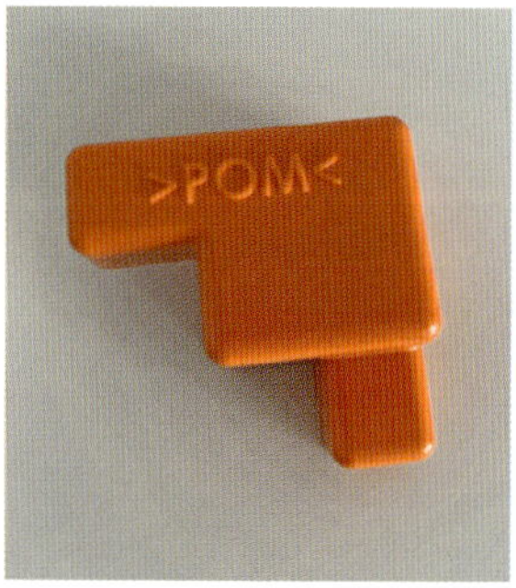

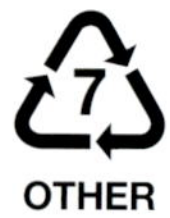

PLASTICS RECYCLING CODE

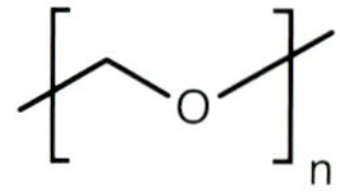

CHEMICAL STRUCTURE

FTIR SPECTRUM

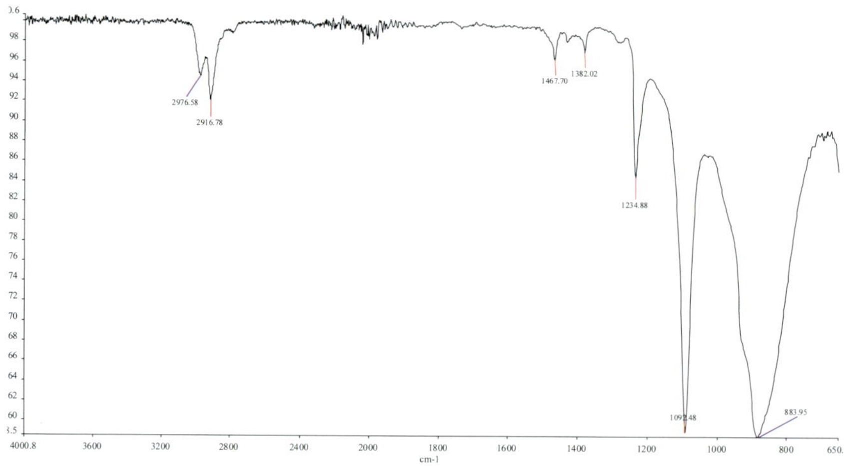

Date of first manufacture
1958 (USA), 1962 (Germany).

Alternate/trade name examples
Delrin, Celcon, Ramtal, Duracon, Kepital, Hostaform, Ultraform.

Polymerization
POM is a highly crystalline material polymerized from anhydrous formaldehyde and stabilized by reaction with acetic anhydride.

Processing
Extrusion, injection molding, blow molding, compression molding, foam molding, rotational molding, transfer molding.

Selected application
Ball bearings, chains, clips, eyeglass frames, furniture sliders, gears, handles, inhalers, knife handles, lock systems, screws, ski bindings, springs, zippers.

Feel/look/sound/smell
Opaque.

Main degradation agents
Light, UV radiation, oxygen, ozone, moisture, temperature, stress, chemicals.

Main signs of degradation and damage
Slow rate of deterioration, damage not commonly detected.

13.15 | Polyoxymethylene (POM) / Polyacetal / Acetal

	PROPERTY	UNITS (SI)	DATA 1947	DATA 1967	DATA 2016
MECHANICAL	**Compressive strength** (ASTM D695)	MPa	—	124	85
	Elongation (ASTM D638)	%	—	—	20–120
	Flexural strength (ASTM D790)	MPa	—	97.4	88–98
	Hardness (ASTM D785)	Rockwell M, R	—	M94–R118	M79–M92 R117–R120
	Impact strength (Izod notched) (ASTM D256)	J/cm	—	—	—
	Tensile modulus (rigidity) (ASTM D638)	GPa	—	2.83	—
	Tensile strength (ASTM D638)	MPa	—	69.0	52–60
PHYSICAL	**Density** (ASTM D792)	g/cm^3	—	1.425	1.35–1.44
	Hildebrand solubility parameter (δ)	$(MPa)^{1/2}$	—	—	—
	Permeability: H_2O vapor O_2 CO_2 (ASTM D1434)	$cm^3 \cdot cm \cdot cm^{-2} \cdot s^{-1} \cdot Pa^{-1} \times 10^{12}$	— — —	— — —	68.3 — —
	Resistance to solvents p = poor, f = fair, g = good	Excellent resistance to practically all solvents			
	Surface tension	mNm^{-1}	—	—	44.6
	Water absorption (equilibrium 24 hours) ⅛-inch (3.175 mm) thickness (ASTM D570)	%	—	0.12	0.8–1.65
THERMAL	**Coefficient of linear thermal expansion** (ASTM D696)	$°C^{-1} (\times 10^{-5})$	—	7.0	10–12
	Glass transition temperature (T_g)	°C	—	—	−60 to −90
	Low-temperature performance / Brittleness temperature (ASTM D746)	°C	—	Good	−60
	Melting point (T_m) (ASTM D1238)	°C	—	—	166–185
	Softening point (Vicat) (ASTM D1525)	°C	—	175	150–160
OPTICAL	**Clarity/transparency**	—	—	Translucent to opaque	—
	Transmittance (ASTM D1003)	%	—	—	—
	Refractive index (23°C) (ASTM D542)	—	—	1.48	1.545–1.553

Data sources:
1947: *Modern Plastics Encyclopedia*, vol. 3, *Charts* (New York: Plastics Catalogue Corporation).
1967: Shell Chemicals, *Properties of Plastics*, rev. metric ed. (London: Shell International Chemical Company and Shell Chemicals UK).
2016: George Wypych, *Handbook of Polymers*, 2nd ed. (Toronto: ChemTec).

13.16 | Polypropylene (PP)

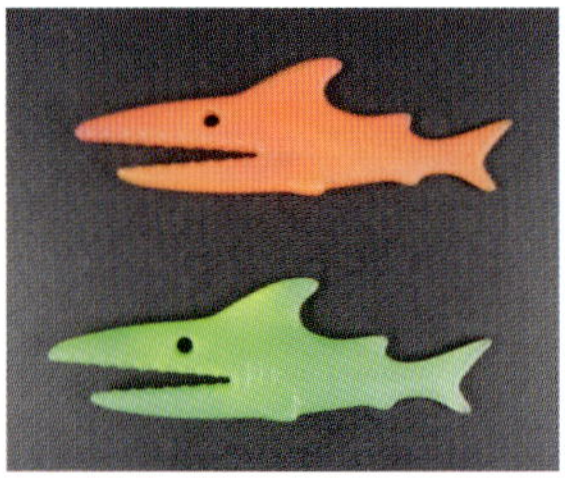

PLASTICS RECYCLING CODE

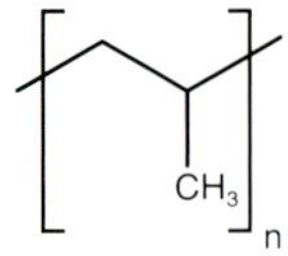

CHEMICAL STRUCTURE

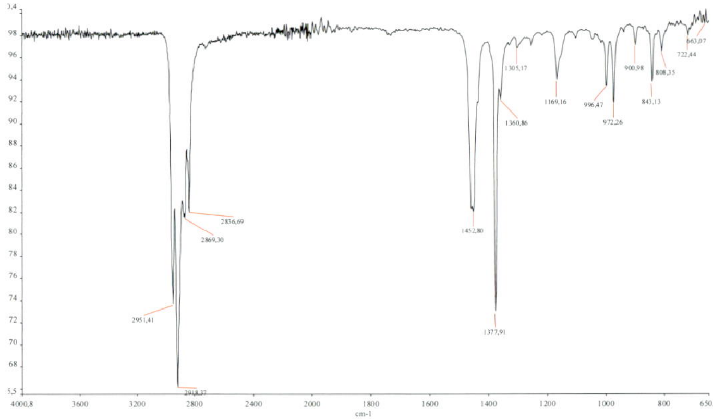

FTIR SPECTRUM

Date of first manufacture
1951 (USA), 1954 (isotactic PP, Italy).

Alternate/trade name examples
Hostalene PP, Moplen, Napryl, Novolen, Propathene.

Polymerization
Polypropylene is made by addition polymerization from propylene using Ziegler-Natta catalysts. Only isotactic PP (with the CH_3 groups arranged along the same side of the chain) is marketed in significant quantities.

Processing
Extrusion, injection molding, blow molding, thermoforming (vacuum forming). Spun into fibers, it has industrial and household textile purposes, and filaments are used in additive manufacturing (rapid prototyping, 3D printing).

Selected applications
Most used plastic. Battery casings, bottles for household and industrial liquids, boxes and suitcases, rope, car bumpers, carpets, chair shells, crates, food containers, food and other packaging, household goods, netting, nonwoven fabrics for diapers and medical applications, outdoor furniture, toys.

Feel/look/sound/smell
Rigid, but flexible. Waxy feel. Transparent or translucent. Smell of wax when aged.

Main degradation agents
Light, UV radiation, oxygen, ozone, moisture, temperature, stress, chemicals.

Main signs of degradation and damage
Yellowing, cracking, brittleness, crumbling, chalky surface, scratches.

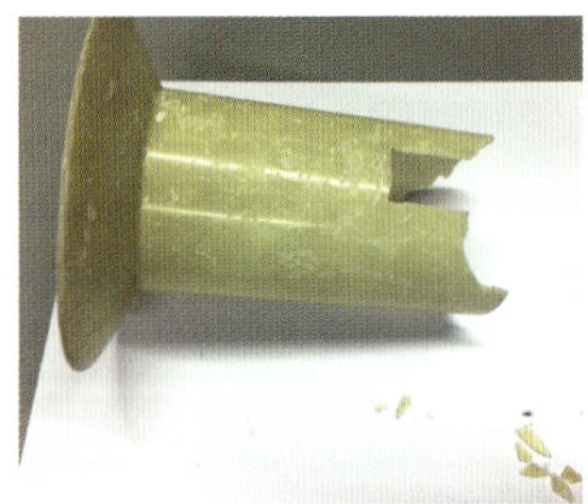

A brittle, crumbling, degraded cylinder.

13.16 | Polypropylene (PP)

	PROPERTY	UNITS (SI)	DATA 1947	DATA 1967	DATA 2016
MECHANICAL	**Compressive strength** (ASTM D695)	MPa	—	58.8-69	40
	Elongation (ASTM D638)	%	—	50-600	10-140
	Flexural strength (ASTM D790)	MPa	—	103.6-172	41
	Hardness (ASTM D785)	Rockwell R	—	R85-R110	R102-R103 R64-R90
	Impact strength (Izod notched) (ASTM D256)	J/cm	—	0.21-3.34	0.18-0.69
	Tensile modulus (rigidity) (ASTM D638)	GPa	—	0.90-1.38	1.2-2.0
	Tensile strength (ASTM D638)	MPa	—	29.0-38.0	26-32
PHYSICAL	**Density** (ASTM D792)	g/cm^3	—	0.90-0.91	—
	Hildebrand solubility parameter (δ)	$(MPa)^{1/2}$	—	16	18.0-19.2
	Permeability: H_2O vapor O_2 CO_2 (ASTM D1434) N.T.P. = normal temperature and pressure	1967: $cc(N{\cdot}T{\cdot}P{\cdot}){\cdot}cm^{-2}{\cdot}sec^{-1}{\cdot}mil^{-1}{\cdot}cm$ Hg^{-1} at 25°C × 10^{-8} 2016: $cm^3{\cdot}cm{\cdot}cm^{-2}{\cdot}s^{-1}{\cdot}Pa^{-1} \times 10^{12}$	— — —	160 4 12	1.58 0.17 —
	Resistance to solvents p = poor, f = fair, g = good	Acetone (p), ethanol (g), isopropanol (g), mineral spirit (p), xylene (p)			
	Surface tension	mNm^{-1}	—	—	20.4
	Water absorption (equilibrium 24 hours) ⅛-inch (3.175 mm) thickness (ASTM D570)	%	—	<0.01	0.02-0.04
THERMAL	**Coefficient of linear thermal expansion** (ASTM D696)	$°C^{-1}$ (× 10^{-5})	—	11	10.5
	Glass transition temperature (T_g)	°C	—	—	−15
	Low-temperature performance / Brittleness temperature (ASTM D746)	°C	—	Fair	−30-+2
	Melting point (T_m) (ASTM D1238)	°C	—	—	120-176
	Softening point (Vicat) (ASTM D1525)	°C	—	150	138-155
OPTICAL	**Clarity/transparency**	—	—	Transparent to translucent	Translucent to white
	Transmittance (ASTM D1003)	%	—	—	—
	Refractive index (23°C) (ASTM D542)	—	—	1.49	1.49-1.51

Data sources:
1947: *Modern Plastics Encyclopedia*, vol. 3, *Charts* (New York: Plastics Catalogue Corporation).
1967: Shell Chemicals, *Properties of Plastics*, rev. metric ed. (London: Shell International Chemical Company and Shell Chemicals UK).
2016: George Wypych, *Handbook of Polymers*, 2nd ed. (Toronto: ChemTec).

13.17 | Polystyrene (PS)

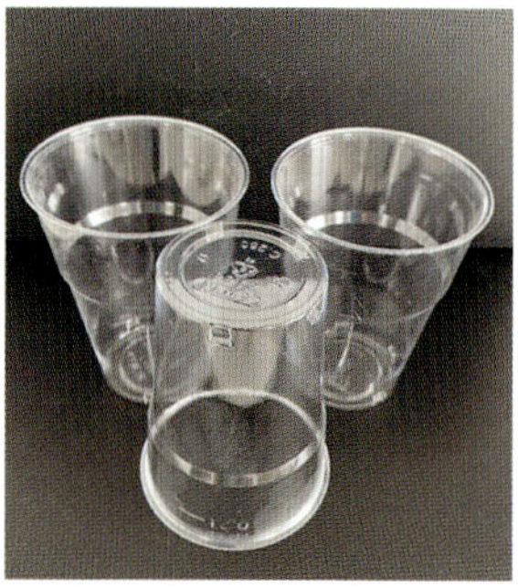

PLASTICS RECYCLING CODE

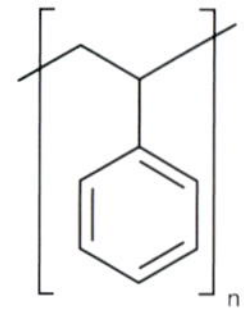

CHEMICAL STRUCTURE

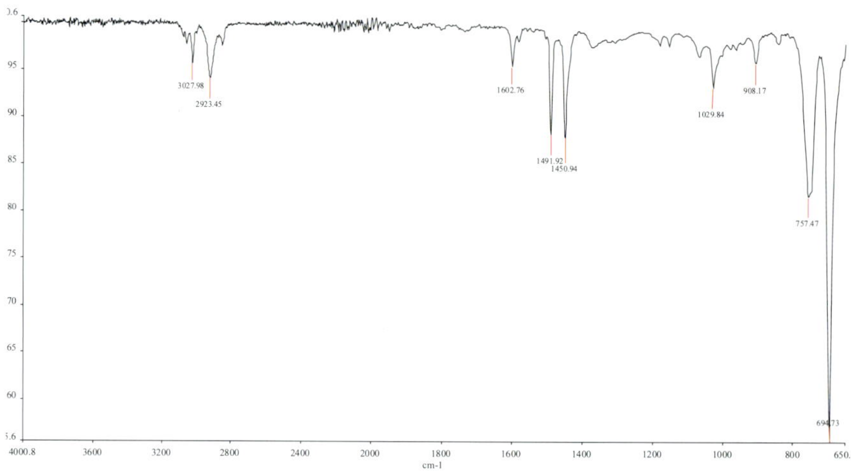

FTIR SPECTRUM

Date of first manufacture
1930 (Germany), 1937 (USA).

Alternate/trade name examples
Abstrex, Bextrene, Carinex, Distrene, Ozolith, Lustrex, Restirolo, Vestyron.

Polymerization
General-purpose polystyrene is synthesized by addition polymerization of ethylene benzene, which is synthesized from benzene and ethylene.

Processing
Extrusion, injection molding, blow molding, foaming, thermoforming (vacuum forming), additive manufacturing (rapid prototyping, 3D printing).

Selected applications
CD cases, coat hangers, coffee cups, disposable cutlery, housewares, insulating foam, office equipment, packaging foam, petri dishes, toothbrushes, toys, yogurt cups.

Feel/look/sound/smell
Transparent; comes in all colors. Glassy sound when tapped.

Main degradation agents
Light, UV radiation, oxygen, ozone, moisture, temperature, stress, chemicals.

Main signs of degradation and damage
Yellowing, brittleness, crumbling, cracking, crazing.

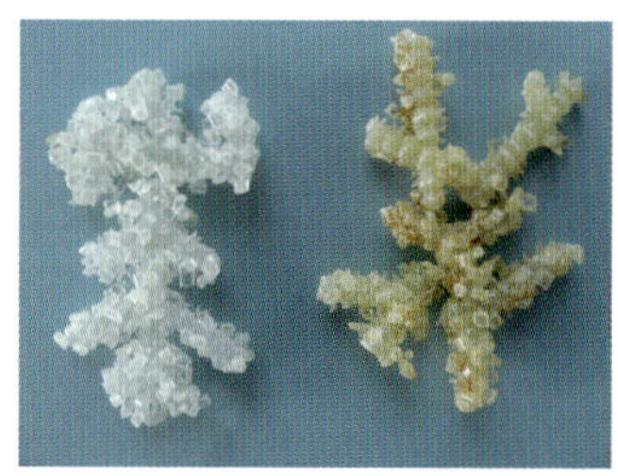

Yellowed decorative snowflakes.

13.17 | Polystyrene (PS)

	PROPERTY	UNITS (SI)	DATA 1947	DATA 1967	DATA 2016
MECHANICAL	**Compressive strength** (ASTM D695)	MPa	79.3-117.2	79.4-110	70
	Elongation (ASTM D638)	%	1.5-3.5	1.0-2.5	2-3
	Flexural strength (ASTM D790)	MPa	33.1-131.0	83.9-117.5	66-95
	Hardness (ASTM D785)	Rockwell M	M55-M95	M65-M90	M58-M64
	Impact strength (Izod notched) (ASTM D256)	J/cm	0.14-0.32	0.10-0.27	0.12-0.20
	Tensile modulus (rigidity) (ASTM D638)	GPa	1.2-3.2	2.76-4.14	3-3.5
	Tensile strength (ASTM D638)	MPa	20.7-58.6	34.5-83.9	40-66
PHYSICAL	**Density** (ASTM D792)	g/cm^3	1.054-1.070	1.04-1.11	1.04-1.06
	Hildebrand solubility parameter (δ)	(MPa)$^{1/2}$	—	—	17.45-23.9
	Permeability: H_2O vapor O_2 CO_2 (ASTM D1434) N.T.P. = normal temperature and pressure	1967: cc(N.T.P.)·cm^{-2}·sec^{-1}·mil^{-1}·cm Hg^{-1} at 25°C × 10^{-8} 2016: cm^3·cm·cm^{-2}·s^{-1}·Pa^{-1} × 10^{12}	— — —	4.700 4 35	135 0.19 —
	Resistance to solvents p = poor, f = fair, g = good	Aromatic hydrocarbons (p), chlorinated hydrocarbons (p), esters (p), ketones (p), higher alcohols (p)			
	Surface tension	mNm^{-1}	—	—	39.3-40.7
	Water absorption (equilibrium 24 hours) ⅛-inch (3.175 mm) thickness (ASTM D570)	%	0.04-0.06	0.03-0.4	0.03-0.1
THERMAL	**Coefficient of linear thermal expansion** (ASTM D696)	°C^{-1} (× 10^{-5})	6-8	6-8	6-8
	Glass transition temperature (T_g)	°C	—	—	85-102
	Low-temperature performance / Brittleness temperature (ASTM D746)	°C	—	Poor	−10
	Melting point (T_m) (ASTM D1238)	°C	—	—	275
	Softening point (Vicat) (ASTM D1525)	°C	—	82-103	74
OPTICAL	**Clarity/transparency**	—	Transparent	Transparent	—
	Transmittance (ASTM D1003)	%	88-92	—	89-90
	Refractive index (23°C) (ASTM D542)	—	1.59	1.59-1.6	1.5894-1.600

Data sources:
1947: *Modern Plastics Encyclopedia*, vol. 3, *Charts* (New York: Plastics Catalogue Corporation).
1967: Shell Chemicals, *Properties of Plastics*, rev. metric ed. (London: Shell International Chemical Company and Shell Chemicals UK).
2016: George Wypych, *Handbook of Polymers*, 2nd ed. (Toronto: ChemTec).

13.18 | Expanded polystyrene (EPS)

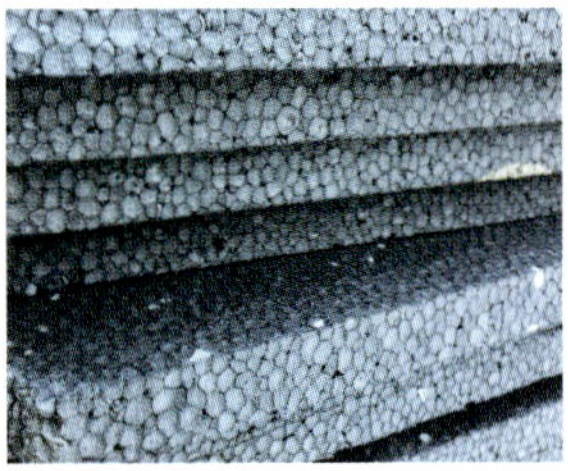

PS

PLASTICS RECYCLING CODE

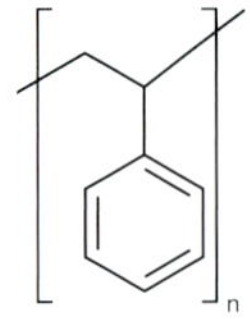

CHEMICAL STRUCTURE

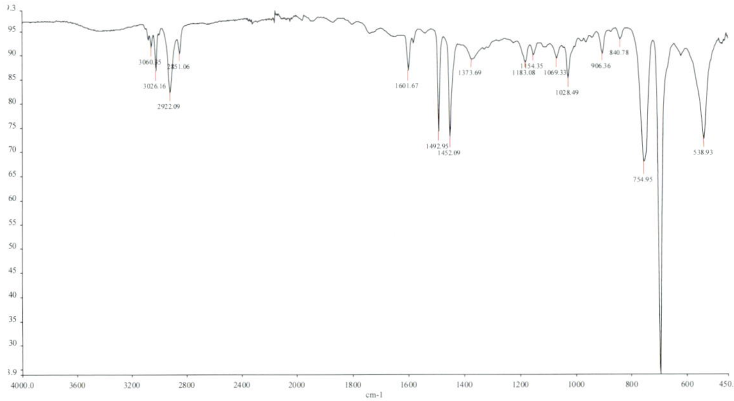

FTIR SPECTRUM

Date of first manufacture
1941 (Germany).

Alternate/trade name examples
Styrofoam, Styropor.

Polymerization
Ethylene benzene is synthesized from benzene and ethylene and polymerized into styrene by addition reaction.

Processing
Polystyrene beads are impregnated with a volatile solvent (pentane) and then heated; while the polystyrene is melting, pentane evaporates and inflates the polystyrene beads. The enlarged beads are reheated and then compressed into blocks. Expanded polystyrene can be processed in various ways to create different densities of material.

Application
Building insulation, buoys, coffee cups, egg cartons, fast food containers, insulation and flotation devices for boats, life preservers, part of artworks (usually inside, as a support), plates and trays, protective packaging (loose or molded "peanuts") for cushioning items inside boxes.

Feel/look/sound/smell
White, rigid lightweight foam with closed cells. Also comes in other colors (mostly blue).

Main degradation agents
Light, UV radiation, oxygen, ozone, moisture, temperature, stress, chemicals.

Main signs of degradation and damage
Yellowing, crumbling, breaks.

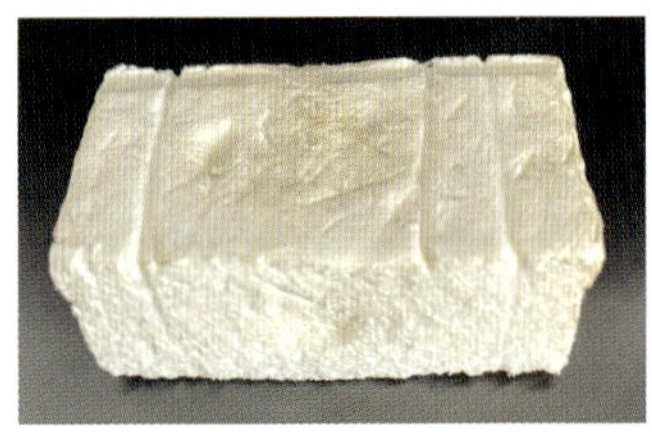

Yellowed Styrofoam.

13.18 | Expanded polystyrene (EPS)

	PROPERTY	UNITS (SI)	DATA 1947	DATA 1967	DATA 2017
MECHANICAL	**Compression set** (ASTM D695)	%	—	—	32.2–45.7
	Elongation (ASTM D638)	%	—	—	5.0–13.4
	Flexural strength (ASTM D790)	MPa	—	—	0.0750–3.00
	Hardness (ASTM D785)	Rockwell	—	—	—
	Impact strength (Izod notched) (ASTM D256)	J/cm	—	—	0.216–0.245
	Tensile modulus (rigidity) (ASTM D638)	GPa	—	—	0.00650–2.65
	Shear strength (ASTM D638)	MPa	—	—	0.0800–1.10
PHYSICAL	**Density** (ASTM D792)	g/cm^3	—	—	0.00310–3.50
	Hildebrand solubility parameter (δ)	$(MPa)^{1/2}$	—	—	—
	Permeability: H_2O vapor O_2 CO_2 (ASTM D1434) N.T.P. = normal temperature and pressure	$cc(N.T.P.)\cdot cm^{-2}\cdot sec^{-1}\cdot mil^{-1}\cdot cm\ Hg^{-1}$ at 25ºC × 10^{-8}	— — —	— — —	— — —
	Resistance to solvents: p = poor, f = fair, g = good	Aromatic hydrocarbons (p), chlorinated hydrocarbons (p), esters (p), ketones (p), higher alcohols (p)			
	Surface tension	mNm^{-1}	—	—	—
	Water absorption (equilibrium 24 hours) ⅛-inch (3.175 mm) thickness (ASTM D570)	%	—	—	0.030–9.0
THERMAL	**Coefficient of linear thermal expansion** (ASTM D696)	°C^{-1} (× 10^{-5})	—	—	6.30–8.00
	Glass transition temperature (T_g)	°C	—	—	—
	Low-temperature performance / Brittleness temperature (ASTM D746)	°C	—	—	—
	Melting point (T_m) (ASTM D1238)	°C	—	—	—
	Softening point (Vicat) (ASTM D1525)	°C	—	—	99.0–104
OPTICAL	**Clarity/transparency**	—	—	—	—
	Transmittance (ASTM D1003)	%	—	—	90
	Refractive index (23°C) (ASTM D542)	—	—	—	—

Data sources:
1947: *Modern Plastics Encyclopedia*, vol. 3, *Charts* (New York: Plastics Catalogue Corporation).
1967: Shell Chemicals, *Properties of Plastics*, rev. metric ed. (London: Shell International Chemical Company and Shell Chemicals UK).
2017: MatWeb, www.matweb.com.

13.19 | High-impact polystyrene (HIPS)

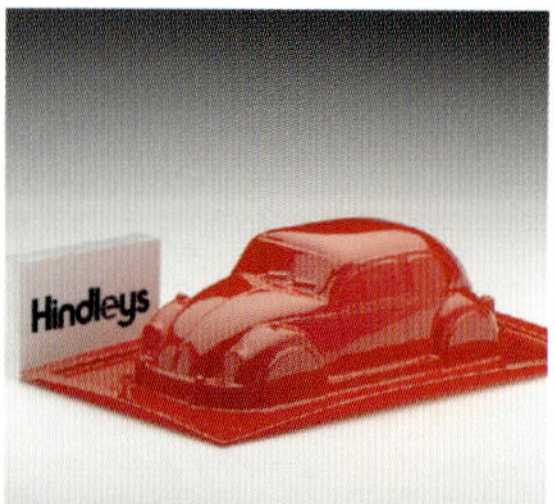

PLASTICS RECYCLING CODE

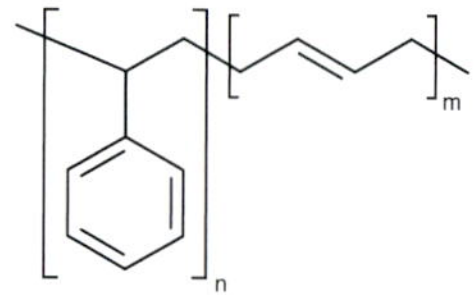

CHEMICAL STRUCTURE

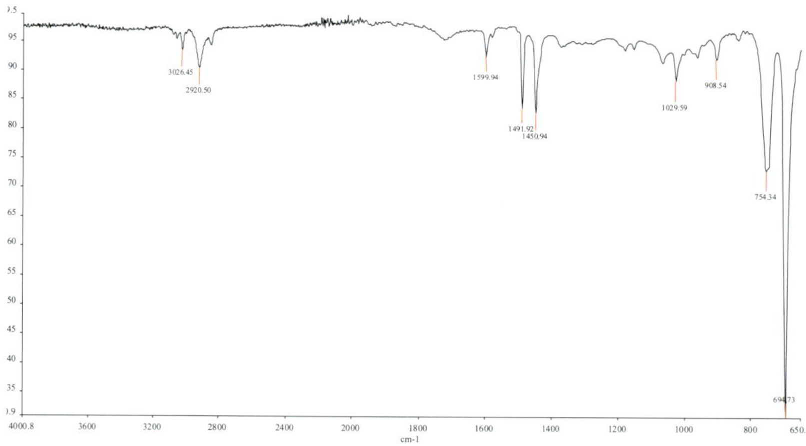

FTIR SPECTRUM

Date of first manufacture
1949.

Alternate/trade name examples
Bextrene, Styron.

Polymerization
Ethylene benzene is synthesized from benzene and ethylene, and polystyrene is synthesized by addition polymerization of ethylene benzene. Polystyrene is grafted with butadiene to improve impact resistance.

Processing
Injection molding, thermoforming (vacuum forming), blow molding, additive manufacturing (rapid prototyping, 3D printing).

Applications
Artworks, cups and lids, canisters, closures, dairy packaging, refrigerator interiors, sheets for packaging and vacuum forming, storage boxes, toys.

Feel/look/sound/smell
Translucent to opaque material. When tapped, HIPS typically has a less glassy sound than polystyrene.

Main degradation agents
Light, UV radiation, oxygen, ozone, moisture, temperature, stress, chemicals.

Main signs of degradation and damage
Yellowing.

13.19 | High-impact polystyrene (HIPS)

	PROPERTY	UNITS (SI)	DATA 1947 (STYRENE ELASTOMER)	DATA 1967	DATA 2017 (IMPACT MODIFIED)
MECHANICAL	**Compressive strength** (ASTM D695)	MPa	—	27.6-62.2	—
	Elongation (ASTM D638)	%	10-300	7-60	1.30-350
	Flexural strength (ASTM D790)	MPa	—	27.6-69	3.9-131
	Hardness (ASTM D785)	Shore D Rockwell R, M	D60-D70	R50-R100 M35-M70	R53-R124 M10-M103
	Impact strength (Izod notched) (ASTM D256)	J/cm	0.8-1.07	0.21-1.61	0.15-9.61
	Tensile modulus (rigidity) (ASTM D638)	GPa	0.3	1.72-3.10	1.1-5.6
	Tensile strength (ASTM D638)	MPa	6.2-8.3	17.2-48.0	12.4-76.0
PHYSICAL	**Density** (ASTM D792)	g/cm^3	0.95-0.97	0.98-1.10	0.848-2.12
	Hildebrand solubility parameter (δ)	(MPa)$^{1/2}$	—	—	—
	Permeability: H_2O vapor O_2 CO_2 (ASTM D1434) N.T.P. = normal temperature and pressure	cc(N.T.P.)·cm−2·sec−1·mil−1·cm Hg^{-1} at 25°C x 10−8	— — —	— 7.2 49.6	— — —
	Resistance to solvents p = poor, f = fair, g = good	aromatic hydrocarbon (p), chlorinated hydrocarbons (p), esters (p), ketones (p), higher alcohols (p)			
	Surface tension	mNm^{-1}	—	—	
	Water absorption (equilibrium 24 hours) ⅛-inch (3.175 mm) thickness (ASTM D570)	%	0.2-0.3	0.1-0.3	0.02-0.12
THERMAL	**Coefficient of linear thermal expansion** (ASTM D696)	°C^{-1} (× 10^{-5})	18-23	3.4-21	0.72-10.0
	Glass transition temperature (T_g)	°C	—	—	—
	Low-temperature performance / Brittleness temperature (ASTM D746)	°C	—	Poor/fair	−20
	Melting point (T_m) (ASTM D1238)	°C	—	130-282	130-282
	Softening point (Vicat) (ASTM D1525)	°C	—	78-100	38.3-155
OPTICAL	**Clarity/transparency**	—	Translucent to opaque	Translucent	—
	Transmittance (ASTM D1003)	%	—	—	82-89
	Refractive index (23°C) (ASTM D542)	—	—	—	—

Data sources:
1947: *Modern Plastics Encyclopedia*, vol. 3, *Charts* (New York: Plastics Catalogue Corporation).
1967: Shell Chemicals, *Properties of Plastics*, rev. metric ed. (London: Shell International Chemical Company and Shell Chemicals UK).
2017: MatWeb, www.matweb.com.

13.20 | Acrylonitrile-butadiene-styrene (ABS)

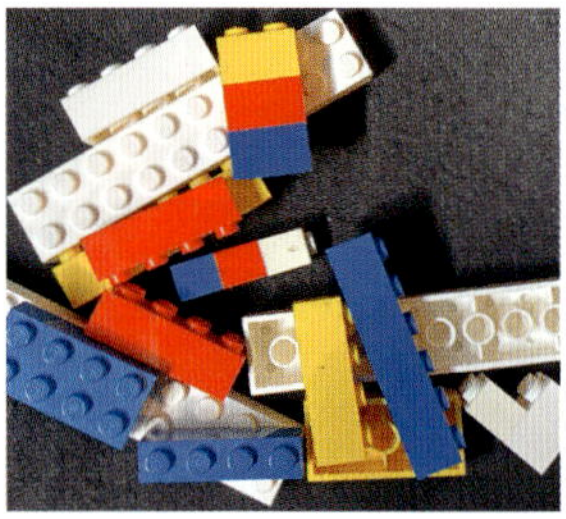

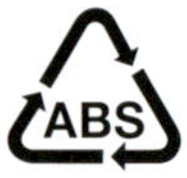

PLASTICS RECYCLING CODE

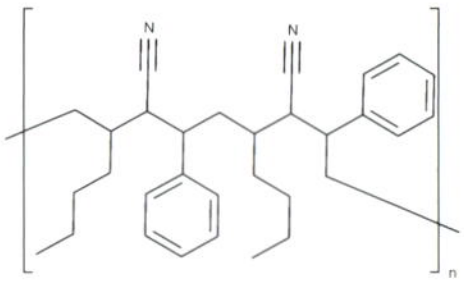

CHEMICAL STRUCTURE

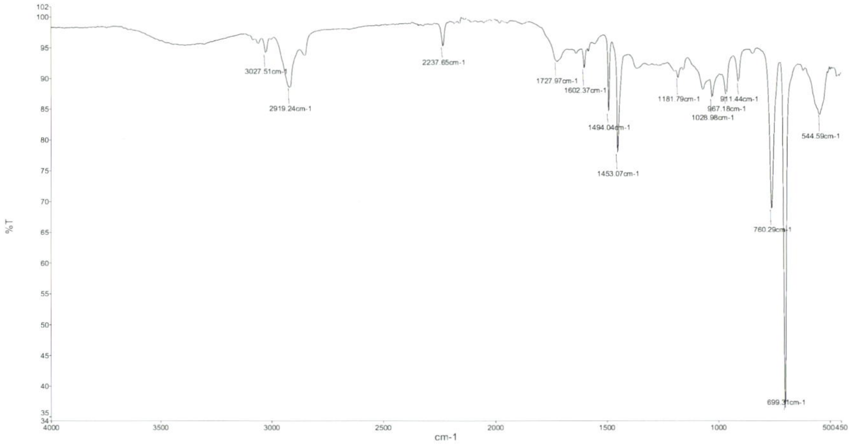

FTIR SPECTRUM

Date of first manufacture
1948 (England), 1954 (USA).

Alternate/trade name examples
Lustran, Terluran, Novodur, Cycolac, Terlux, MAGNUM.

Polymerization
Acrylonitrile-butadiene-styrene is an amorphous terpolymer made by addition polymerization of acrylonitrile with styrene, with butadiene grafted onto the polymer backbone.

Processing
Extrusion, injection molding, casting, film lamination, rotational molding, thermoforming (vacuum forming), additive manufacturing (3D printing, rapid prototyping).

Selected applications
Boats, computer housings, domestic appliances (kitchen appliance housings, refrigerator liners, vacuum cleaners), furniture, hot tubs, lawn mower covers, luggage, lunch boxes and toolboxes, medical applications, office equipment housings, packaging, pipes and fittings, power tools, safety helmets, telephone handsets, toys.

Feel/look/sound/smell
Soft feel. Sounds softer than glassy polystyrene sounds.

Degradation agents
Light, UV radiation, oxygen, ozone, moisture, temperature, stress, chemicals.

Main signs of degradation and damage
Discoloration, yellowing.

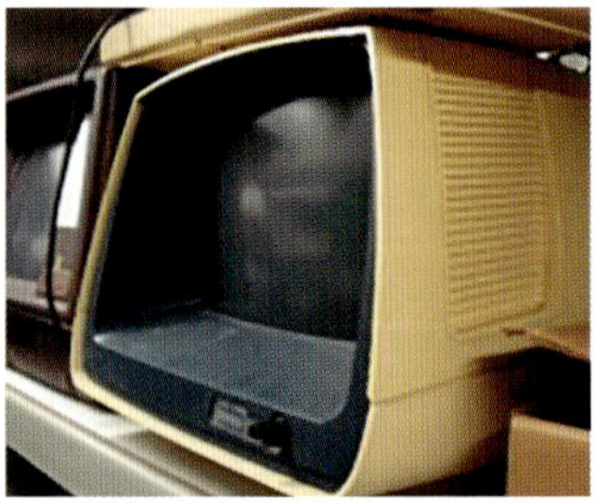

Yellowed computer housing.

13.20 | Acrylonitrile-butadiene-styrene (ABS)

	PROPERTY	UNITS (SI)	DATA 1947	DATA 1967	DATA 2016
MECHANICAL	**Compressive strength** (ASTM D695)	MPa	—	17.2-76	65-86
	Elongation (ASTM D638)	%	—	10-140	8-20
	Flexural strength (ASTM D790)	MPa	—	24.8-93.2	55-125
	Hardness (ASTM D785)	Rockwell R	—	R30-R118	R101-R124
	Impact strength (Izod notched) (ASTM D256)	J/cm	—	1.60-6.41	0.3-4.5
	Tensile modulus (rigidity) (ASTM D638)	GPa	—	0.69-2.83	1.21-1.94
	Tensile strength (ASTM D638)	MPa	—	17.2-62.0	25-65
PHYSICAL	**Density** (ASTM D792)	g/cm^3	—	0.99-1.10	1.03-1.09
	Hildebrand solubility parameter (δ)	$(MPa)^{1/2}$	—	—	19.2
	Permeability: H_2O vapor O_2 CO_2 (ASTM D1434)	$cm^3 \cdot cm \cdot cm^{-2} \cdot s^{-1} \cdot Pa^{-1}$	— — —	— — —	— — —
	Resistance to solvents p = poor, f = fair, g = good	Alcohols (f), aliphatic hydrocarbons (f), esters (p), ketones (p)			
	Surface tension	mNm^{-1}	—	—	—
	Water absorption (equilibrium 24 hours) ⅛-inch (3.175 mm) thickness (ASTM D570)	%	—	0.1-0.3	0.7-1.03
THERMAL	**Coefficient of linear thermal expansion** (ASTM D696)	$°C^{-1}$ ($\times 10^{-5}$)	—	6-13	6-11
	Glass transition temperature (T_g)	°C	—	—	102-107
	Low-temperature performance / Brittleness temperature (ASTM D746)	°C	—	Poor	−40
	Melting point (T_m) (ASTM D1238)	°C	—	—	220-260
	Softening point (Vicat) (ASTM D1525)	°C	—	85	90-112
OPTICAL	**Clarity/transparency**	—	—	Translucent or opaque	—
	Transmittance (ASTM D1003)	%	—	—	80-90
	Refractive index (23°C) (ASTM D542)	—	—	—	1.540

Data sources:
1947: *Modern Plastics Encyclopedia*, vol. 3, *Charts* (New York: Plastics Catalogue Corporation).
1967: Shell Chemicals, *Properties of Plastics*, rev. metric ed. (London: Shell International Chemical Company and Shell Chemicals UK).
2016: George Wypych, *Handbook of Polymers*, 2nd ed. (Toronto: ChemTec).

13.21 | Acrylonitrile-styrene-acrylate (ASA)

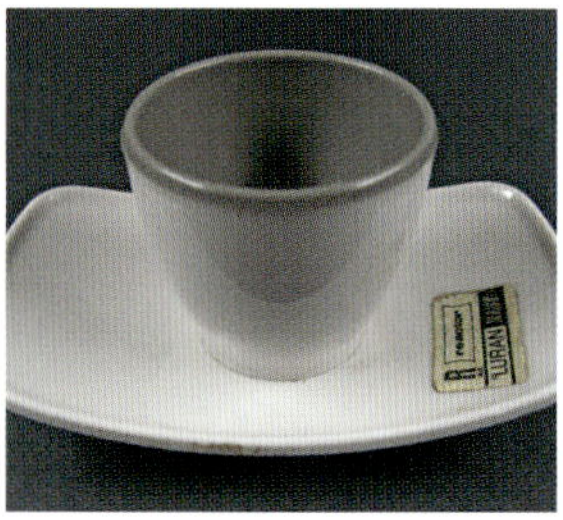

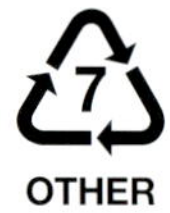

PLASTICS RECYCLING CODE

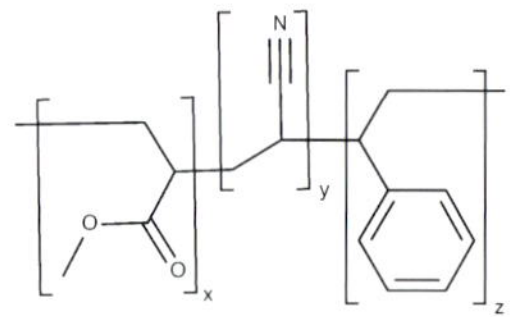

CHEMICAL STRUCTURE

FTIR SPECTRUM

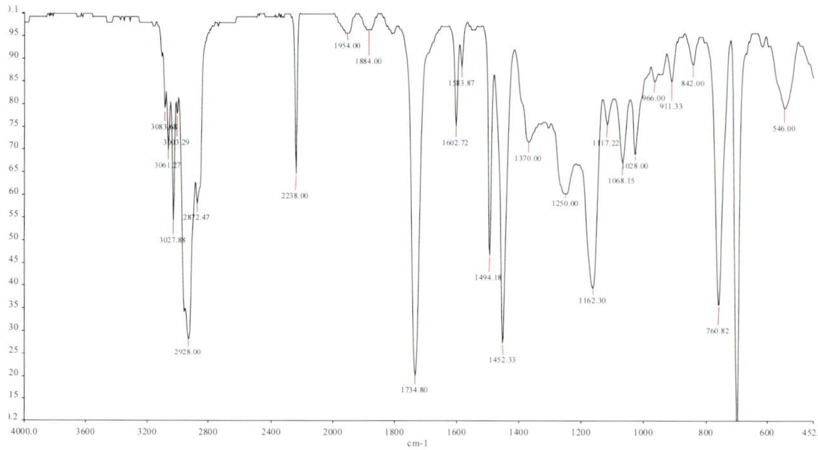

Date of first manufacture
1940s (USA), 1970 (Germany).

Alternate/trade name examples
Luran S, Dialac.

Polymerization
ASA is a two-phase thermoplastic consisting of a styrene-acrylonitrile (SAN) copolymer onto which is grafted a polybutyl acrylate elastomer.

Processing
Extrusion, injection molding, thermoforming (vacuum forming), additive manufacturing (rapid prototyping, 3D printing).

Selected applications
Automotive parts, cups and saucers, household items, kitchen accessories, marine applications, outdoor structural parts, pool accessories, ski bindings, skylights.

Feel/look/sound/smell
Sounds softer than glassy polystyrene sounds.

Main degradation agents
Light, UV radiation, oxygen, ozone, moisture, temperature, stress, chemicals.

Main signs of degradation and damage
Slow rate of deterioration, damage not commonly detected.

13.21 | Acrylonitrile-styrene-acrylate (ASA)

	PROPERTY	UNITS (SI)	DATA 1947	DATA 1967	DATA 2016
MECHANICAL	**Compressive strength** (ASTM D695)	MPa	—	—	35-65
	Elongation (ASTM D638)	%	—	—	7-37
	Flexural strength (ASTM D790)	MPa	—	—	56-75
	Hardness (ASTM D785)	Rockwell R	—	—	R100-R103
	Impact strength (Izod notched) (ASTM D256)	J/cm	—	—	0.65-2.10
	Tensile modulus (rigidity) (ASTM D638)	GPa	—	—	1.9-2.6
	Tensile strength (ASTM D638)	MPa	—	—	29-51
PHYSICAL	**Density** (ASTM D792)	g/cm^3	—	—	1.06-1.1
	Hildebrand solubility parameter (δ)	$(MPa)^{1/2}$	—	—	—
	Permeability: H_2O vapor O_2 CO_2 (ASTM D1434)	g/m^2·day $cm^3/$ m^2·day·bar	— — —	— — —	30-35 150-560 —
	Resistance to solvents p = poor, f = fair, g = good	Alcohols (f), aliphatic hydrocarbons (g), aromatic hydrocarbons (p), esters (p), ketones (p)			
	Surface tension	mNm^{-1}	—	—	—
	Water absorption (equilibrium 24 hours) ⅛-inch (3.175 mm) thickness (ASTM D570)	%	—	—	0.55-1.65
THERMAL	**Coefficient of linear thermal expansion** (ASTM D696)	$°C^{-1}$ ($\times 10^{-5}$)	—	—	9.5-12
	Glass transition temperature (T_g)	°C	—	—	103-127
	Low-temperature performance / Brittleness temperature (ASTM D746)	°C	—	—	—
	Melting point (T_m) (ASTM D1238)	°C	—	—	180-200
	Softening point (Vicat) (ASTM D1525)	°C	—	—	85-100
OPTICAL	**Clarity/transparency**	—	—	—	—
	Transmittance (ASTM D1003)	%	—	—	—
	Refractive index (23°C) (ASTM D542)	—	—	—	1.57-1.677

Data sources:
1947: *Modern Plastics Encyclopedia*, vol. 3, *Charts* (New York: Plastics Catalogue Corporation).
1967: Shell Chemicals, *Properties of Plastics*, rev. metric ed. (London: Shell International Chemical Company and Shell Chemicals UK).
2016: George Wypych, *Handbook of Polymers*, 2nd ed. (Toronto: ChemTec).

13.22 | Styrene-acrylonitrile (SAN)

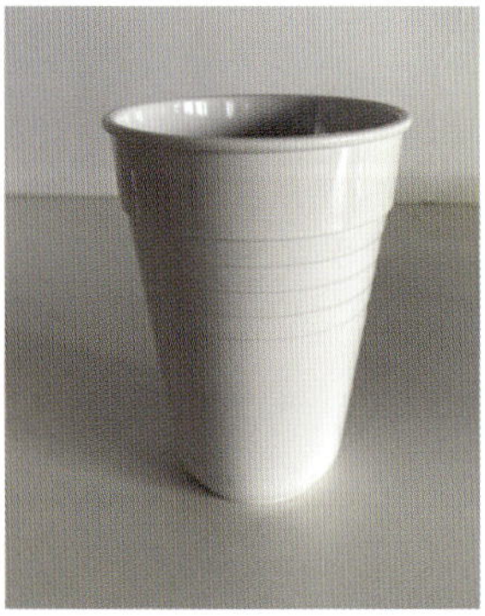

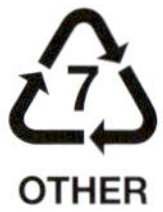

PLASTICS RECYCLING CODE

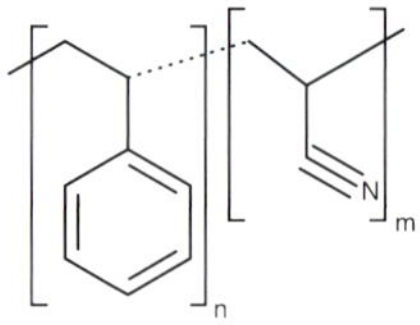

CHEMICAL STRUCTURE

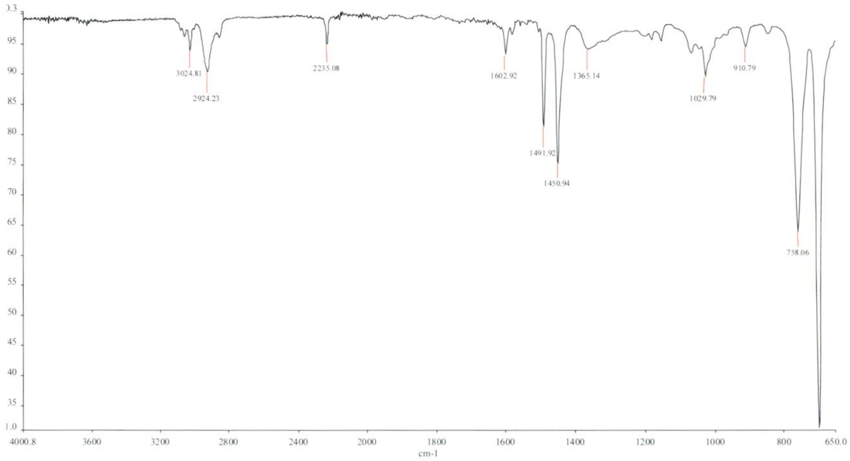

FTIR SPECTRUM

Date of first manufacture
1938 (UK).

Alternate/trade name examples
Tyril, Luran.

Polymerization
Styrene-acrylonitrile is made by addition polymerization using free-radical initiators from styrene and acrylonitrile.

Processing
Extrusion, injection molding, blow molding, casting, thermoforming.

Selected applications
Battery cases, bottles, cups, food containers, instrument lenses, kitchenware.

Feel/look/sound/smell
Outstanding transparency, sounds softer than glassy polystyrene sounds.

Main degradation agents
Light, UV radiation, oxygen, ozone, moisture, temperature, stress, chemicals.

Main signs of degradation and damage
Slow rate of deterioration, damage not commonly detected.

13.22 | Styrene-acrylonitrile (SAN)

	PROPERTY	UNITS (SI)	DATA 1947	DATA 1967	DATA 2016
MECHANICAL	**Compressive strength** (ASTM D695)	MPa	—	96.6-117	—
	Elongation (ASTM D638)	%	—	1.5-3.5	2.5-4.0
	Flexural strength (ASTM D790)	MPa	—	96.6 -131	—
	Hardness (ASTM D785)	Rockwell M	—	M80-M83	—
	Impact strength (Izod notched) (ASTM D256)	J/cm	—	0.19-0.27	—
	Tensile modulus (rigidity) (ASTM D638)	GPa	—	2.76-3.87	2.2-17.1
	Tensile strength (ASTM D638)	MPa	—	65.7-83.9	61-79
PHYSICAL	**Density** (ASTM D792)	g/cm^3	—	1.075-1.10	1.07-1.09
	Hildebrand solubility parameter (δ)	$(MPa)^{1/2}$	—	—	—
	Permeability: H_2O vapor O_2 CO_2 (ASTM D1434)	$cm^3 \cdot cm \cdot cm^{-2} \cdot s^{-1} \cdot Pa^{-1}$	— — —	— — —	— — —
	Resistance to solvents p = poor, f = fair, g = good	Ethyl acetate (p), ketones (p), alcohols (g), aliphatic hydrocarbons (g)			
	Surface tension	mNm^{-1}	—	—	—
	Water absorption (equilibrium 24 hours) ⅛-inch (3.175 mm) thickness (ASTM D570)	%	—	0.1-0.3	0.25-0.35
THERMAL	**Coefficient of linear thermal expansion** (ASTM D696)	$°C^{-1}$ ($\times 10^{-5}$)	—	6-8	7
	Glass transition temperature (T_g)	°C	—	—	103-112
	Low-temperature performance / Brittleness temperature (ASTM D746)	°C	—	Poor	−20
	Melting point (T_m) (ASTM D1238)	°C	—	—	—
	Softening point (Vicat) (ASTM D1525)	°C	—	85-103	105-120
OPTICAL	**Clarity/transparency**	—	—	Transparent	—
	Transmittance (ASTM D1003)	%	—	—	85-92
	Refractive index (23°C) (ASTM D542)	—	—	1.57	1.5700

Data sources:
1947: *Modern Plastics Encyclopedia*, vol. 3, *Charts* (New York: Plastics Catalogue Corporation).
1967: Shell Chemicals, *Properties of Plastics*, rev. metric ed. (London: Shell International Chemical Company and Shell Chemicals UK).
2016: George Wypych, *Handbook of Polymers*, 2nd ed. (Toronto: ChemTec).

13.23 | Polyvinyl chloride, plasticized (PVC-P)

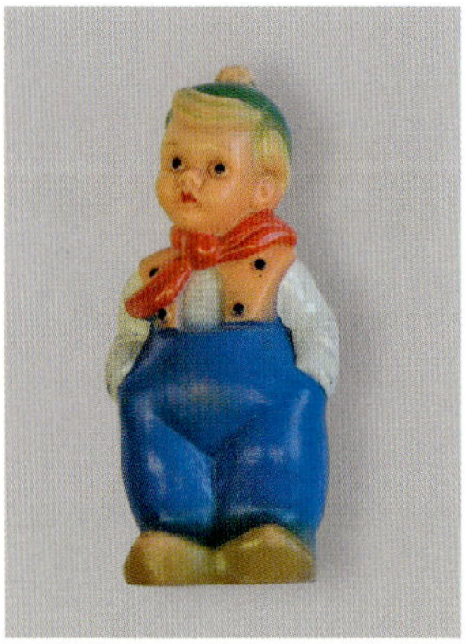

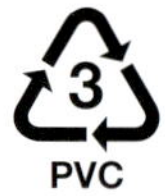

PLASTICS RECYCLING CODE

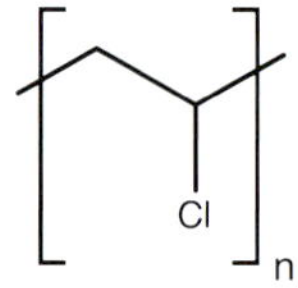

CHEMICAL STRUCTURE

FTIR SPECTRUM

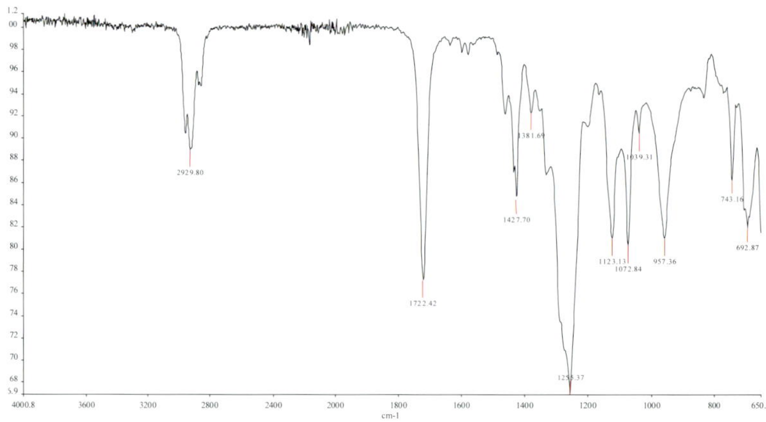

Date of first manufacture
1926 (USA), 1940 (UK).

Alternate/trade name examples
Vintec, Tygon, Vinylite.

Polymerization
Polyvinyl chloride is made by addition polymerization of vinyl chloride, which is synthesized from ethylene and chlorine. PVC can be mixed with plasticizers, which are sometimes added in concentrations as high as 50%. The most commonly used plasticizer is di(2-ethylhexyl) phthalate (DEHP), also known as dioctyl phthalate (DOP).

Processing
Extrusion, injection molding, blow molding, thermoforming (vacuum forming), calendering, plastisol coating (dipping), rotational molding.

Selected applications
Films, flooring, foils, garden and other hoses, gloves, imitation leather upholstery, jewelry, packaging, raincoats, shower curtains, tablecloths, toys (Barbie dolls).

Feel/look/sound/smell
Flexible. Transparent or opaque. Smell of plasticizers.

Main degradation agents
Light, UV radiation, oxygen, ozone, moisture, temperature, stress, chemicals.

Main signs of degradation and damage
Loss of plasticizers: exudation of liquids, efflorescence of solid white materials; stiffening, discoloration from yellow to brown to black; loss of flexibility (hardening), tears, breaks.

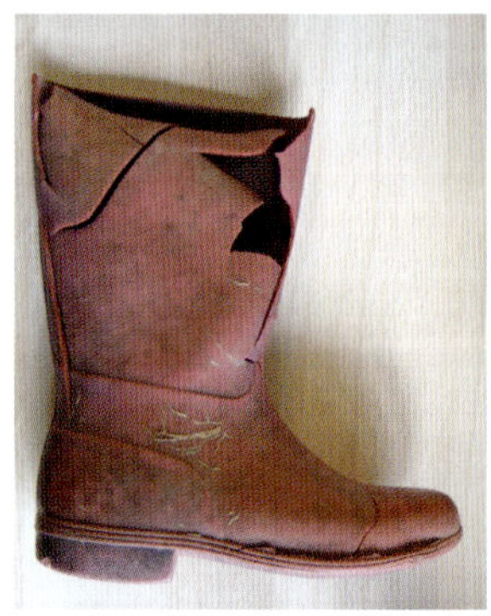

Stiff, hardened, cracked rain boot.

13.23 | Polyvinyl chloride, plasticized (PVC-P)

	PROPERTY	UNITS (SI)	DATA 1947	DATA 1967	DATA 2016
MECHANICAL	**Compressive strength** (ASTM D695)	MPa	—	89.7	—
	Elongation (ASTM D638)	%	—	—	3.3-430
	Flexural strength (ASTM D790)	MPa	—	83.9	67-107
	Hardness (ASTM D785)	Rockwell M Shore A, D	—	M105	M66-M69 D22-D25 A30-A96
	Impact strength (Izod notched) (ASTM D256)	J/cm	—	8.03	0.33-13.02
	Tensile modulus (rigidity) (ASTM D638)	GPa	—	4.14	2.4-4.0
	Tensile strength (ASTM D638)	MPa	—	—	7.1-68.9
PHYSICAL	**Density** (ASTM D792)	g/cm^3	—	1.35	1.37-1.43
	Hildebrand solubility parameter (δ)	(MPa)$^{1/2}$	—	—	19.19-20.1
	Permeability: H_2O vapor O_2 CO_2 (ASTM D1434)	cm^3·cm·cm^{-2}·s^{-1}·Pa^{-1} × 10^{-9}	— — —	— — —	0.12 0.034 —
	Resistance to solvents p = poor, f = fair, g = good	Acetone (p), esters (p), ethanol (g), isopropanol (g), mineral spirit (p), xylene (p)			
	Surface tension	mNm^{-1}	—	—	32-46
	Water absorption (equilibrium 24 hours) ⅛-inch (3.175 mm) thickness (ASTM D570)	%	—	0.10	0.04-0.4
THERMAL	**Coefficient of linear thermal expansion** (ASTM D696)	°C^{-1} (× 10^{-5})	—	5	3.5-7.1
	Glass transition temperature (T_g)	°C	—	—	82-87
	Low-temperature performance / Brittleness temperature (ASTM D746)	°C	—	Good	−20-0
	Melting point (T_m) (ASTM D1238)	°C	—	—	103-230
	Softening point (Vicat) (ASTM D1525)	°C	—	78	—
OPTICAL	**Clarity/transparency**	—	—	—	—
	Transmittance (ASTM D1003)	%	—	—	—
	Refractive index (23°C) (ASTM D542)	—	—	—	1.532-1.548

Data sources:
1947: *Modern Plastics Encyclopedia*, vol. 3, *Charts* (New York: Plastics Catalogue Corporation).
1967: Shell Chemicals, *Properties of Plastics*, rev. metric ed. (London: Shell International Chemical Company and Shell Chemicals UK).
2016: George Wypych, *Handbook of Polymers*, 2nd ed. (Toronto: ChemTec).

13.24 | Polyvinyl chloride, rigid, unplasticized (PVC-U)

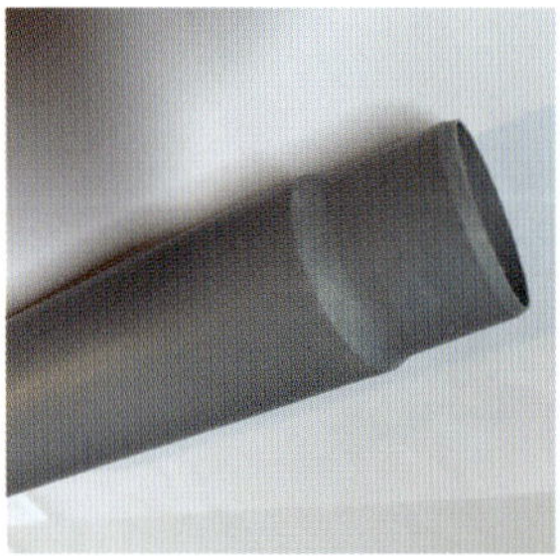

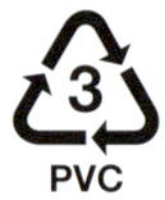

PLASTICS RECYCLING CODE

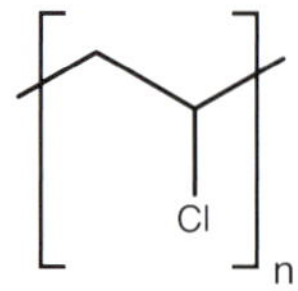

CHEMICAL STRUCTURE

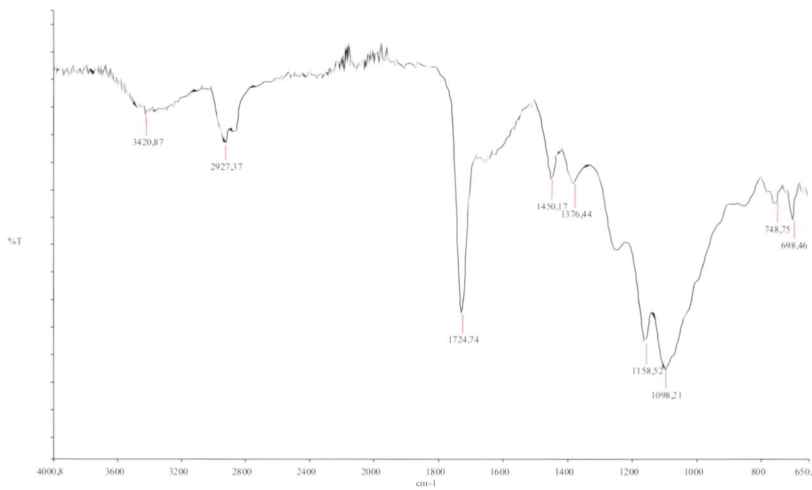

FTIR SPECTRUM

Date of first manufacture
1912 (patented Germany); commercially available since 1930.

Alternate/trade name examples
Vestolit, Dural, Exelon.

Polymerization
Polyvinyl chloride is made by addition polymerization of vinyl chloride, Vinyl chloride is synthesized from ethylene and chlorine.

Processing
Extrusion, injection molding, blow molding.

Selected applications
Gutters, plumbing pipe, transparent bottles, ventilators, window frames.

Feel/look/sound/smell
Rigid. Translucent or opaque.

Main degradation agents
Light, UV radiation, oxygen, ozone, moisture, temperature, stress, chemicals.

Main signs of degradation and damage
Yellowing, chalky surface, brittleness, cracking, breaks.

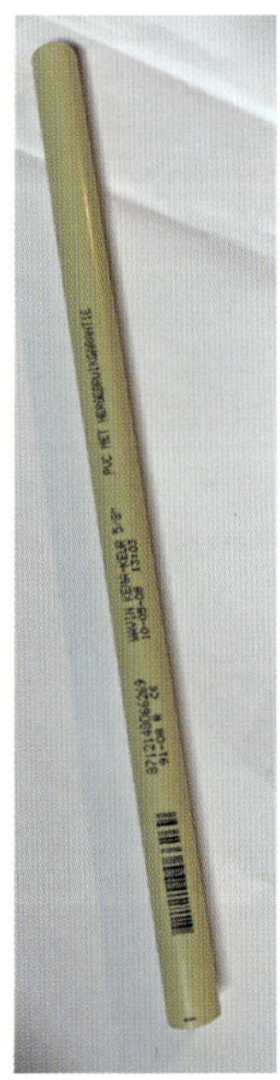

Yellowed electricity pipe.

13.24 | Polyvinyl chloride, rigid, unplasticized (PVC-U)

	PROPERTY	UNITS (SI)	DATA 1947	DATA 1967	DATA 2017
MECHANICAL	**Compressive strength** (ASTM D695)	MPa	—	55.3	—
	Elongation (ASTM D638)	%	2–550	2–40	2–330
	Flexural strength (ASTM D790)	MPa	—	93.2	50.7–104
	Hardness (ASTM D785)	Rockwell R Shore D	—	R110	R100–R111 D66–D88
	Impact strength (Izod notched) (ASTM D256)	J/cm	—	0.54–1.61	0.16–14.0
	Tensile modulus (rigidity) (ASTM D638)	GPa	—	2.42	1.82–7.03
	Tensile strength (ASTM D638)	MPa	6.89–62.1	58.8	23.8–68.9
PHYSICAL	**Density** (ASTM D792)	g/cm^3	1.2–1.6	1.38–1.4	1.16–1.65
	Hildebrand solubility parameter (δ)	$(MPa)^{1/2}$	—	—	—
	Permeability: H_2O vapor O_2 CO_2 (ASTM D1434) N.T.P. = normal temperature and pressure	$cc(N.T.P.)\cdot cm^{-2}\cdot sec^{-1}\cdot mil^{-1}\cdot cm\ Hg^{-1}$ at 25°C × 10^{-8}	— — —	630 0.47 1.52	— — —
	Resistance to solvents p = poor, f = fair, g = good	Acetone (p), esters (p), ethanol (g), isopropanol (g), mineral spirit (p), xylene (p)			
	Surface tension	mNm^{-1}	—	—	—
	Water absorption (equilibrium 24 hours) ⅛-inch (3.175 mm) thickness (ASTM D570)	%	0.1–0.6	0.05	0.2
THERMAL	**Coefficient of linear thermal expansion** (ASTM D696)	$°C^{-1}$ (× 10^{-5})	—	5	2.7–14.0
	Glass transition temperature (T_g)	°C	—	—	—
	Low-temperature performance / Brittleness temperature (ASTM D746)	°C	—	—	−30–15.0
	Melting point (T_m) (ASTM D1238)	°C	—	—	177–216
	Softening point (Vicat) (ASTM D1525)	°C	—	82	51.0–97.8
OPTICAL	**Clarity/transparency**	—	—	Transparent to opaque	—
	Transmittance (ASTM D1003)	%	—	—	7.5–90
	Refractive index (23°C) (ASTM D542)	—	—	1.52	—

Data sources:
1947: *Modern Plastics Encyclopedia*, vol. 3, *Charts* (New York: Plastics Catalogue Corporation).
1967: Shell Chemicals, *Properties of Plastics*, rev. metric ed. (London: Shell International Chemical Company and Shell Chemicals UK).
2017: MatWeb, www.matweb.com.

13.25 | Polyvinyl chloride–polyvinyl acetate copolymer (PVC/PVAC)

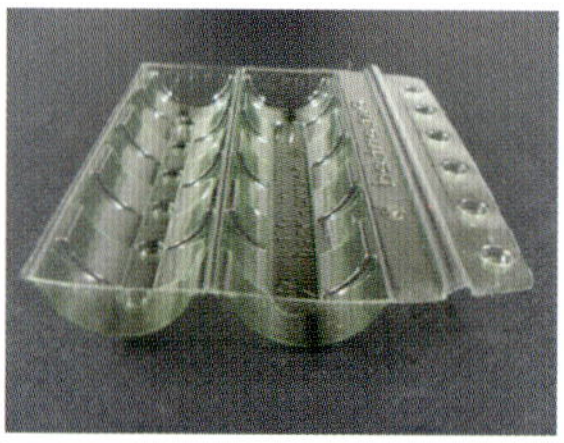

7
OTHER

PLASTICS RECYCLING CODE

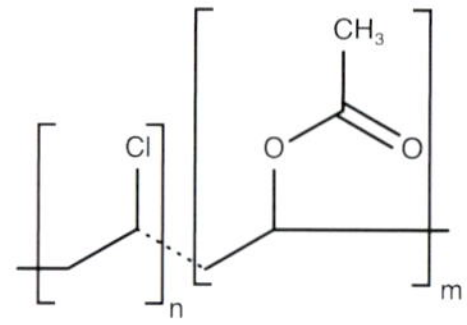

CHEMICAL STRUCTURE

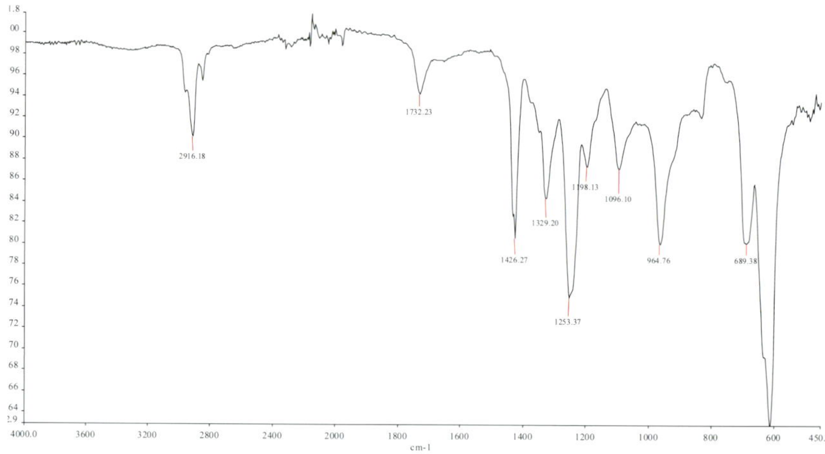

FTIR SPECTRUM

Date of first manufacture
1948 (for records).

Alternate/trade name examples
N/A

Polymerization
PVC/PVAC copolymer is polymerized from vinyl acetate and vinyl chloride.

Processing
Extrusion, calendaring, vacuum forming, thermoforming, compression molding.

Selected applications
Aluminum foil coating, blister packs, credit cards, flooring, paints, printing and other inks, records, rigid sheet.

Feel/look/sound/smell
Rigid.

Main degradation agents
Light, UV radiation, oxygen, ozone, moisture, temperature, stress, chemicals.

Main signs of degradation and damage
Yellowing and breaking.

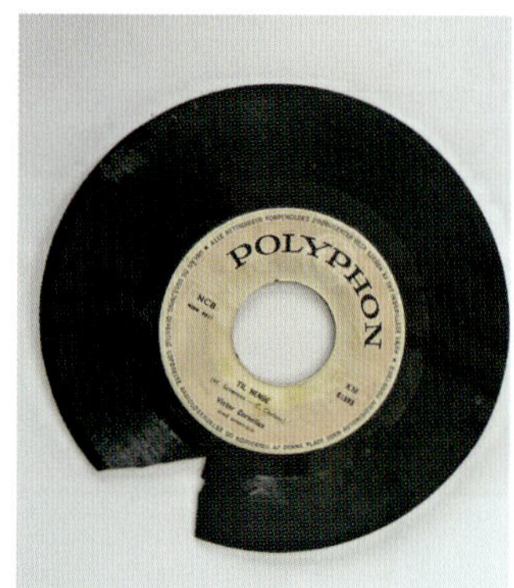

Broken 45 rpm record.

13.25 | Polyvinyl chloride-polyvinyl acetate copolymer (PVC/PVAC)

	PROPERTY	UNITS (SI)	DATA 1947	DATA 1967	DATA 2016
MECHANICAL	**Compressive strength** (ASTM D695)	MPa	—	55.3-89.7	—
	Elongation (ASTM D638)	%	—	200-450	290
	Flexural strength (ASTM D790)	MPa	51.7-96.5	83.9	—
	Hardness (ASTM D785)	Rockwell	—	—	—
	Impact strength (Izod notched) (ASTM D256)	J/cm	0.13-0.27	0.27-0.54	—
	Tensile modulus (rigidity) (ASTM D638)	GPa	2.4-2.8	2.42-4.14	1.7
	Tensile strength (ASTM D638)	MPa	31.0-55.2	51.9-58.8	3.27-4.86
PHYSICAL	**Density** (ASTM D792)	g/cm^3	1.34-1.45	1.37-1.45	1.30-1.37
	Hildebrand solubility parameter (δ)	$(MPa)^{1/2}$	—	—	—
	Permeability: H_2O vapor O_2 CO_2 (ASTM D1434)	$cm^3 \cdot cm \cdot cm^{-2} \cdot s^{-1} \cdot Pa^{-1}$	— — —	— — —	— — —
	Resistance to solvents p = poor, f = fair, g = good	Acetone (p), esters(p), ethanol (g), isopropanol (g), mineral spirit (p), xylene (p)			
	Surface tension	mNm^{-1}	—	—	—
	Water absorption (equilibrium 24 hours) ⅛-inch (3.175 mm) thickness (ASTM D570)	%	0.05-0.15	0.08	—
THERMAL	**Coefficient of linear thermal expansion** (ASTM D696)	$°C^{-1}$ ($\times 10^{-5}$)	6.9	7	—
	Glass transition temperature (T_g)	°C	—	—	58-75
	Low-temperature performance / Brittleness temperature (ASTM D746)	°C	—	Good	—
	Melting point (T_m) (ASTM D1238)	°C	—	—	100-110
	Softening point (Vicat) (ASTM D1525)	°C	—	—	38
OPTICAL	**Clarity/transparency**	—	Transparent or opaque	Transparent or opaque	—
	Transmittance (ASTM D1003)	%	—	—	—
	Refractive index (23°C) (ASTM D542)	—	1.52-1.53	1.55	1.45-1.47

Data sources:
1947: *Modern Plastics Encyclopedia*, vol. 3, *Charts* (New York: Plastics Catalogue Corporation).
1967: Shell Chemicals, *Properties of Plastics*, rev. metric ed. (London: Shell International Chemical Company and Shell Chemicals UK).
2016: George Wypych, *Handbook of Polymers*, 2nd ed. (Toronto: ChemTec).

13.26 | Polyvinylidene chloride (PVDC)

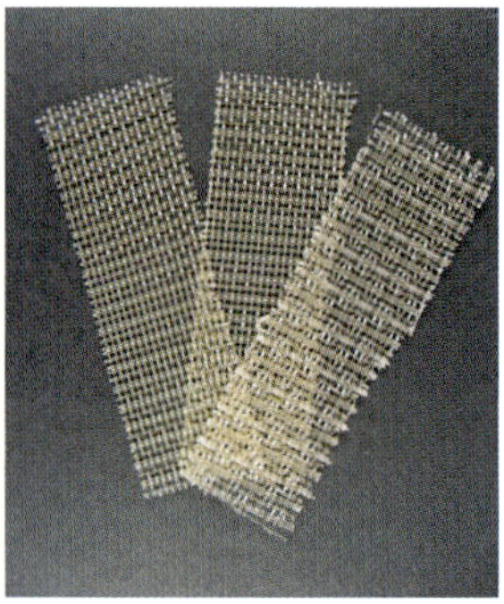

7
OTHER

PLASTICS RECYCLING CODE

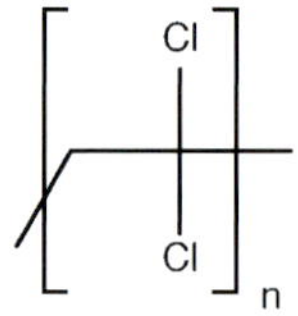

CHEMICAL STRUCTURE

FTIR SPECTRUM

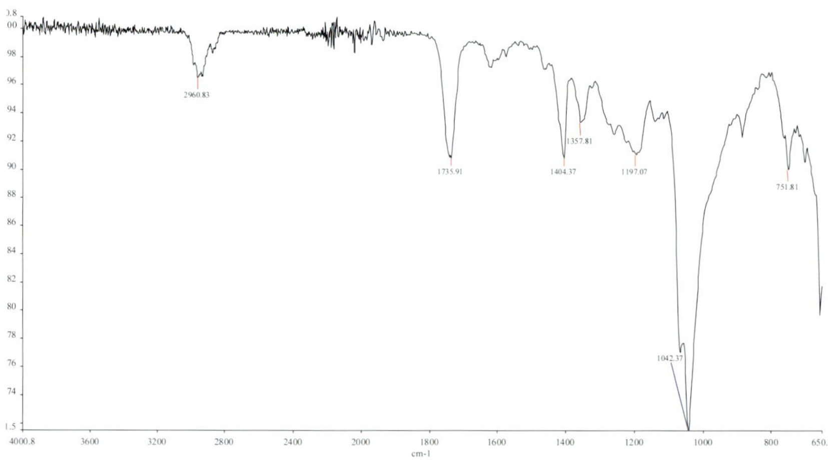

Date of first manufacture
1933 (USA), 1949 (cling wrap).

Alternate/trade name examples
Saran, Diofan D, Sconatex.

Polymerization
Polyvinylidene chloride is made by addition polymerization from vinylidene chloride.

Processing
Extrusion, film extrusion, thermoforming (vacuum forming), fiber spinning.

Selected applications
Barrier films for food packaging, blister packaging, cling wrap, coatings, fibers, pharmaceutical applications, tape.

Feel/look/sound/smell
Transparent.

Main degradation agents
Light, UV radiation, oxygen, ozone, moisture, temperature, stress, chemicals.

Main signs of degradation and damage
Yellowing. Brittleness.

13.26 | Polyvinylidene chloride (PVDC)

	PROPERTY	UNITS (SI)	DATA 1947	DATA 1967	DATA 2016
MECHANICAL	**Compressive strength** (ASTM D695)	MPa	—	—	55
	Elongation (ASTM D638)	%	—	—	10–20 15–40 (oriented)
	Flexural strength (ASTM D790)	MPa	—	—	40
	Hardness (ASTM D785)	Rockwell R	—	—	R55
	Impact strength (Izod notched) (ASTM D256)	J/cm	—	—	—
	Tensile modulus (rigidity) (ASTM D638)	GPa	—	—	—
	Tensile strength (ASTM D638)	MPa	—	—	24–69 207–414 (oriented)
PHYSICAL	**Density** (ASTM D792)	g/cm^3	—	—	1.70–1.775
	Hildebrand solubility parameter (δ)	$(MPa)^{1/2}$	—	—	23.1; 25.0
	Permeability: H_2O vapor (ASTM D1434)	$cm^3 \cdot cm \cdot cm^{-2} \cdot s^{-1} \cdot Pa^{-1} \times 10^{12}$	—	—	—
	O_2		—	—	0.000383
	CO_2		—	—	—
	Resistance to solvents p = poor, f = fair, g = good	Alcohols (g), aliphatic hydrocarbons (g), aromatic hydrocarbons (p), esters (f), ketones (f)			
	Surface tension	mNm^{-1}	—	—	45.4
	Water absorption (equilibrium 24 hours) ⅛-inch (3.175 mm) thickness (ASTM D570)	%	—	—	—
THERMAL	**Coefficient of linear thermal expansion** (ASTM D696)	$°C^{-1}$ ($\times 10^{-5}$)	—	—	—
	Glass transition temperature (T_g)	°C	—	—	−17 to −20
	Low-temperature performance / Brittleness temperature (ASTM D746)	°C	—	—	−10–+10
	Melting point (T_m) (ASTM D1238)	°C	—	—	158–208
	Softening point (Vicat) (ASTM D1525)	°C	—	—	100–150
OPTICAL	**Clarity/transparency**	—	—	—	—
	Transmittance (ASTM D1003)	%	—	—	—
	Refractive index (23°C) (ASTM D542)	—	—	—	1.60–1.63

Data sources:
1947: *Modern Plastics Encyclopedia*, vol. 3, *Charts* (New York: Plastics Catalogue Corporation).
1967: Shell Chemicals, *Properties of Plastics*, rev. metric ed. (London: Shell International Chemical Company and Shell Chemicals UK).
2016: George Wypych, *Handbook of Polymers*, 2nd ed. (Toronto: ChemTec).

13.27 | Cellophane (regenerated cellulose)

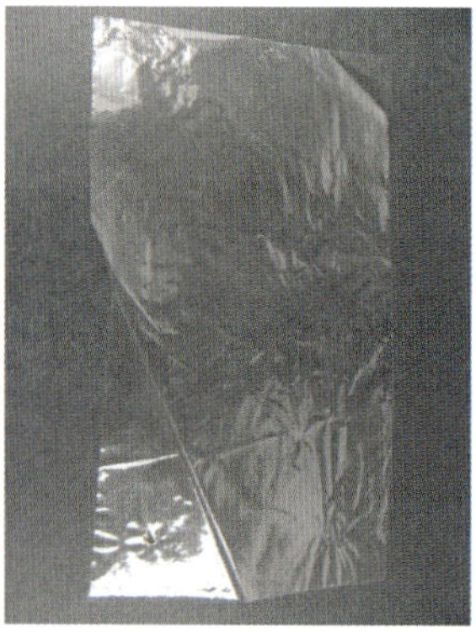

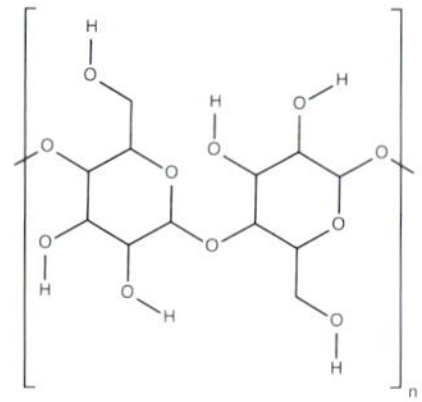

CHEMICAL STRUCTURE

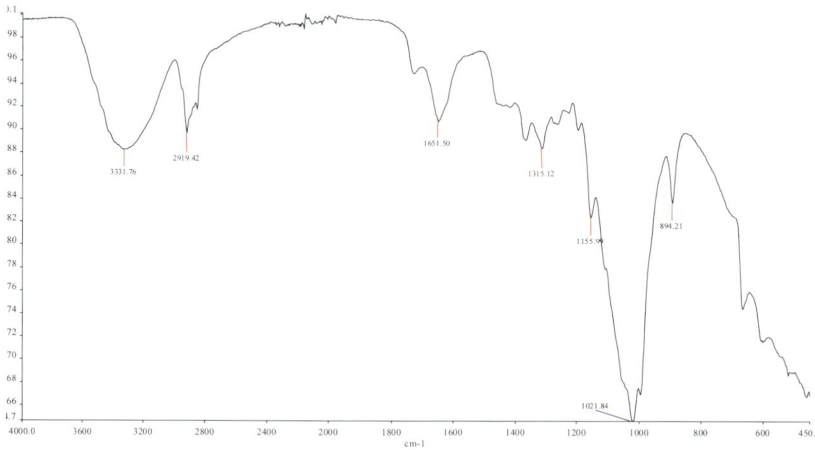

FTIR SPECTRUM

Date of first manufacture
1892 (Switzerland), viscose patented 1911 (Switzerland).

Alternate/trade name examples
N/A

Polymerization
Cellulose fibers from wood or cotton are dissolved into an alkaline solution, which forms a viscous solution. The addition of sulfuric acid and sodium sulfate causes the cellulose to re-form, creating regenerated cellulose. Ethylene glycol or glycerol is added as a plasticizer.

Processing
Extrusion, film extrusion. Like paper, cellophane is stronger in the casting (longitudinal) direction than in the transverse direction.

Selected applications
Cellophane is biodegradable. It has a high moisture vapor transmission rate and is suitable for packaging cigarettes and cigars and food.

Feel/look/sound/smell
Crisp, transparent. Comes in many bright colors.

Main degradation agents
Light, UV radiation, oxygen, ozone, moisture, temperature, stress, chemicals, microbes.

Main signs of degradation and damage
Yellowing, distortion, brittleness, tears, breaks.

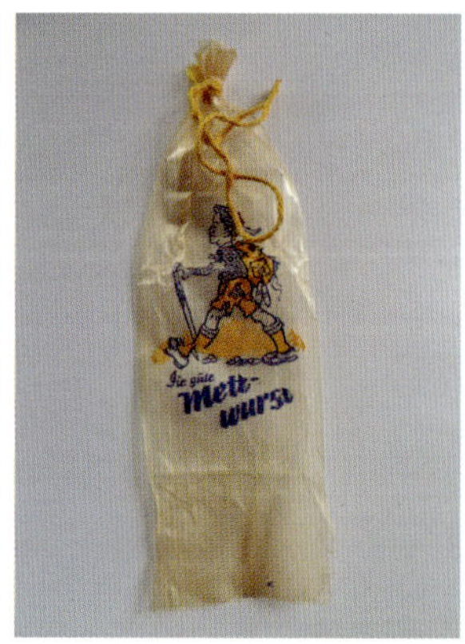

Yellowed and brittle grocery packaging.

13.27 | Cellophane (regenerated cellulose)

	PROPERTY	UNITS (SI)	DATA 1947	DATA 1967	DATA 2017
MECHANICAL	**Compressive strength** (ASTM D695)	MPa	—	—	—
	Elongation (ASTM D638)	%	—	—	18 transverse 55 longitudinal
	Flexural strength (ASTM D790)	MPa	—	—	—
	Hardness (ASTM D785)	Rockwell	—	—	—
	Impact strength (Izod notched) (ASTM D256)	J/cm	—	—	—
	Tensile modulus (rigidity) (ASTM D638)	GPa	—	—	3-5
	Tensile strength (ASTM D638)	MPa	—	—	55 transverse 120 longitudinal
PHYSICAL	**Density** (ASTM D792)	g/cm^3	—	—	1.44
	Hildebrand solubility parameter (δ)	$(MPa)^{1/2}$	—	—	—
	Permeability: H_2O vapor O_2 CO_2 (ASTM D1434)	$cm^3 \cdot cm \cdot cm^{-2} \cdot s^{-1} \cdot Pa^{-1} \times 10^{-13}$	—	—	20000 0.004-0.04 0.004-0.05
	Resistance to solvents p = poor, f = fair, g = good	Alcohols (g), aliphatic hydrocarbons (g), aromatic hydrocarbons (g), esters (g), ketones (g)			
	Surface tension	mNm^{-1}	—	—	—
	Water absorption (equilibrium 24 hours) ⅛-inch (3.175 mm) thickness (ASTM D570)	%	—	—	50
THERMAL	**Coefficient of linear thermal expansion** (ASTM D696)	$°C^{-1}$ ($\times 10^{-5}$)	—	—	—
	Glass transition temperature (T_g)	°C	—	—	—
	Low-temperature performance / Brittleness temperature (ASTM D746)	°C	—	—	81-100
	Melting point (T_m) (ASTM D1238)	°C	—	—	—
	Softening point (Vicat) (ASTM D1525)	°C	—	—	166-180
OPTICAL	**Clarity/transparency**	—	—	—	—
	Transmittance (ASTM D1003)	%	—	—	90
	Refractive index (23°C) (ASTM D542)	—	—	—	—

Data sources:
1947: *Modern Plastics Encyclopedia*, vol. 3, *Charts* (New York: Plastics Catalogue Corporation).
1967: Shell Chemicals, *Properties of Plastics*, rev. metric ed. (London: Shell International Chemical Company and Shell Chemicals UK).
2017: MatWeb, www.matweb.com.

13.28 | Polycaprolactone (PCL)

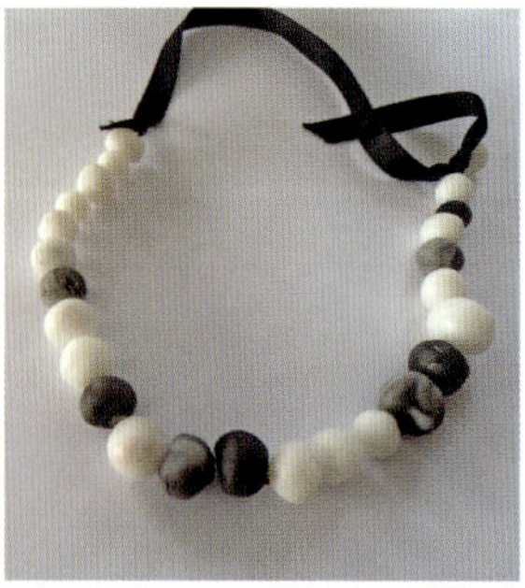

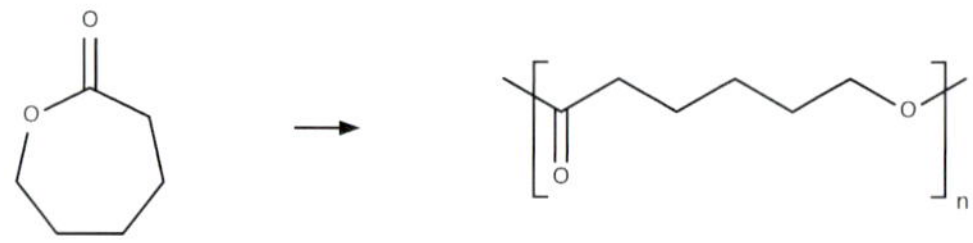

CHEMICAL STRUCTURE

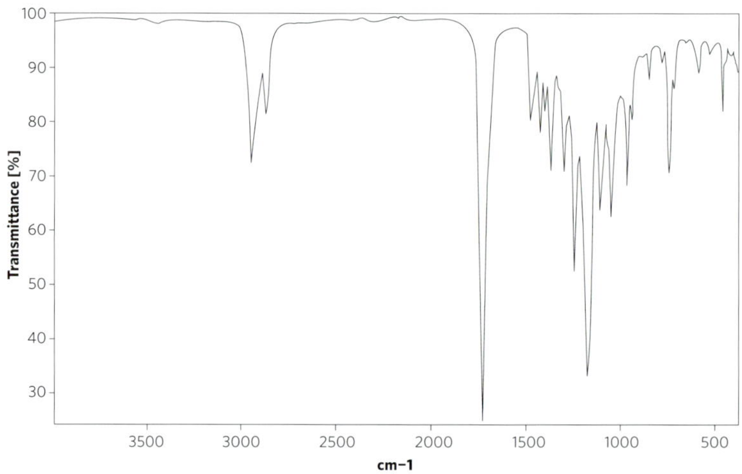

FTIR SPECTRUM

Date of first manufacture
1930 (invented but not used), 1965 (application). New interest arose in the 1990s due to the material's biodegradability.

Alternate/trade name examples
ShapeLock, Polymorph, TechTack.

Polymerization
Polycaprolactone is polymerized by the ring-opening addition reaction of ε-caprolactone.

Processing
Extrusion, additive manufacturing (rapid prototyping, 3D printing).

Selected applications
PCL is a biodegradable polyester used by hobbyists as a modeling material and in works of art. In medicine, it is used as an implantable biomaterial and for splinting.

Feel/look/sound/smell
Whitish, opaque. Waxy feel.

Main degradation agents
Light, UV radiation, oxygen, ozone, moisture, temperature, stress, chemicals, microbes.

Main signs of degradation and damage
Yellowing.

13.28 | Polycaprolactone (PCL)

	PROPERTY	UNITS (SI)	DATA 1947	DATA 1967	DATA 2016
MECHANICAL	**Compressive strength** (ASTM D695)	MPa	—	—	—
	Elongation (ASTM D638)	%	—	—	300–600
	Flexural strength (ASTM D790)	MPa	—	—	200–500
	Hardness (ASTM D785)	Rockwell	—	—	—
	Impact strength (Izod notched) (ASTM D256)	J/cm	—	—	1.20–3.75
	Tensile modulus (rigidity) (ASTM D638)	GPa	—	—	0.21–0.44
	Tensile strength (ASTM D638)	MPa	—	—	7.6–58
PHYSICAL	**Density** (ASTM D792)	g/cm^3	—	—	1.07–1.20
	Hildebrand solubility parameter (δ)	$(MPa)^{1/2}$	—	—	—
	Permeability: H_2O vapor O_2 CO_2 (ASTM D1434)	$cm^3 \cdot cm \cdot cm^{-2} \cdot s^{-1} \cdot Pa^{-1}$	— — —	— — —	— — —
	Resistance to solvents p = poor, f = fair, g = good	Methanol (f), organic solvents (p), alcohols (g)			
	Surface tension	mNm^{-1}	—	—	—
	Water absorption (equilibrium 24 hours) ⅛-inch (3.175 mm) thickness (ASTM D570)	%	—	—	—
THERMAL	**Coefficient of linear thermal expansion** (ASTM D696)	$°C^{-1}$ ($\times 10^{-5}$)	—	—	—
	Glass transition temperature (T_g)	°C	—	—	−60 to −72
	Low-temperature performance / Brittleness temperature (ASTM D746)	°C	—	—	—
	Melting point (T_m) (ASTM D1238)	°C	—	—	58–63
	Softening point (Vicat) (ASTM D1525)	°C	—	—	—
OPTICAL	**Clarity/transparency**	—	—	—	—
	Transmittance (ASTM D1003)	%	—	—	—
	Refractive index (23°C) (ASTM D542)	—	—	—	—

Data sources:
1947: *Modern Plastics Encyclopedia*, vol. 3, *Charts* (New York: Plastics Catalogue Corporation).
1967: Shell Chemicals, *Properties of Plastics*, rev. metric ed. (London: Shell International Chemical Company and Shell Chemicals UK).
2016: George Wypych, *Handbook of Polymers*, 2nd ed. (Toronto: ChemTec).

13.29 | Polyhydroxybutyrate (PHB)

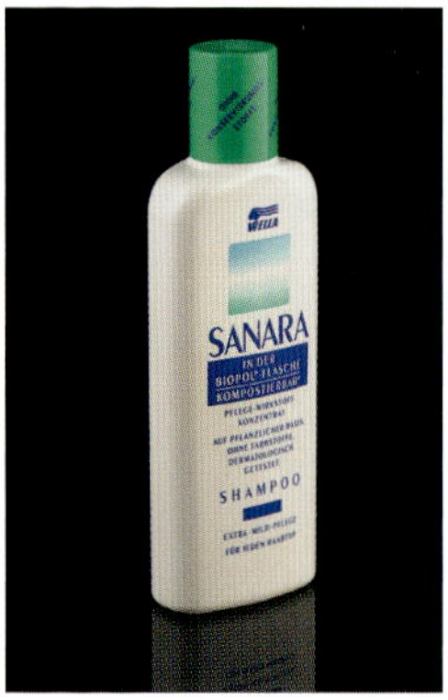

CHEMICAL STRUCTURE

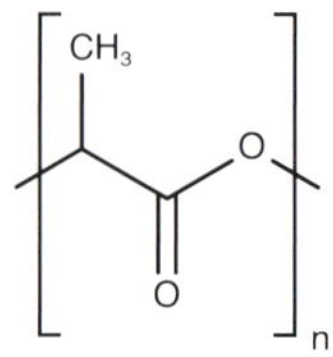

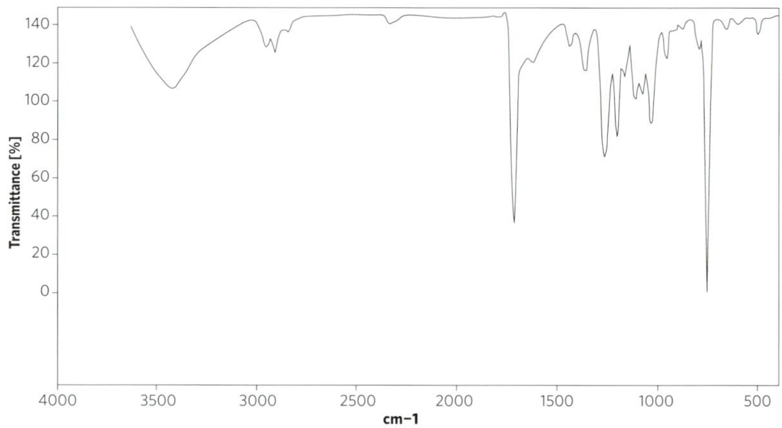

FTIR SPECTRUM

Date of first manufacture
1926 (discovered); in 1992, renewed interest arose through the need for biodegradable plastics.

Alternate/trade name examples
Biopol.

Polymerization
Polyhydroxybutyrate belongs to the linear polyester family of hydroxy alkanoates, all of which are produced during fermentation of sugars.

Processing
Injection molding, injection blow molding, film casting.

Selected applications
Biodegradable bottles and packaging film. Used for medical applications such as controlled-release drug packaging.

Feel/look/sound/smell
White or yellow.

Main degradation agents
Light, UV radiation, oxygen, ozone, moisture, temperature, stress, chemicals, microbes.

Main signs of degradation and damage
Brittleness to total disintegration.

13.29 | Polyhydroxybutyrate (PHB)

	PROPERTY	UNITS (SI)	DATA 1947	DATA 1967	DATA 2016
MECHANICAL	**Compressive strength** (ASTM D695)	MPa	—	—	—
	Elongation (ASTM D638)	%	—	—	5-58
	Flexural strength (ASTM D790)	MPa	—	—	—
	Hardness (ASTM D785)	Rockwell	—	—	—
	Impact strength (Izod notched) (ASTM D256)	J/cm	—	—	—
	Tensile modulus (rigidity) (ASTM D638)	GPa	—	—	3.5
	Tensile strength (ASTM D638)	MPa	—	—	40-62
PHYSICAL	**Density** (ASTM D792)	g/cm^3	—	—	1.17-1.25
	Hildebrand solubility parameter (δ)	$(MPa)^{1/2}$	—	—	19.2
	Permeability: H_2O vapor O_2 CO_2 (ASTM D1434)	$cm^3 \cdot cm \cdot cm^{-2} \cdot s^{-1} \cdot Pa^{-1}$	— — —	— — —	— — —
	Resistance to solvents p = poor, f = fair, g = good	Alcohols (g), esters (p), ketones (g)			
	Surface tension	mNm^{-1}	—	—	—
	Water absorption (equilibrium 24 hours) ⅛-inch (3.175 mm) thickness (ASTM D570)	%	—	—	—
THERMAL	**Coefficient of linear thermal expansion** (ASTM D696)	$°C^{-1}$ ($\times 10^{-5}$)	—	—	—
	Glass transition temperature (T_g)	°C	—	—	−4–+2.4
	Low-temperature performance / Brittleness temperature (ASTM D746)	°C	—	—	—
	Melting point (T_m) (ASTM D1238)	°C	—	—	166-185
	Softening point (Vicat) (ASTM D1525)	°C	—	—	53-96
OPTICAL	**Clarity/transparency**	—	—	—	—
	Transmittance (ASTM D1003)	%	—	—	—
	Refractive index (23°C) (ASTM D542)	—	—	—	—

Data sources:
1947: *Modern Plastics Encyclopedia*, vol. 3, *Charts* (New York: Plastics Catalogue Corporation).
1967: Shell Chemicals, *Properties of Plastics*, rev. metric ed. (London: Shell International Chemical Company and Shell Chemicals UK).
2016: George Wypych, *Handbook of Polymers*, 2nd ed. (Toronto: ChemTec).

13.30 | Polylactic acid (PLA)

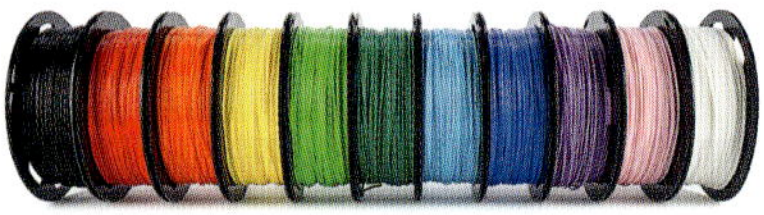

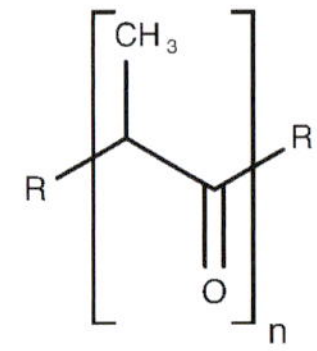

CHEMICAL STRUCTURE

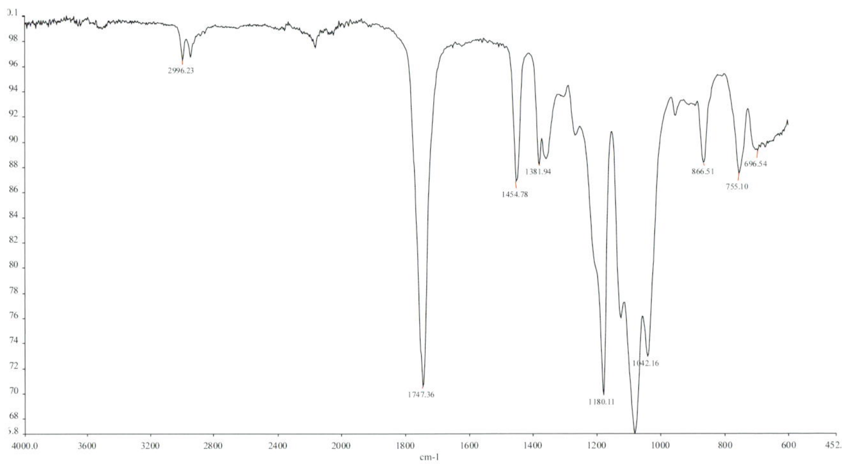

FTIR SPECTRUM

Date of first manufacture
1920 (USA); in 1989, new interest arose due to the need for biodegradable plastics.

Alternate/trade name examples
Lactron, Ecogehr.

Polymerization
Lactic acid from renewable sources such as sugar cane or corn is heated, and with the addition of enzymes the lactic acid polymerizes into polylactic acid.

Processing
Extrusion, injection molding, foaming, melt-spinning fibers, filaments for fused deposition modeling.

Selected applications
Bottles, clips, cups, disposable tableware, fibers, film, garbage bags, sheet, packaging, shopping bags. PLA is compostable.

Feel/look/sound/smell
Rigid to flexible.

Main degradation agents
Light, UV radiation, oxygen, ozone, moisture, temperature, stress, chemicals, microbes.

Main signs of degradation and damage
Total disintegration.

13.30 | Polylactic acid (PLA)

	PROPERTY	UNITS (SI)	DATA 1947	DATA 1967	DATA 2016
MECHANICAL	**Compressive strength** (ASTM D695)	MPa	—	—	—
	Elongation (ASTM D638)	%	—	—	4-6
	Flexural strength (ASTM D790)	MPa	—	—	83
	Hardness (ASTM D785)	Rockwell	—	—	—
	Impact strength (Izod notched) (ASTM D256)	J/cm	—	—	13-24.6
	Tensile modulus (rigidity) (ASTM D638)	GPa	—	—	3.7-4.1
	Tensile strength (ASTM D638)	MPa	—	—	52-72
PHYSICAL	**Density** (ASTM D792)	g/cm^3	—	—	1.21-1.29
	Hildebrand solubility parameter (δ)	$(MPa)^{1/2}$	—	—	19.0-21.0
	Permeability: H_2O vapor O_2 CO_2 (ASTM D1434)	$cm^3 \cdot cm^{-3} \cdot cm \cdot Hg^{-1}$ $\times 10^4$	— — —	— — —	110 2.2-4.9 —
	Resistance to solvents p = poor, f = fair, g = good	Alcohols (p), aliphatic hydrocarbons (p), aromatic hydrocarbons (p), esters (p), ketones (p)			
	Surface tension	mNm^{-1}	—	—	—
	Water absorption (equilibrium 24 hours) ⅛-inch (3.175 mm) thickness (ASTM D570)	%	—	—	0.5
THERMAL	**Coefficient of linear thermal expansion** (ASTM D696)	$°C^{-1}$ ($\times 10^{-5}$)	—	—	—
	Glass transition temperature (T_g)	°C	—	—	55-75
	Low-temperature performance / Brittleness temperature (ASTM D746)	°C	—	—	—
	Melting point (T_m) (ASTM D1238)	°C	—	—	164-178
	Softening point (Vicat) (ASTM D1525)	°C	—	—	55-60
OPTICAL	**Clarity/transparency**	—	—	—	—
	Transmittance (ASTM D1003)	%	—	—	2.2
	Refractive index (23°C) (ASTM D542)	—	—	—	1.35-1.45

Data sources:
1947: *Modern Plastics Encyclopedia*, vol. 3, *Charts* (New York: Plastics Catalogue Corporation).
1967: Shell Chemicals, *Properties of Plastics*, rev. metric ed. (London: Shell International Chemical Company and Shell Chemicals UK).
2016: George Wypych, *Handbook of Polymers*, 2nd ed. (Toronto: ChemTec).

13.31 | Thermoplastic starch (TPS)

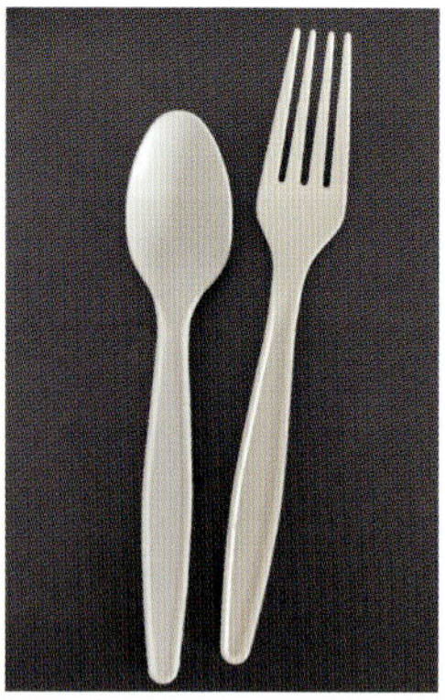

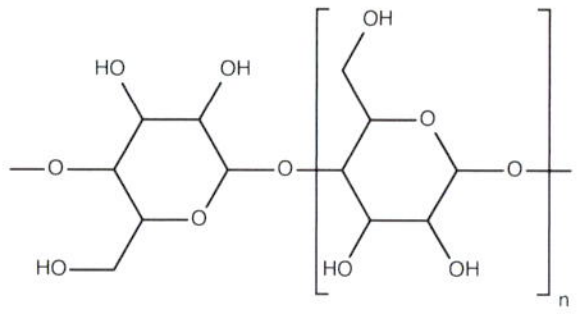

CHEMICAL STRUCTURE

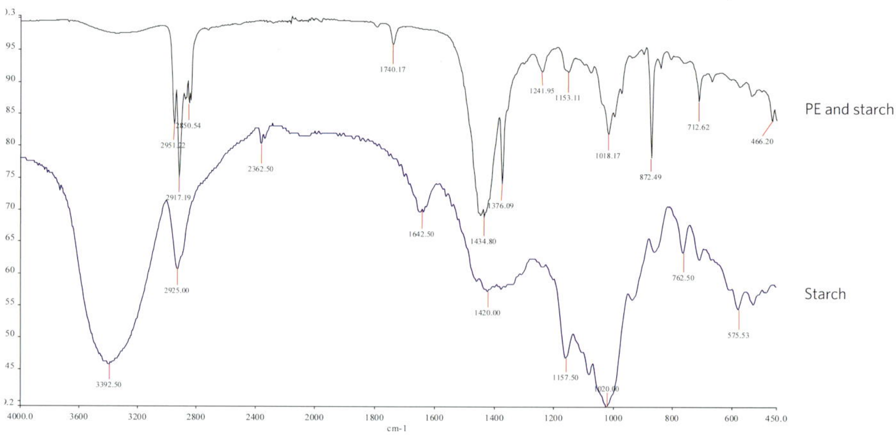

FTIR SPECTRUM

Date of first manufacture
1993.

Alternate/trade name examples
BIOPAR.

Polymerization
Thermoplastic starch is made by heating starch. The resulting polymerization of amylose and amylopectin gives the thermoplastic starch its properties. At polymerization, a biodegradable plasticizer such as glycerol is added.

Processing
Extrusion, injection molding, foam processing. Thermoplastic starch is typically blended with other biodegradable polymers rather than used solely.

Selected applications
Disposable tableware and cutlery, packing "peanuts." Blended with polyethylene, TPS is used to produce "green" plastic grocery bags.

Feel/look/sound/smell
Soft, waxy feel.

Main degradation agents
Light, UV radiation, oxygen, ozone, moisture, temperature, stress, chemicals, microbes.

Main signs of degradation and damage
Brittleness to total disintegration.

13.31 | Thermoplastic starch (TPS)

	PROPERTY	UNITS (SI)	DATA 1947	DATA 1967	DATA 2016
MECHANICAL	**Compressive strength** (ASTM D695)	MPa	—	—	—
	Elongation (ASTM D638)	%	—	—	10
	Flexural strength (ASTM D790)	MPa	—	—	—
	Hardness (ASTM D785)	Rockwell	—	—	—
	Impact strength (Izod notched) (ASTM D256)	J/cm	—	—	—
	Tensile modulus (rigidity) (ASTM D638)	GPa	—	—	1.50
	Tensile strength (ASTM D638)	MPa	—	—	30.0
PHYSICAL	**Density** (ASTM D792)	g/cm^3	—	—	1.40
	Hildebrand solubility parameter (δ)	$(MPa)^{1/2}$	—	—	—
	Permeability: H_2O vapor O_2 CO_2 (ASTM D1434)	$cm^3 \cdot cm \cdot cm^{-2} \cdot s^{-1} \cdot Pa^{-1}$	— — —	— — —	— — —
	Resistance to solvents p = poor, f = fair, g = good	Not commonly mentioned			
	Surface tension	mNm^{-1}	—	—	—
	Water absorption (equilibrium 24 hours) ⅛-inch (3.175 mm) thickness (ASTM D570)	%	—	—	—
THERMAL	**Coefficient of linear thermal expansion** (ASTM D696)	$°C^{-1}$ ($\times 10^{-5}$)	—	—	—
	Glass transition temperature (T_g)	°C	—	—	—
	Low-temperature performance / Brittleness temperature (ASTM D746)	°C	—	—	81–100
	Melting point (T_m) (ASTM D1238)	°C	—	—	—
	Softening point (Vicat) (ASTM D1525)	°C	—	—	166–180
OPTICAL	**Clarity/transparency**	—	—	—	—
	Transmittance (ASTM D1003)	%	—	—	—
	Refractive index (23°C) (ASTM D542)	—	—	—	—

Data sources:
1947: *Modern Plastics Encyclopedia*, vol. 3, *Charts* (New York: Plastics Catalogue Corporation).
1967: Shell Chemicals, *Properties of Plastics*, rev. metric ed. (London: Shell International Chemical Company and Shell Chemicals UK).
2016: George Wypych, *Handbook of Polymers*, 2nd ed. (Toronto: ChemTec).

14.1 | Casein-formaldehyde (CS)

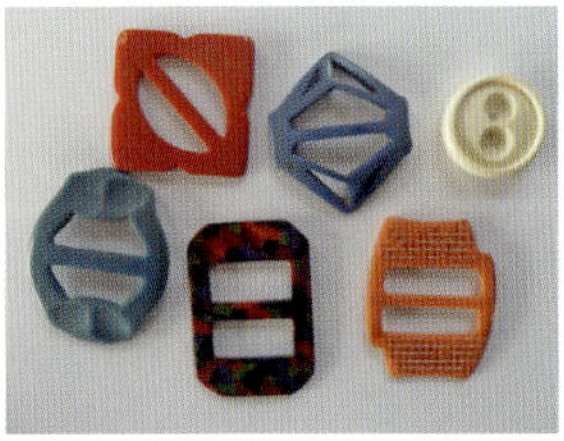

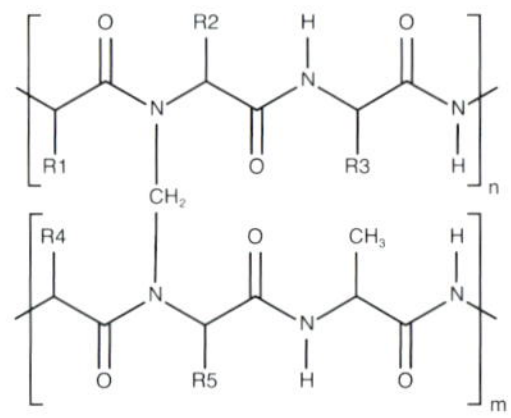

CHEMICAL STRUCTURE

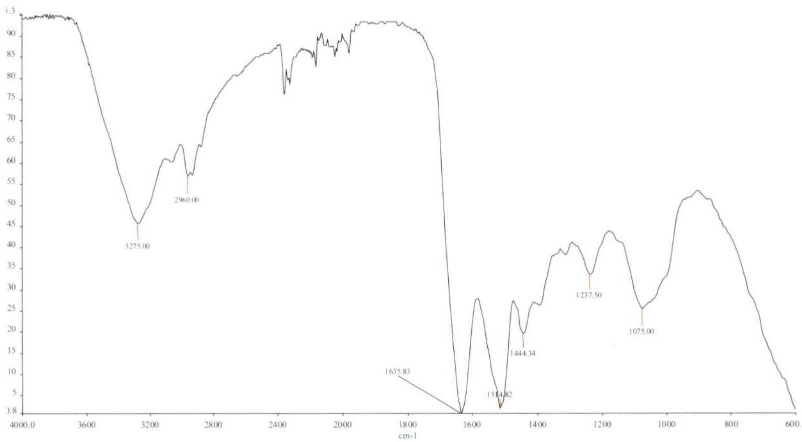

FTIR SPECTRUM

Date of first manufacture
1897 (Germany), 1904 (England).

Alternate/trade name examples
Galalith, Lactoid, Erinoid, Galorn, Ameroid.

Polymerization
Casein-formaldehyde is made from milk protein (casein) by condensation polymerization of the acid groups of the diamino acids with the amino groups of other amino acids under elimination of water. The formaldehyde has the effect of hardening the casein molecules, building up three-dimensional networks.

Processing
Extrusion into pipes, sheet, and rods. Can be drilled, cut, embossed, and dyed.

Selected applications
Adding pigments to casein formaldehyde made it suitable for imitation tortoiseshell and ivory objects like buckles, buttons, combs, knife handles, knitting needles, toiletry sets.

Feel/look/sound/smell
Rigid. Resembles ivory or horn; translucent or opaque.

Main degradation agents
Moisture (changes in relative humidity), light, UV radiation, oxygen, ozone, temperature, stress, chemicals.

Main signs of degradation and damage
Warping, shrinkage, cracking, crazing, discoloration.

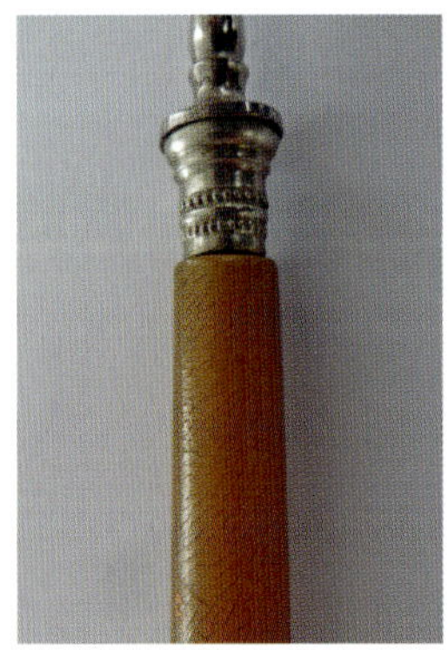

Fork handle with crazing.

14.1 | Casein-formaldehyde (CS)

	PROPERTY	UNITS (SI)	DATA 1947	DATA 1967	DATA 2016
MECHANICAL	**Compressive strength** (ASTM D695)	MPa	186-365	—	—
	Elongation (ASTM D638)	%	2.5	—	—
	Flexural strength (ASTM D790)	MPa	68.9-124.1	—	—
	Hardness (ASTM D785)	Rockwell M	M70-M100	—	—
	Impact strength (Izod notched) (ASTM D256)	J/cm	0.53	—	—
	Tensile modulus (rigidity) (ASTM D638)	GPa	3.51-3.93	—	—
	Tensile strength (ASTM D638)	MPa	68.9	—	—
PHYSICAL	**Density** (ASTM D792)	g/cm^3	1.35	—	—
	Hildebrand solubility parameter (δ)	$(MPa)^{1/2}$	—	—	—
	Permeability: H_2O vapor O_2 CO_2 (ASTM D1434)	$cm^3 \cdot cm \cdot cm^{-2} \cdot s^{-1} \cdot Pa^{-1}$	— — —	— — —	— — —
	Resistance to solvents p = poor, f = fair, g = good	Resistant to solvents			
	Surface tension	mNm^{-1}	—	—	—
	Water absorption (equilibrium 24 hours) ⅛-inch (3.175 mm) thickness (ASTM D570)	%	7-14	—	—
THERMAL	**Coefficient of linear thermal expansion** (ASTM D696)	$°C^{-1}$ ($\times 10^{-5}$)	4.1-6.8	—	—
	Glass transition temperature (T_g)	°C	—	—	—
	Low-temperature performance / Brittleness temperature (ASTM D746)	°C	—	—	—
	Melting point (T_m) (ASTM D1238)	°C	—	—	—
	Softening point (Vicat) (ASTM D1525)	°C	148.8	—	—
OPTICAL	**Clarity/transparency**	—	Translucent or opaque	—	—
	Transmittance (ASTM D1003)	%	—	—	—
	Refractive index (23°C) (ASTM D542)	—	—	—	—

Data sources:
1947: *Modern Plastics Encyclopedia*, vol. 3, *Charts* (New York: Plastics Catalogue Corporation).
1967: Shell Chemicals, *Properties of Plastics*, rev. metric ed. (London: Shell International Chemical Company and Shell Chemicals UK).
2016: George Wypych, *Handbook of Polymers*, 2nd ed. (Toronto: ChemTec).

14.2 | Epoxy (EP)

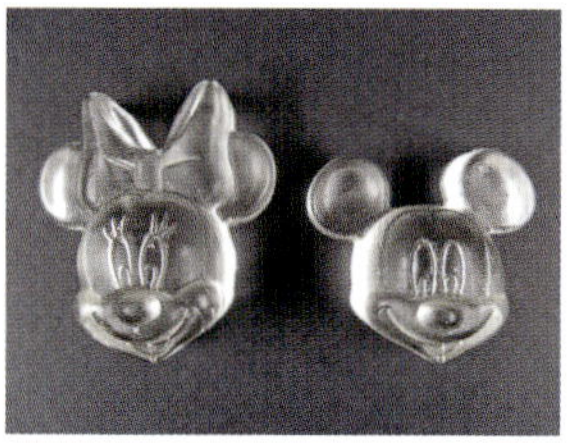

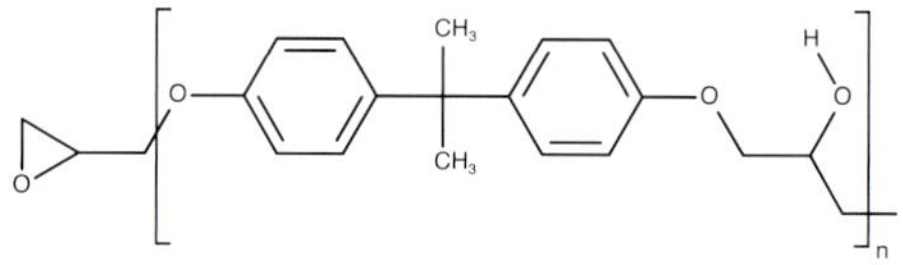

CHEMICAL STRUCTURE

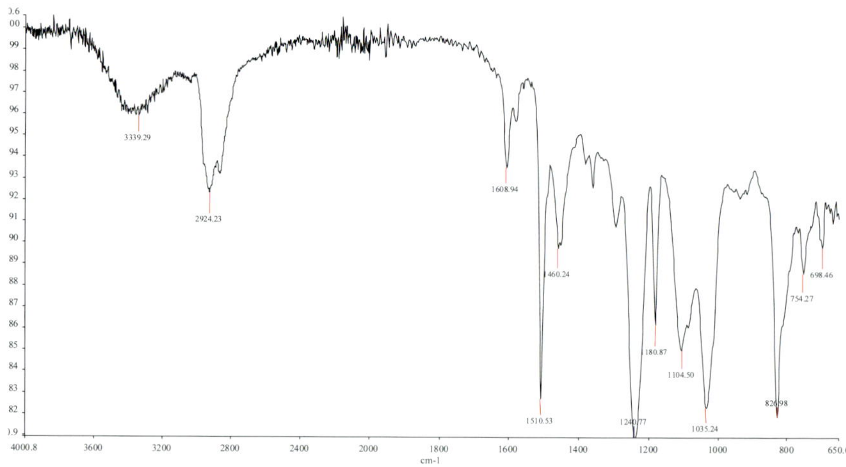

FTIR SPECTRUM

Date of first manufacture
1934 (Germany), 1943 (Switzerland), 1946 (USA).

Alternate/trade name examples
Epikote, Epodite, Epon, Lekutherm.

Polymerization
Epoxy is synthesized by reacting bisphenol A with epichlorohydrin in the presence of a catalyst. Curing is achieved with amines as hardener. Recent developments include epoxies modified to contain thermoplastic processing properties.

Processing
Casting, pultrusion, spraying, coatings, lamination.

Selected applications
Adhesives for protecting floors, boats, and works of art; industrial paint; stove enamels.

Feel/look/sound/smell
Transparent, translucent to opaque.

Main degradation agents
Light, UV radiation, oxygen, ozone, moisture, temperature, stress, chemicals.

Main signs of degradation and damage
Yellowing.

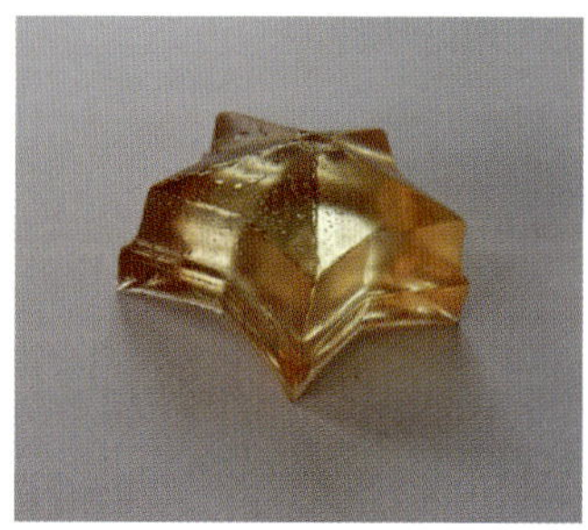

Yellowed decorative star.

14.2 | Epoxy (EP)

	PROPERTY	UNITS (SI)	DATA 1947	DATA 1967	DATA 2016
MECHANICAL	**Compressive strength** (ASTM D695)	MPa	—	104-203	116-404
	Elongation (ASTM D638)	%	—	5-10	1.3-705 thermoplastic 3-5 adhesive
	Flexural strength (ASTM D790)	MPa	—	69-138	74-325
	Hardness (ASTM D785)	Shore D	—	—	D62-D95
	Impact strength (Izod notched) (ASTM D256)	J/cm	—	0.16-0.49	0.25-12.46
	Tensile modulus (rigidity) (ASTM D638)	GPa	—	1.38-4.14	3.6-4.3
	Tensile strength (ASTM D638)	MPa	—	—	12-24
PHYSICAL	**Density** (ASTM D792)	g/cm^3	—	—	—
	Hildebrand solubility parameter (δ)	$(MPa)^{1/2}$	—	1.0-3.2	1.15-1.3
	Permeability: H_2O vapor O_2 CO_2 (ASTM D1434)	$cm^3 \cdot cm \cdot cm^{-2} \cdot s^{-1} \cdot Pa^{-1}$	— — —	— — —	— — —
	Resistance to solvents p = poor, f = fair, g = good	Ketones (p), hydrocarbons (g), alcohols (g), aromatics (p)			
	Surface tension	mNm^{-1}	—	—	—
	Water absorption (equilibrium 24 hours) ⅛-inch (3.175 mm) thickness (ASTM D570)	%	—	—	39.1-51.2
THERMAL	**Coefficient of linear thermal expansion** (ASTM D696)	$°C^{-1}$ ($\times 10^{-5}$)	—	—	—
	Glass transition temperature (T_g)	°C	—	5-9	1.7-6.7
	Low-temperature performance / Brittleness temperature (ASTM D746)	°C	—	—	37-127 thermoplastic 103-193 adhesive
	Melting point (T_m) (ASTM D1238)	°C	—	Good	—
	Softening point (Vicat) (ASTM D1525)	°C	—	—	90-245
OPTICAL	**Clarity/transparency**	—	—	Transparent to opaque	—
	Transmittance (ASTM D1003)	%	—	—	—
	Refractive index (23°C) (ASTM D542)	—	—	1.6	1.51-1.58

Data sources:
1947: *Modern Plastics Encyclopedia*, vol. 3, *Charts* (New York: Plastics Catalogue Corporation).
1967: Shell Chemicals, *Properties of Plastics*, rev. metric ed. (London: Shell International Chemical Company and Shell Chemicals UK).
2016: George Wypych, *Handbook of Polymers*, 2nd ed. (Toronto: ChemTec).

14.3 | Melamine-formaldehyde (MF)

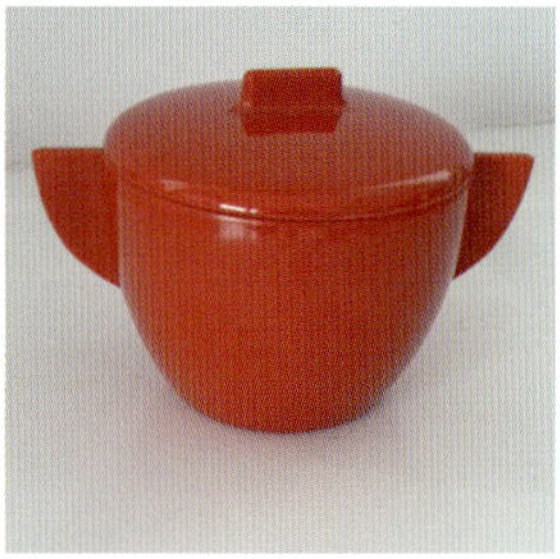

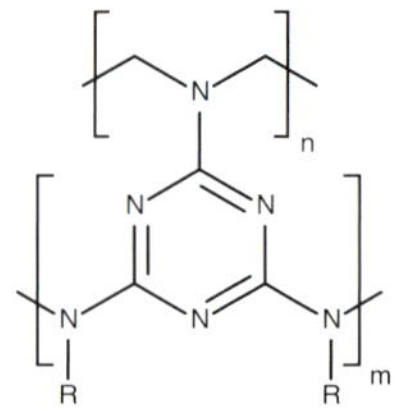

CHEMICAL STRUCTURE

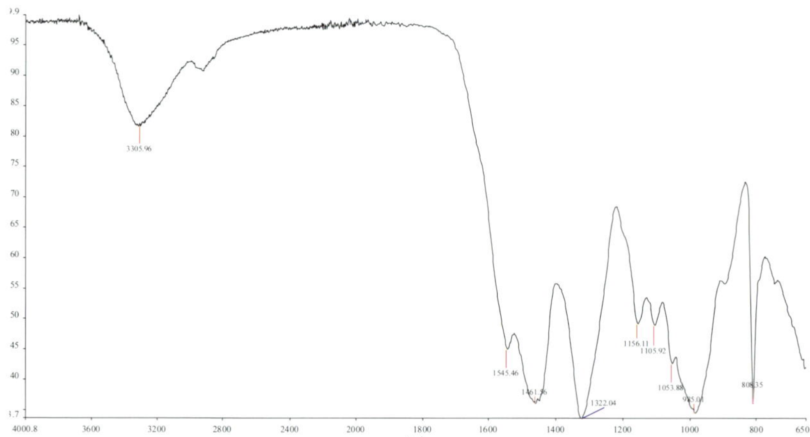

FTIR SPECTRUM

Date of first manufacture
1933 (Germany), 1935 (USA).

Alternate/trade name examples
Melamine, Formica, Argosy, Gaydon; Melaware; Melmex, Melmac, Plaskon, Resimene, Trespa.

Polymerization
Melamine-formaldehyde is formed by condensation polymerization of formaldehyde and melamine.

Processing
Compression molding, laminating.

Selected applications
Melamine-formaldehyde is more resistant to shock, scratching, and chemicals than phenol-formaldehyde and urea-formaldehyde, so it became more popular than those materials. Melamine-formaldehyde is used for camping equipment, dinnerware, insulation, kitchen utensils, and paper laminates. The laminate form, commonly known as Formica, is used on cabinet fronts and countertops, over chipboard, and on furniture.

Feel/look/sound/smell
Rigid, hard. Bright, high-gloss colors.

Main degradation agents
Light, UV radiation, oxygen, ozone, moisture, temperature, stress, chemicals.

Main signs of degradation and damages
Slow rate of deterioration, damage not commonly detected.

14.3 | Melamine-formaldehyde (MF)

	PROPERTY	UNITS (SI)	DATA 1947 (A-CELLULOSE FILLED)	DATA 1967 (A-CELLULOSE FILLED)	DATA 2016
MECHANICAL	**Compressive strength** (ASTM D695)	MPa	186.2-296.5	172-297	—
	Elongation (ASTM D638)	%	—	0.6-0.9	—
	Flexural strength (ASTM D790)	MPa	62.1-110.3	69-110.5	80
	Hardness (ASTM D785)	Rockwell	R111-R122	R110-R125	—
	Impact strength (Izod notched) (ASTM D256)	J/cm	0.13-0.19	0.13-0.19	—
	Tensile modulus (rigidity) (ASTM D638)	GPa	6.9	8.97	—
	Tensile strength (ASTM D638)	MPa	48.3-89.6	48.0-89.7	45-52
PHYSICAL	**Density** (ASTM D792)	g/cm^3	1.45-1.55	1.47-1.52	1-1.14
	Hildebrand solubility parameter (δ)	$(MPa)^{1/2}$	—	—	—
	Permeability: H_2O vapor O_2 CO_2 (ASTM D1434)	$cm^3 \cdot cm \cdot cm^{-2} \cdot s^{-1} \cdot Pa^{-1}$	— — —	— — —	— — —
	Resistance to solvents p = poor, f = fair, g = good	Resistant to solvents			
	Surface tension	mNm^{-1}	—	—	—
	Water absorption (equilibrium 24 hours) ⅛-inch (3.175 mm) thickness (ASTM D570)	%	0.1-0.6	0.1-0.6	—
THERMAL	**Coefficient of linear thermal expansion** (ASTM D696)	$°C^{-1}$ ($\times 10^{-5}$)	2.5-4.5	4.0	—
	Glass transition temperature (T_g)	°C	—	—	—
	Low-temperature performance / Brittleness temperature (ASTM D746)	°C	—	Good	—
	Melting point (T_m) (ASTM D1238)	°C	—	—	90-101
	Softening point (Vicat) (ASTM D1525)	°C	—	—	—
OPTICAL	**Clarity/transparency**	—	—	—	—
	Transmittance (ASTM D1003)	%	—	—	—
	Refractive index (23°C) (ASTM D542)	—	—	—	—

Data sources:
1947: *Modern Plastics Encyclopedia*, vol. 3, *Charts* (New York: Plastics Catalogue Corporation).
1967: Shell Chemicals, *Properties of Plastics*, rev. metric ed. (London: Shell International Chemical Company and Shell Chemicals UK).
2016: George Wypych, *Handbook of Polymers*, 2nd ed. (Toronto: ChemTec).

14.4 | Phenol-formaldehyde / Cast phenolic (PF)

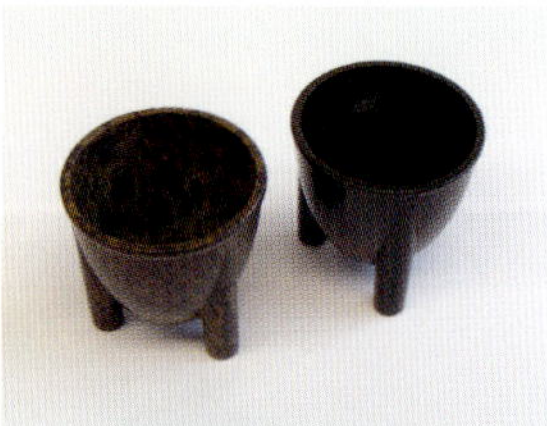

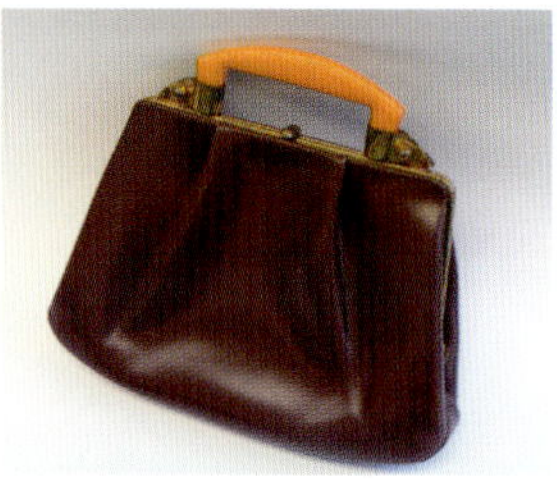

CHEMICAL STRUCTURE

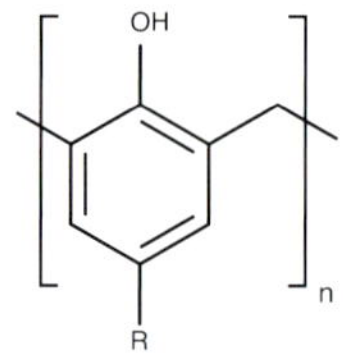

FTIR SPECTRUM

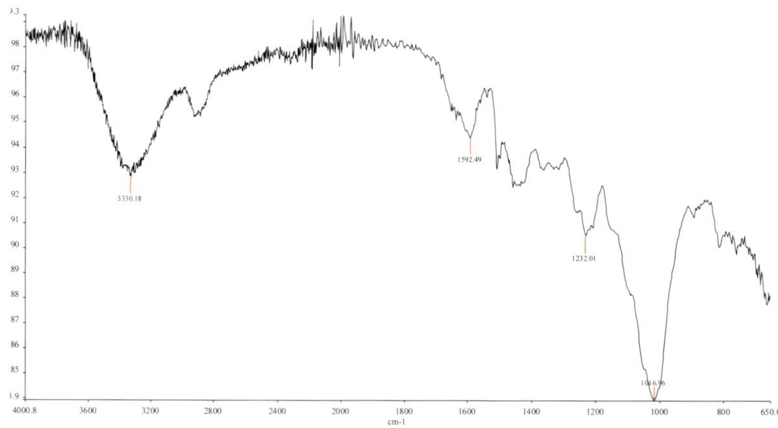

Date of first manufacture
1907 (USA); objects manufactured until the 1950s, when it was replaced by melamine-formaldehyde, which has better properties (durability). Currently used only in laminated composites for industrial applications.

Alternate/trade name examples
Bakelite, Trolithan, Catalin, Durez, Resinox, Textolite.

Polymerization
Phenol-formaldehyde is polymerized from phenol with formaldehyde. Acid catalysis with an excess of formaldehyde during polymerization led to cast phenolic.

Processing
Compression molding, casting, laminating.

Selected applications
Ashtrays; door handles; electrical equipment cases for hairdryers, plugs, radios, razors, switches, and telephones; records. Dyes were added to cast phenolics to color products such as cutlery handles, jewelry, and napkin rings.

Feel/look/sound/smell
Hard, rigid. Color is typically dark brown, red, or green; often appears mottled. Cast phenolic without colorants can be transparent to translucent and comes in various colors. When the object is rubbed, a carbolic acid (phenolic, "hospital") smell can be noticed.

Main degradation agents
Chemicals, light, moisture, oxygen, ozone, stress, temperature, UV radiation.

Main signs of degradation and damage
Discoloration, dulling, breaks.

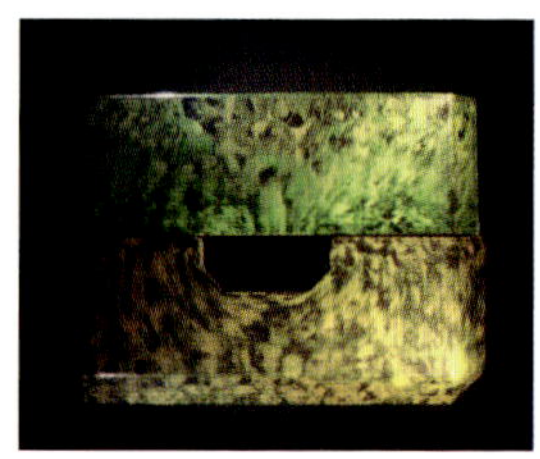

Discolored soap box.

14.4 | Phenol-formaldehyde / Cast phenolic (PF)

	PROPERTY	UNITS (SI)	DATA 1947	DATA 1967	DATA 2016
MECHANICAL	**Compressive strength** (ASTM D695)	MPa	69-207	69-203	69-200
	Elongation (ASTM D638)	%	1.0-1.5	1.0-1.5	1.5-2.2
	Flexural strength (ASTM D790)	MPa	82.7-117.2	83.9-103.6	75-120
	Hardness (ASTM D785)	Rockwell M	M120-M130	M124-M128	M93-M128
	Impact strength (Izod notched) (ASTM D256)	J/cm	0.11-0.19	0.10-0.19	8.7
	Tensile modulus (rigidity) (ASTM D638)	GPa	4.8-6.9	5.18-6.90	7.58
	Tensile strength (ASTM D638)	MPa	48.3-68.9	48.0-55.3	34.4-62
PHYSICAL	**Density** (ASTM D792)	g/cm^3	1.25-1.3	1.25-1.3	1.24-1.32
	Hildebrand solubility parameter (δ)	$(MPa)^{1/2}$	—	—	—
	Permeability: H_2O vapor O_2 CO_2 (ASTM D1434)	$cm^3 \cdot cm \cdot cm^{-2} \cdot s^{-1} \cdot Pa^{-1}$	— — —	— — —	— — —
	Resistance to solvents p = poor, f = fair, g = good	Resistant to solvents			
	Surface tension	mNm^{-1}	—	—	—
	Water absorption (equilibrium 24 hours) ⅛-inch (3.175 mm) thickness (ASTM D570)	%	0.1-0.2	0.1-0.2	0.1-0.4
THERMAL	**Coefficient of linear thermal expansion** (ASTM D696)	$°C^{-1}$ ($\times 10^{-5}$)	2.5-6	2.5-6	1.7-2.2
	Glass transition temperature (T_g)	°C	—	—	200
	Low-temperature performance / Brittleness temperature (ASTM D746)	°C	—	—	—
	Melting point (T_m) (ASTM D1238)	°C	—	—	90-107
	Softening point (Vicat) (ASTM D1525)	°C	—	—	—
OPTICAL	**Clarity/transparency**	—	Transparent to translucent	—	—
	Transmittance (ASTM D1003)	%	—	—	—
	Refractive index (23°C) (ASTM D542)	—	1.5-1.7	1.5-1.7	1.7

Data sources:
1947: *Modern Plastics Encyclopedia*, vol. 3, *Charts* (New York: Plastics Catalogue Corporation).
1967: Shell Chemicals, *Properties of Plastics*, rev. metric ed. (London: Shell International Chemical Company and Shell Chemicals UK).
2016: George Wypych, *Handbook of Polymers*, 2nd ed. (Toronto: ChemTec).

14.5 | Unsaturated polyester (UP)
Glass fiber-reinforced unsaturated polyester (GRP)

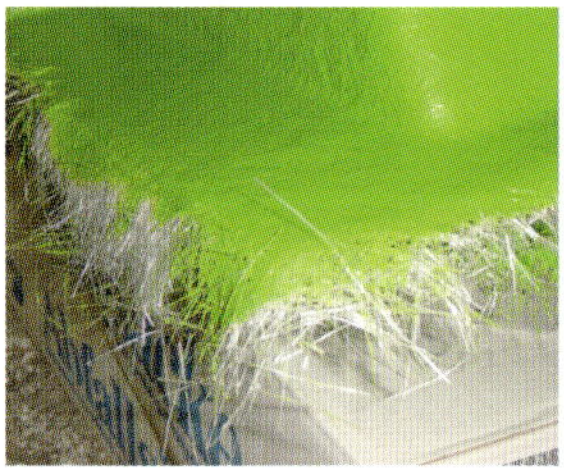

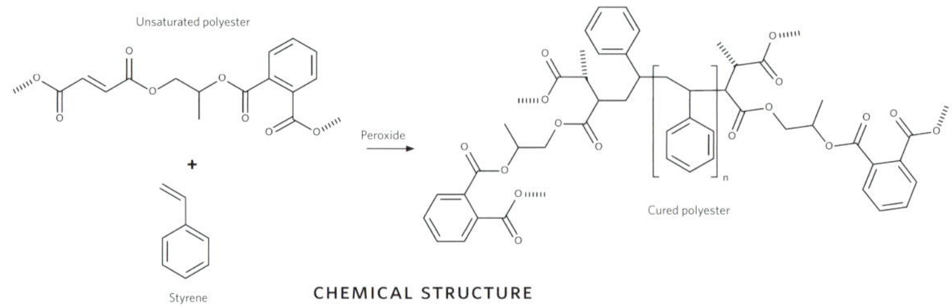

CHEMICAL STRUCTURE

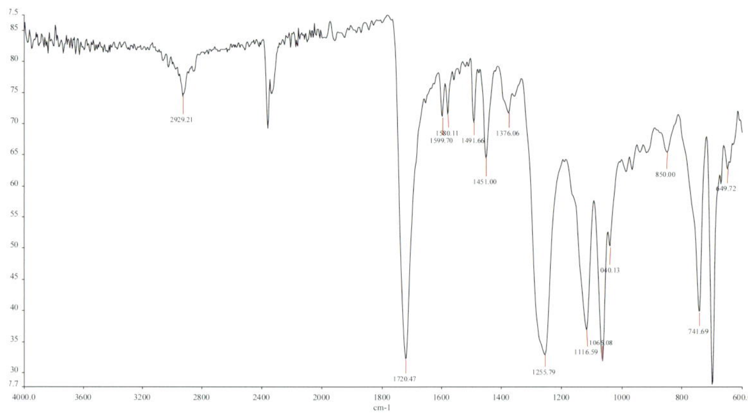

FTIR SPECTRUM

Date of first manufacture
1933 (USA), 1941 (fiber-reinforced, USA).

Alternate/trade name examples
GRP (glass fiber-reinforced unsaturated polyester), fiberglass (FRP).

Polymerization
Unsaturated polyesters are synthesized using condensation polymerization from glycol (neopentyl glycol) and dicarboxylic acids (phthalic acids, isophthalic acid, maleic anhydride) using peroxide as a curing agent.

Processing
Molding, compression molding, casting, pultrusion, sheet molding, reinforcing by hand layup and spray layup in an open mold. Materials for reinforced unsaturated polyesters (GRP) can be glass (chopped fibers, rovings, mats, balls) and aramid or carbon fibers.

Selected applications
Airplanes, bathrooms, beetles in entomology collections, boats, cars, chairs, embedded biological specimens, gel coats, laminates, mobile homes, paperweights, pipe, sculptures, shower stalls, tanks and other very large containers, wind turbine blades.

Feel/look/sound/smell
Rigid, hard. Transparent when no colorants or fillers are added. Freshly made products have smell of styrene.

Main degradation agents
Light, UV radiation, oxygen, ozone, moisture, temperature, stress, chemicals.

Main signs of degradation and damage
Yellowing, brittleness, chalky surfaces (mostly on glass fiber-reinforced objects), cracking, breakage.

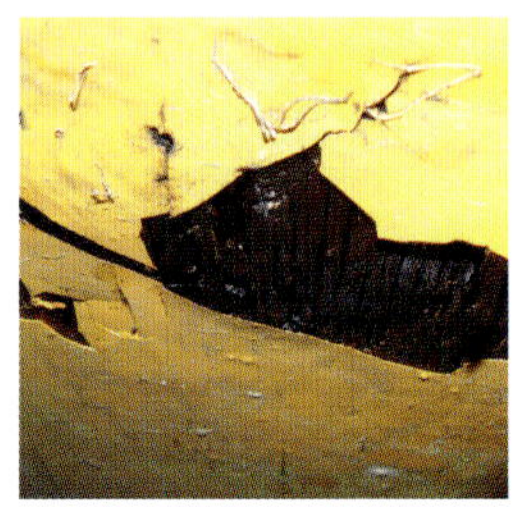

Detail of fig. 4.36b with cracking caused by water absorption.

14.5 | Unsaturated polyester (UP)
Glass fiber-reinforced unsaturated polyester (GRP)

	PROPERTY	UNITS (SI)	DATA 1947	DATA 1967	DATA 2016
MECHANICAL	**Compressive strength** (ASTM D695)	MPa	—	—	104–131
	Elongation (ASTM D638)	%	—	—	1.2–5.0
	Flexural strength (ASTM D790)	MPa	—	—	3200–3900
	Hardness (ASTM D785)	Rockwell	—	—	—
	Impact strength (Izod notched) (ASTM D256)	J/cm	—	—	—
	Tensile modulus (rigidity) (ASTM D638)	GPa	—	—	1.97
	Tensile strength (ASTM D638)	MPa	—	—	22–85
PHYSICAL	**Density** (ASTM D792)	g/cm^3	—	—	1.1–1.12
	Hildebrand solubility parameter (δ)	$(MPa)^{1/2}$	—	—	—
	Permeability: H_2O vapor O_2 CO_2 (ASTM D1434)	$cm^3 \cdot cm \cdot cm^{-2} \cdot s^{-1} \cdot Pa^{-1}$	— — —	— — —	— — —
	Resistance to solvents p = poor, f = fair, g = good	Alcohols (g), aliphatic hydrocarbons (p), aromatic hydrocarbons (p), halogenated hydrocarbons (p)			
	Surface tension	mNm^{-1}	—	—	—
	Water absorption (equilibrium 24 hours) ⅛-inch (3.175 mm) thickness (ASTM D570)	%	—	—	0.2
THERMAL	**Coefficient of linear thermal expansion** (ASTM D696)	$°C^{-1}$ ($\times 10^{-5}$)	—	—	3.1
	Glass transition temperature (Tg)	°C	—	—	−61 before cure 94–125 after cure
	Low-temperature performance / Brittleness temperature (ASTM D746)	°C	—	—	—
	Melting point (T_m) (ASTM D1238)	°C	—	—	—
	Softening point (Vicat) (ASTM D1525)	°C	—	—	—
OPTICAL	**Clarity/transparency**	—	—	—	—
	Transmittance (ASTM D1003)	%	—	—	—
	Refractive index (23°C) (ASTM D542)	—	—	—	1.54–1.57*

Data sources:
1947: *Modern Plastics Encyclopedia*, vol. 3, *Charts* (New York: Plastics Catalogue Corporation).
1967: Shell Chemicals, *Properties of Plastics*, rev. metric ed. (London: Shell International Chemical Company and Shell Chemicals UK).
2016: George Wypych, *Handbook of Polymers*, 2nd ed. (Toronto: ChemTec).
*2011: Anna Laganà and Thea B. van Oosten, "Back to Transparency, Back to Life: Research into the Restoration of Broken Transparent Unsaturated Polyester and Poly(methyl methacrylate) Works of Art," in *ICOM-CC 16th Triennial Conference Preprints, Lisbon, September 19–23, 2011*, ed. J. Bridgland (Almada, Portugal: Critério Produção Gráfica, Lda), 1–9.

14.6 | Urea-formaldehyde (UF)

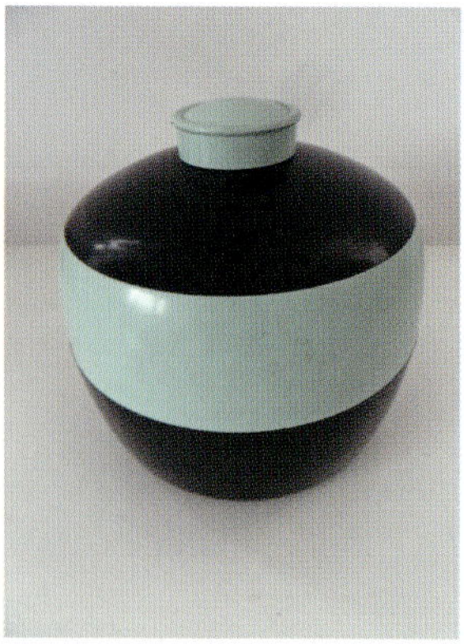

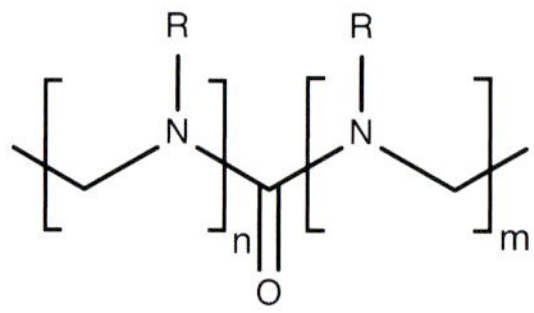

CHEMICAL STRUCTURE

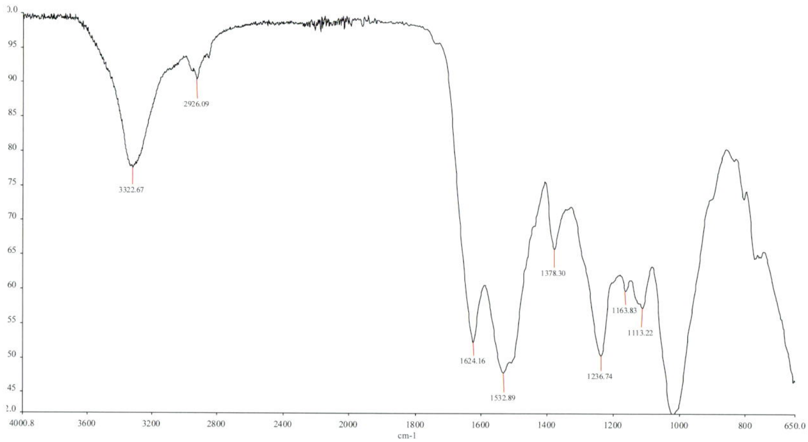

FTIR SPECTRUM

Date of first manufacture
Early 1920s (UK, Germany), 1925 (UK), 1927 (USA). Replaced by melamine-formaldehyde plastics after the 1950s.

Alternate/trade name examples
Beetle, Bandalasta, Linga Longa, Beatl, Plaskon, Scarab, Uformite.

Polymerization
Urea-formaldehyde is formed by the condensation reaction of formaldehyde and urea.

Processing
Compression molding, injection molding.

Selected applications
Cigarette boxes, electrical fittings, jewelry, lamp shades, picnic sets and tableware, radio and other electrical housings, telephone handsets.

Feel/look/sound/smell
Glossy, white, or pastel colors and speckled, mottled, or marbled effects. Faint fishy smell when rubbed.

Main degradation agents
Light, UV radiation, oxygen, ozone, moisture, temperature, stress, chemicals.

Main signs of degradation and damage
Discoloration, yellowing. Dull surface, cracking, scratches.

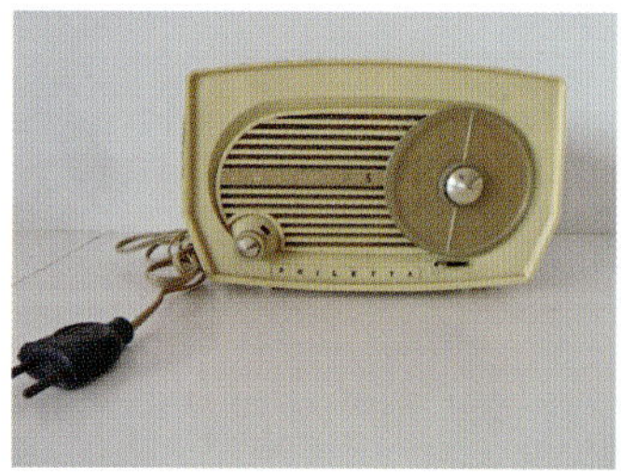

Yellowed radio housing.

14.6 | Urea-formaldehyde (UF)

	PROPERTY	UNITS (SI)	DATA 1947	DATA 1967 (A-CELLULOSE FILLED)	DATA 2016
MECHANICAL	**Compressive strength** (ASTM D695)	MPa	186.2-296.5	172-241	—
	Elongation (ASTM D638)	%	—	0.5-1.0	—
	Flexural strength (ASTM D790)	MPa	62.1-110.3	69-110.5	80-170
	Hardness (ASTM D785)	Rockwell M	—	M115-M120	—
	Impact strength (Izod notched) (ASTM D256)	J/cm	0.13-0.19	0.13-0.19	—
	Tensile modulus (rigidity) (ASTM D638)	GPa	6.9	10.4	—
	Tensile strength (ASTM D638)	MPa	48.3-89.6	41.4-89.7	55
PHYSICAL	**Density** (ASTM D792)	g/cm^3	1.45-1.55	1.47-1.52	1.2-1.31
	Hildebrand solubility parameter (δ)	$(MPa)^{1/2}$	—	—	25.74
	Permeability: H_2O vapor O_2 CO_2 (ASTM D1434)	$cm^3 \cdot cm \cdot cm^{-2} \cdot s^{-1} \cdot Pa^{-1}$	— — —	— — —	— — —
	Resistance to solvents p = poor, f = fair, g = good	Resistant to solvents			
	Surface tension	mNm^{-1}	—	—	—
	Water absorption (equilibrium 24 hours) ⅛-inch (3.175 mm) thickness (ASTM D570)	%	0.75-2.0	0.4-0.8	3
THERMAL	**Coefficient of linear thermal expansion** (ASTM D696)	$°C^{-1}$ ($\times 10^{-5}$)	2.4-4.5	2.7	—
	Glass transition temperature (T_g)	°C	—	—	—
	Low-temperature performance / Brittleness temperature (ASTM D746)	°C	—	Good	—
	Melting point (T_m) (ASTM D1238)	°C	—	—	119
	Softening point (Vicat) (ASTM D1525)	°C	—	—	—
OPTICAL	**Clarity/transparency**	—	Translucent or opaque	—	—
	Transmittance (ASTM D1003)	%	—	—	—
	Refractive index (23°C) (ASTM D542)	—	1.54-1.56	1.54-1.56	1.43

Data sources:
1947: *Modern Plastics Encyclopedia*, vol. 3, *Charts* (New York: Plastics Catalogue Corporation).
1967: Shell Chemicals, *Properties of Plastics*, rev. metric ed. (London: Shell International Chemical Company and Shell Chemicals UK).
2016: George Wypych, *Handbook of Polymers*, 2nd ed. (Toronto: ChemTec).

15.1 | Balata rubber*

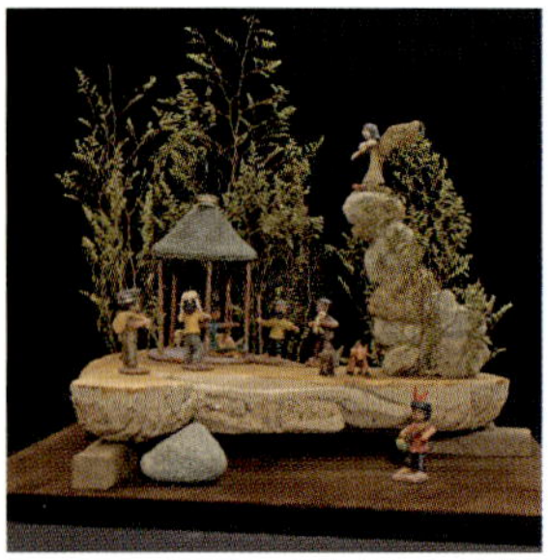

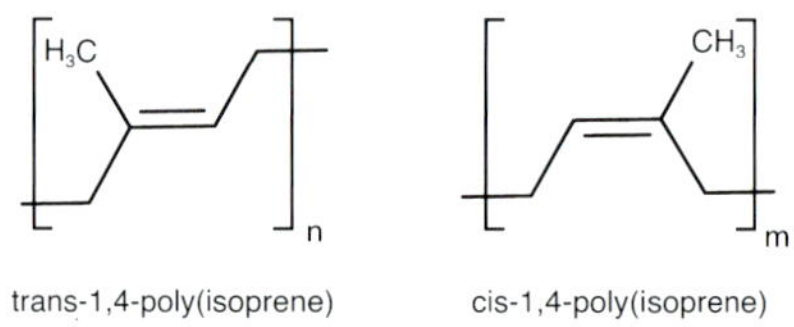

CHEMICAL STRUCTURE

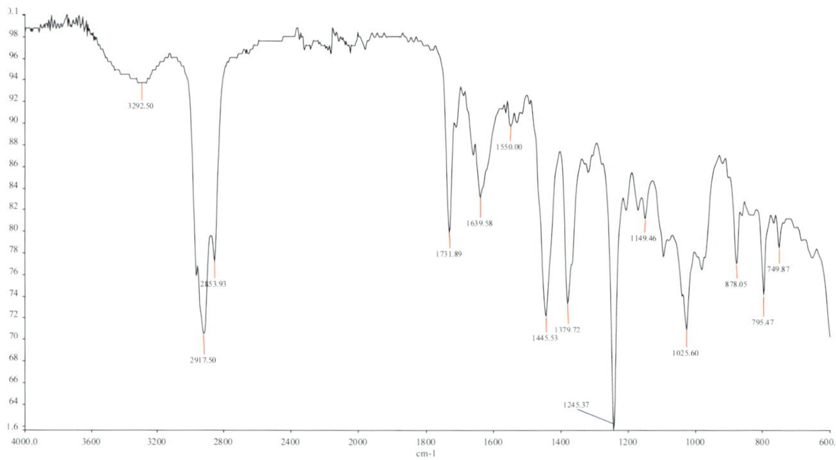

FTIR SPECTRUM

Date of first manufacture
1850s (USA, UK); long prior use by Indigenous peoples.

Alternate/trade name examples
N/A

Processing
Balata is the evaporated milky juice from *Manilkara* and other trees in the Sapotaceae family in North and South America. Trees are tapped, the sap collected, and the water evaporated. Raw balata sheets are brown and leathery and become moldable when heated to a temperature of 50°C. Processing involves mastication, blending, molding, and vulcanization.

Selected applications
Cable insulation (19th century), golf balls, impregnation of canvas and transmission belts, souvenir items.

Feel/look/sound/smell
Balata is pliable but not as elastic as natural rubber.

Main degradation agents
Oxygen, ozone, light, UV radiation, moisture, temperature, stress, chemicals.

Main signs of degradation and damage
Hardening (losing elasticity), cracking, crumbling.

15.1 | Balata rubber*

	PROPERTY	UNITS (SI)	DATA 1947	DATA 1967	DATA 2016 CIS	DATA 2016 TRANS
MECHANICAL	**Compression set** (ASTM D395)	%	—	—	—	—
	Elongation (ASTM D638)	%	—	—	100-800	250-400 (vulcanized)
	Flexural strength (ASTM D790)	MPa	—	—	—	—
	Hardness (ASTM D785)	Shore A	—	—	A30-A90 (unvulcanized) A30-A45 (vulcanized)	A90-A95 (vulcanized)
	Impact strength (Izod notched) (ASTM D256)	J/cm	—	—	—	—
	Flexural modulus (stiffness) (ASTM D790)	GPa	—	—	—	—
	Tear strength (ASTM D624)	kNm^{-1}	—	—	151-159	—
	Tensile strength (ASTM D638)	MPa	—	—	28 (vulcanized)	19.6-36.7 (vulcanized)
PHYSICAL	**Density** (ASTM D792)	g/cm^3	—	—	0.906-0.93 0.950 (vulcanized)	0.90-0.95
	Hildebrand solubility parameter (δ)	$(MPa)^{1/2}$	—	—	17-18.4	—
	Permeability: H_2O vapor O_2 CO_2 (ASTM D1434)	$cm^3 \cdot cm \cdot cm^{-2} \cdot s^{-1} \cdot Pa^{-1} \times 10^{12}$	— — —	— — —	— 1.76 —	172 — —
	Resistance to solvents p = poor, f = fair, g = good	Aliphatic hydrocarbons (p), esters (p), fats/oils (p), ketones (f), alcohols (g)				
	Surface tension	mNm^{-1}	—	—	32	31
	Water absorption (equilibrium 24 hours) ⅛-inch (3.175 mm) thickness (ASTM D570)	%	—	—	—	—
THERMAL	**Coefficient of linear thermal expansion** (ASTM D696)	$°C^{-1}$ ($\times 10^{-5}$)	—	—	0.67	—
	Glass transition temperature (T_g)	°C	—	—	−75 (natural rubber) −70 to −72 (synthetic polyisoprene)	−63 to −68
	Low-temperature performance / Brittleness temperature (ASTM D746)	°C	—	—	—	—
	Melting point (T_m) (ASTM D1238)	°C	—	—	—	—
	Softening point (Vicat) (ASTM D1525)	°C	—	—	—	—
OPTICAL	**Clarity/transparency**	—	—	—	—	—
	Transmittance (ASTM D1003)	%	—	—	—	—
	Refractive index (23°C) (ASTM D542)	—	—	—	1.5191-1.52	—

Data sources:
1947: *Modern Plastics Encyclopedia*, vol. 3, *Charts* (New York: Plastics Catalogue Corporation).
1967: Shell Chemicals, *Properties of Plastics*, rev. metric ed. (London: Shell International Chemical Company and Shell Chemicals UK).
2016: George Wypych, *Handbook of Polymers*, 2nd ed. (Toronto: ChemTec).
*Balata is a blend of cis- and trans-1,4-polyisoprene.

15.2 | Butadiene rubber (BR)

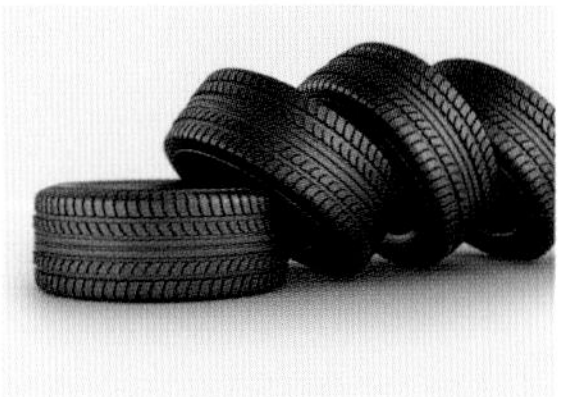

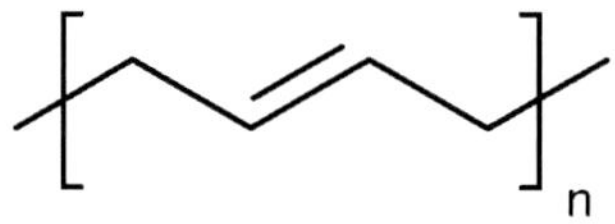

CHEMICAL STRUCTURE

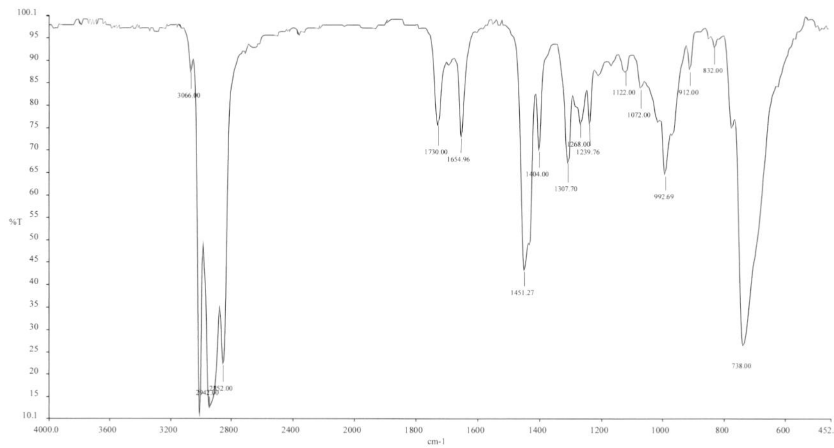

FTIR SPECTRUM

Date of first manufacture
1926.

Alternate/trade name examples
Budene, Buna, diene, polybutadiene, Taktene.

Polymerization
Butadiene is a synthetic rubber that is polymerized from the monomer 1,3-butadiene. It can polymerize in three different configurations: cis, trans, and vinyl. A mixture of configurations is mostly used for manufacturing products.

Processing
Mixing, mastication, molding, vulcanization.

Selected applications
Golf balls, tires. Used as a toughening agent in polystyrene to form high-impact polystyrene and in acrylonitrile-butadiene-styrene (ABS).

Feel/look/sound/smell
Rubber smell.

Main degradation agents
Oxygen, ozone, light, UV radiation, moisture, temperature, stress, chemicals.

Main signs of degradation and damage
No degradation phenomena. Butadiene also has high resistance to wear and abrasion.

15.2 | Butadiene rubber (BR)

	PROPERTY	UNITS (SI)	DATA 1947	DATA 1967	DATA 2016 TRANS	DATA 2016 CIS
MECHANICAL	**Compression set** (ASTM D395)	%	—	—	—	13–15
	Elongation (ASTM D638)	%	—	—	400	450–620
	Flexural strength (ASTM D790)	MPa	—	—	—	—
	Hardness (ASTM D785)	Shore A	—	—	—	A64–A88
	Impact strength (Izod notched) (ASTM D256)	J/cm	—	—	—	—
	Tensile modulus (rigidity) (ASTM D638)	GPa	—	—	15–20	—
	Tear strength (ASTM D624)	kNm^{-1}	—	—	—	36.3–696
	Tensile strength (ASTM D638)	MPa	—	—	250–400	16.2–20
PHYSICAL	**Density** (ASTM D792)	g/cm^3	—	—	0.91	0.890–0.915
	Hildebrand solubility parameter (δ)	$(MPa)^{1/2}$	—	—	17.09–17.6	16.02–18.0
	Permeability: H_2O vapor	$cm^3 \cdot cm \cdot cm^{-2} \cdot s^{-1} \cdot Pa^{-1}$	—	—	—	—
	O_2		—	—	—	—
	CO_2 (ASTM D1434)		—	—	—	—
	Resistance to solvents p = poor, f = fair, g = good	Oils (p), aromatic hydrocarbons (p), water (f), ketones (f), aliphatic hydrocarbons (p)				
	Surface tension	mNm^{-1}	—	—	—	32.0
	Water absorption (equilibrium 24 hours) ⅛-inch (3.175 mm) thickness (ASTM D570)	%	—	—	—	—
THERMAL	**Coefficient of linear thermal expansion** (ASTM D696)	$°C^{-1}$ ($\times 10^{-5}$)	—	—	—	—
	Glass transition temperature (T_g)	°C	—	—	−72 to −87	−102
	Low-temperature performance / Brittleness temperature (ASTM D746)	°C	—	—	—	−100
	Melting point (T_m) (ASTM D1238)	°C	—	—	80–145	1
	Softening point (Vicat) (ASTM D1525)	°C	—	—	—	—
OPTICAL	**Clarity/transparency**	—	—	—	—	—
	Transmittance (ASTM D1003)	%	—	—	—	—
	Refractive index (23°C) (ASTM D542)	—	—	—	1.516–1.605	1.516–1.605

Data sources:
1947: *Modern Plastics Encyclopedia*, vol. 3, *Charts* (New York: Plastics Catalogue Corporation).
1967: Shell Chemicals, *Properties of Plastics*, rev. metric ed. (London: Shell International Chemical Company and Shell Chemicals UK).
2016: George Wypych, *Handbook of Polymers*, 2nd ed. (Toronto: ChemTec).

15.3 | Butyl rubber / Isobutylene-isoprene rubber (IIR)

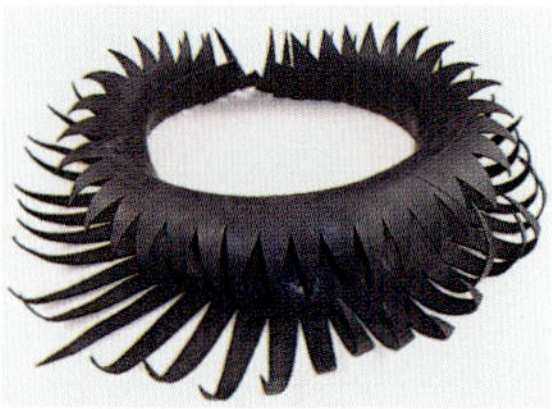

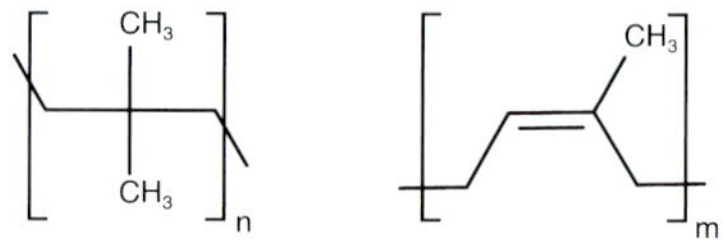

isobutylene isoprene

CHEMICAL STRUCTURE

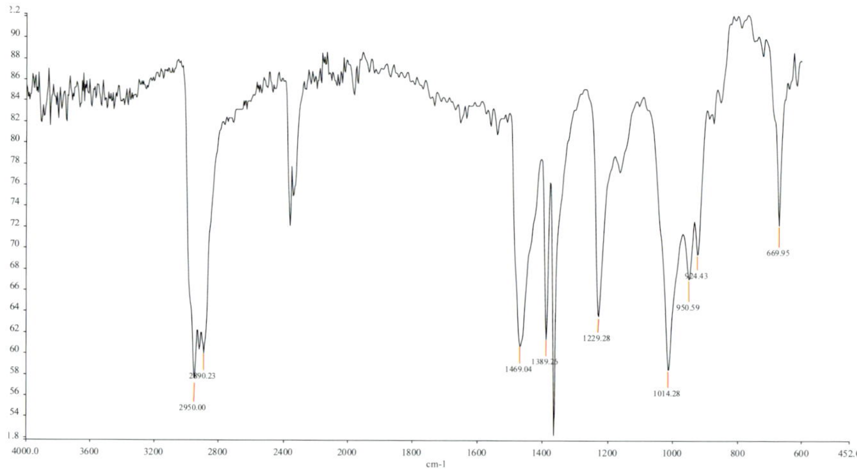

FTIR SPECTRUM

Date of first manufacture
1937 (USA).

Alternate/trade name examples
Esso Butyl.

Polymerization
Butyl rubber, a synthetic elastomer, is synthesized by addition polymerization from isobutylene and isoprene.

Processing
Blending, molding, vulcanization.

Selected applications
Commonly used for adhesives, belting, bladders, O-rings, protective clothing, roof coatings, sealants. Also used for the production of inner tubes due to its very low permeability to gases and impermeability to oxygen.

Feel/look/sound/smell
Mostly black. Rubber smell.

Main degradation agents
Oxygen, ozone, light, UV radiation, moisture, temperature, stress, chemicals.

Main signs of degradation and damages
Slow rate of deterioration, damage not commonly detected.

15.3 | Butyl rubber / Isobutylene-isoprene rubber (IIR)

	PROPERTY	UNITS (SI)	DATA 1947	DATA 1967	DATA 2016
MECHANICAL	**Compression set** (ASTM D395)	%	—	—	72
	Elongation (ASTM D638)	%	≤750	—	670-800
	Flexural strength (ASTM D790)	MPa	—	—	—
	Hardness (ASTM D785)	Shore A	—	—	A36-A90
	Impact strength (Izod notched) (ASTM D256)	J/cm	—	—	—
	Flexural modulus (stiffness) (ASTM D790)	GPa	—	—	—
	Tear strength (ASTM D624)	kNm^{-1}	—	—	31-50
	Tensile strength (ASTM D638)	MPa	≤19.3	—	8.4-19
PHYSICAL	**Density** (ASTM D792)	g/cm^3	0.91	—	0.917-0.92
	Hildebrand solubility parameter (δ)	$(MPa)^{1/2}$	—	—	15.9-16.47
	Permeability: H_2O vapor O_2 CO_2 (ASTM D1434)	$cm^3 \cdot cm \cdot cm^{-2} \cdot s^{-1} \cdot Pa^{-1}$ $\times 10^{12}$	— — —	— — —	36.7 0.0977 —
	Resistance to solvents p = poor, f = fair, g = good	Aliphatic hydrocarbons (p), aromatic hydrocarbons (p), fats and oils (p), ketones (p), esters (g), alcohols (g)			
	Surface tension	mNm^{-1}	—	—	33.6
	Water absorption (equilibrium 24 hours) ⅛-inch (3.175 mm) thickness (ASTM D570)	%	—	—	—
THERMAL	**Coefficient of linear thermal expansion** (ASTM D696)	$°C^{-1}$ ($\times 10^{-5}$)	—	—	13-75
	Glass transition temperature (T_g)	°C	—	—	−71 to −65
	Low-temperature performance / Brittleness temperature (ASTM D746)	°C	—	—	-60
	Melting point (T_m) (ASTM D1238)	°C	—	—	1.5
	Softening point (Vicat) (ASTM D1525)	°C	—	—	—
OPTICAL	**Clarity/transparency**	—	—	—	—
	Transmittance (ASTM D1003)	%	—	—	—
	Refractive index (23°C) (ASTM D542)	—	—	—	1.5081-1.5092

Data sources:
1947: *Modern Plastics Encyclopedia*, vol. 3, *Charts* (New York: Plastics Catalogue Corporation).
1967: Shell Chemicals, *Properties of Plastics*, rev. metric ed. (London: Shell International Chemical Company and Shell Chemicals UK).
2016: George Wypych, *Handbook of Polymers*, 2nd ed. (Toronto: ChemTec).

15.4 | Chloroprene rubber (CR)

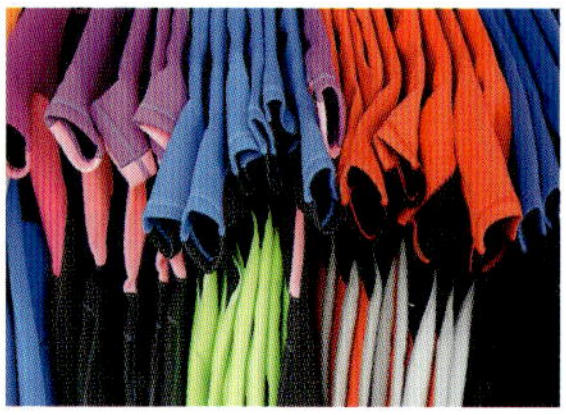

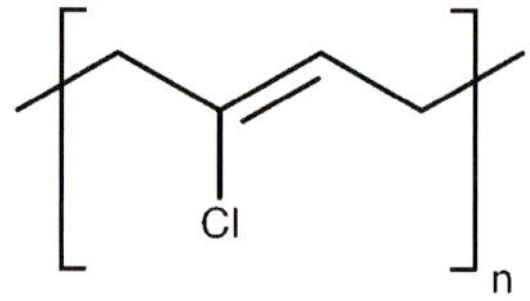

CHEMICAL STRUCTURE

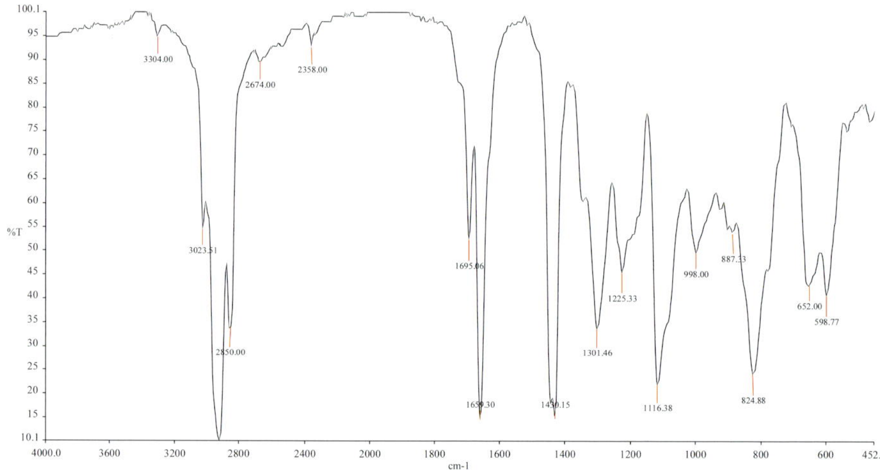

FTIR SPECTRUM

Date of first manufacture
1930 (USA), 1950 (Germany).

Alternate/trade name examples
Neoprene, Baypren, Butaclor, Denka.

Polymerization
Polychloroprene, a synthetic rubber, is polymerized from the monomer chloroprene (2-chlorobuta-1,3-diene).

Processing
Extrusion, compression molding, injection molding, dipping, foaming, calendering.

Selected applications
Adhesives, automotive belts, cable insulation, gaskets, hoses. CR is also produced as foam rubber sheets for knee and wrist braces, laptop computer covers, tubing, and wet suits.

Feel/look/sound/smell
Nearly odorless.

Main degradation agents
Oxygen, ozone, light, UV radiation, moisture, temperature, stress, chemicals.

Main signs of degradation and damage
Hardening.

15.4 | Chloroprene rubber (CR)

	PROPERTY	UNITS (SI)	DATA 1947	DATA 1967	DATA 2016
MECHANICAL	**Compression set** (ASTM D395)	%	—	—	10-32
	Elongation (ASTM D638)	%	0.5-2.2	—	380-955
	Flexural strength (ASTM D790)	MPa	62.1-110.3	—	—
	Hardness (ASTM D785)	Rockwell M Shore A	M80-M110	—	A42-A85
	Impact strength (Izod notched) (ASTM D256)	J/cm	0.08-1.23	—	—
	Flexural modulus (stiffness) (ASTM D790)	GPa	0.68-4.14	—	—
	Tear strength (ASTM D624)	kNm^{-1}	—	—	8.8-50
	Tensile strength (ASTM D638)	MPa	18.6-34.5	—	10.3-20.9
PHYSICAL	**Density** (ASTM D792)	g/cm^3	1.64	—	1.22-1.25
	Hildebrand solubility parameter (δ)	$(MPa)^{1/2}$	—	—	17.6-19.1
	Permeability: H_2O vapor O_2 CO_2 (ASTM D1434)	$cm^3 \cdot cm \cdot cm^{-2} \cdot s^{-1} \cdot Pa^{-1}$ $\times 10^{12}$	— — —	— — —	68.3 0.296 —
	Resistance to solvents p = poor, f = fair, g = good	Esters (p), ketones (p), chlorinated solvents (p), aromatic hydrocarbons (p); resistant to oils/fats/grease and ozone.			
	Surface tension	mNm^{-1}	—	—	43.8
	Water absorption (equilibrium 24 hours) ⅛-inch (3.175 mm) thickness (ASTM D570)	%	0.1-0.3	—	0.9
THERMAL	**Coefficient of linear thermal expansion** (ASTM D696)	$°C^{-1}$ ($\times 10^{-5}$)	12-13	—	—
	Glass transition temperature (T_g)	°C	—	—	−25 to −46
	Low-temperature performance / Brittleness temperature (ASTM D746)	°C	—	—	−35 to −55
	Melting point (T_m) (ASTM D1238)	°C	—	—	45-92
	Softening point (Vicat) (ASTM D1525)	°C	—	—	—
OPTICAL	**Clarity/transparency**	—	Translucent to opaque	—	—
	Transmittance (ASTM D1003)	%	—	—	—
	Refractive index (23°C) (ASTM D542)	—	1.56	—	1.552-1.558

Data sources:
1947: *Modern Plastics Encyclopedia*, vol. 3, *Charts* (New York: Plastics Catalogue Corporation).
1967: Shell Chemicals, *Properties of Plastics*, rev. metric ed. (London: Shell International Chemical Company and Shell Chemicals UK).
2016: George Wypych, *Handbook of Polymers*, 2nd ed. (Toronto: ChemTec).

15.5 | Ethylene-propylene-diene-monomer rubber (EPDM)

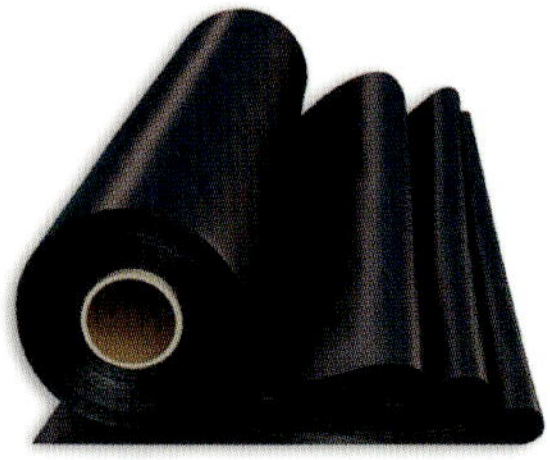

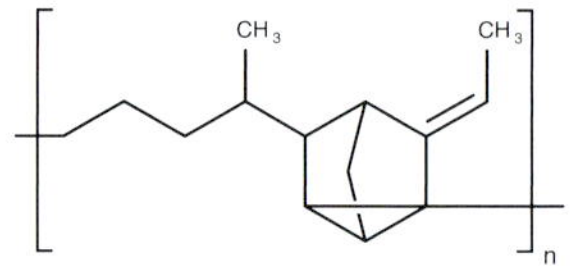

CHEMICAL STRUCTURE

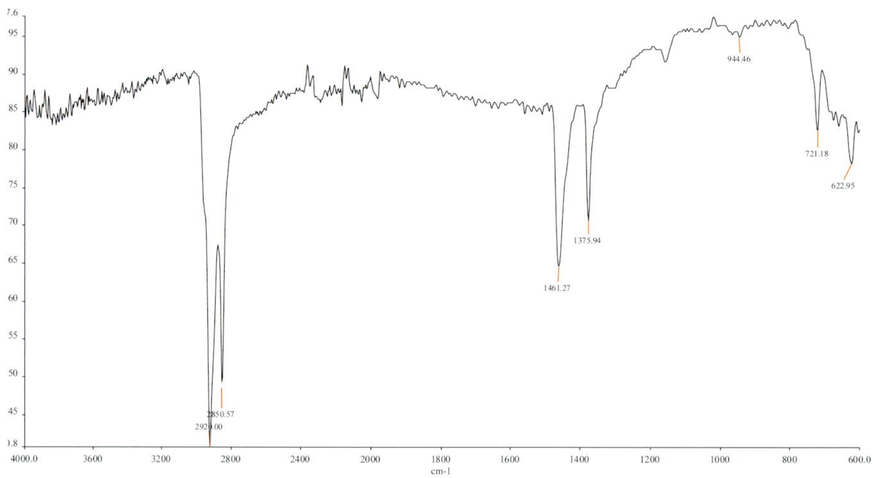

FTIR SPECTRUM

Date of first manufacture
1960s.

Alternate/trade name examples
Dutral, Nordel, Vistalon, Royalene.

Polymerization
EPDM elastomer is a terpolymer of ethylene, propylene, and a diene-component polymerized by addition polymerization.

Processing
Extrusion, calendering, coating, molding. Plasticizers such as paraffin oil, dibutyl phthalate, dioctyl phthalate, and oxidized vegetable oils are added during processing.

Selected applications
Flexible covers, hoses, inner tubes, profiles for window frames and automotive manufacturing, roofing, shoes.

Feel/look/sound/smell
Mild smell to none at all.

Main degradation agents
Light, UV radiation, oxygen, ozone, moisture, temperature, stress, chemicals.

Main signs of degradation and damage
Slow rate of deterioration indoors, fast rate of deterioration outdoors.

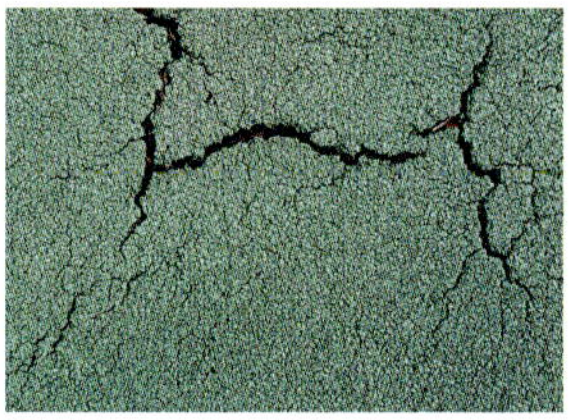

Degraded EPDM rubber tile.

15.5 | Ethylene-propylene-diene-monomer rubber (EPDM)

	PROPERTY	UNITS (SI)	DATA 1947	DATA 1967	DATA 2016
MECHANICAL	**Compression set** (ASTM D395)	%	—	—	50
	Elongation (ASTM D638)	%	—	—	250-760
	Flexural strength (ASTM D790)	MPa	—	—	18-23
	Hardness (ASTM D785)	Shore A	—	—	A50-A90
	Impact strength (Izod notched) (ASTM D256)	J/cm	—	—	540-680
	Flexural modulus (stiffness) (ASTM D790)	GPa	—	—	0.58-0.77
	Tear strength (ASTM D624)	kNm^{-1}	—	—	114-142
	Tensile strength (ASTM D638)	MPa	—	—	8.8-25
PHYSICAL	**Density** (ASTM D792)	g/cm^3	—	—	0.85-0.90
	Hildebrand solubility parameter (δ)	$(MPa)^{1/2}$	—	—	16.0-16.5
	Permeability: H_2O vapor O_2 CO_2 (ASTM D1434)	$cm^3 \cdot cm \cdot cm^{-2} \cdot s^{-1} \cdot Pa^{-1}$	— — —	— — —	— — —
	Resistance to solvents p = poor, f = fair, g = good	Alcohols (g), aliphatic hydrocarbons (g), aromatic hydrocarbons (p), esters (g), fats/oils (p), ketones (g)			
	Surface tension	mNm^{-1}	—	—	34.5
	Water absorption (equilibrium 24 hours) ⅛-inch (3.175 mm) thickness (ASTM D570)	%	—	—	—
THERMAL	**Coefficient of linear thermal expansion** (ASTM D696)	$°C^{-1}$ ($\times 10^{-5}$)	—	—	—
	Glass transition temperature (T_g)	°C	—	—	−48 to −69
	Low-temperature performance / Brittleness temperature (ASTM D746)	°C	—	—	—
	Melting point (T_m) (ASTM D1238)	°C	—	—	—
	Softening point (Vicat) (ASTM D1525)	°C	—	—	—
OPTICAL	**Clarity/transparency**	—	—	—	—
	Transmittance (ASTM D1003)	%	—	—	—
	Refractive index (23°C) (ASTM D542)	—	—	—	1.48

Data sources:
1947: *Modern Plastics Encyclopedia*, vol. 3, *Charts* (New York: Plastics Catalogue Corporation).
1967: Shell Chemicals, *Properties of Plastics*, rev. metric ed. (London: Shell International Chemical Company and Shell Chemicals UK).
2016: George Wypych, *Handbook of Polymers*, 2nd ed. (Toronto: ChemTec).

15.6 | Ethylene-vinyl acetate elastomer (EVAC)

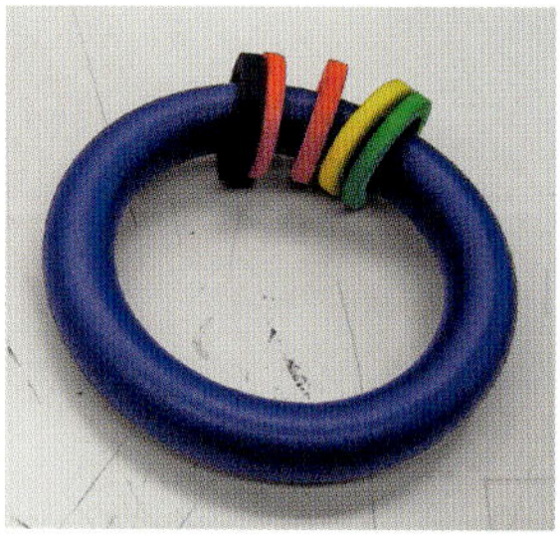

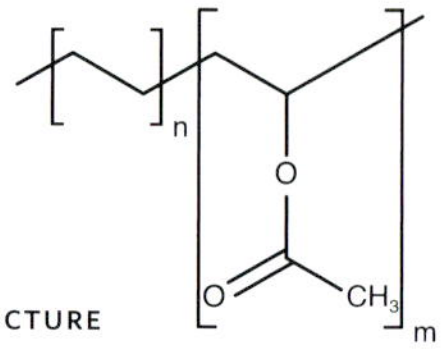

CHEMICAL STRUCTURE

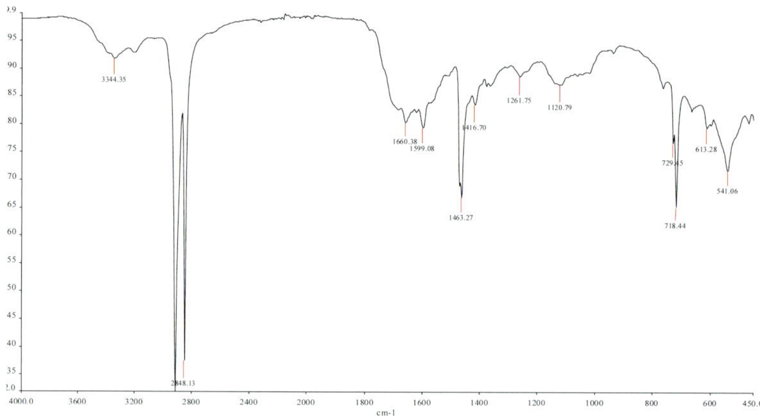

FTIR SPECTRUM

Date of first manufacture
1970s.

Alternate/trade name examples
Levapren 450, Evazote, Elvax, Escorene; Ultrathene.

Polymerization
EVAC is a copolymer of ethylene and vinyl acetate synthesized by addition polymerization and Ziegler-Natta catalysts. By using a blowing agent at polymerization, EVAC as foam rubber is processed. The amount of vinyl acetate copolymerized with ethylene determines the product's properties:
- 0%–4% vinyl acetate gives the properties of polyethylene.
- 4%–40% vinyl acetate gives thermoplastic elastomeric properties.
- 11% vinyl acetate is a hot melt adhesive.
- 40% and greater vinyl acetate produces a thermoplastic elastomer.

Processing
Extrusion, injection molding, reaction injection molding. Cellular foam is produced using injection or compression molding with the addition of chemical foaming agents.

Selected applications
Adhesives; craft foam sheets; padding in a variety of sports boots, gloves, and helmets; sandals.

Feel/look/sound/smell
Soft feel. Good clarity and gloss. Distinctive vinegar-like odor.

Main degradation agents
Light, UV radiation, oxygen, ozone, moisture, temperature, stress, chemicals.

Main signs of degradation and damage
Cracking, crumbling, breaks, tearing.

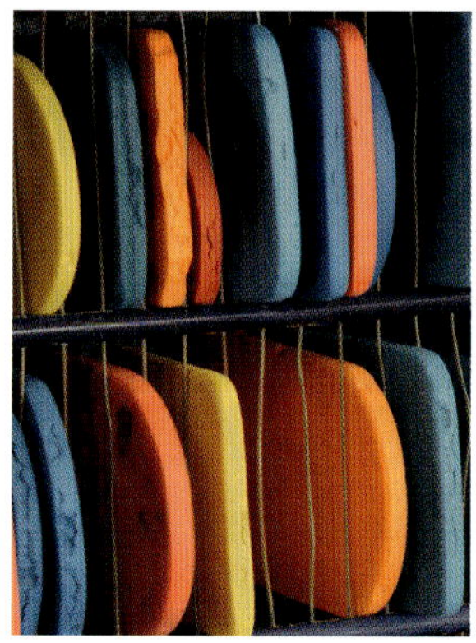

Swim kickboards with breakage and tears at the edges.

15.6 | Ethylene-vinyl acetate elastomer (EVAC)

	PROPERTY	UNITS (SI)	DATA 1947	DATA 1967	DATA 2016
MECHANICAL	**Compression set** (ASTM D395)	%	—	—	23
	Elongation (ASTM D638)	%	—	—	300–860
	Flexural strength (ASTM D790)	MPa	—	—	—
	Hardness (ASTM D785)	Shore A Shore D	—	—	A65–A96 D15–D43
	Impact strength (Izod notched) (ASTM D256)	J/cm	—	—	—
	Flexural modulus (stiffness) (ASTM D790)	MPa	—	—	28–121
	Tear strength (ASTM D624)	kNm^{-1}	—	—	—
	Tensile strength (ASTM D638)	MPa	—	—	2–41
PHYSICAL	**Density** (ASTM D792)	g/cm^3	—	—	0.92–0.98
	Hildebrand solubility parameter (δ)	$(MPa)^{1/2}$	—	—	—
	Permeability: H_2O vapor O_2 CO_2 (ASTM D1434)	$cm^3 \cdot cm \cdot cm^{-2} \cdot s^{-1} \cdot Pa^{-1}$	— — —	— — —	— — —
	Resistance to solvents p = poor, f = fair, g = good	Aliphatic hydrocarbons (p), esters (p), fats/oils (p), ketones (f), alcohols (g)			
	Surface tension	mNm^{-1}	—	—	—
	Water absorption (equilibrium 24 hours) ⅛-inch (3.175 mm) thickness (ASTM D570)	%	—	—	0.005–0.13
THERMAL	**Coefficient of linear thermal expansion** (ASTM D696)	$°C^{-1}$ ($\times 10^{-5}$)	—	—	0.16–0.25
	Glass transition temperature (T_g)	°C	—	—	−38 to −42
	Low-temperature performance / Brittleness temperature (ASTM D746)	°C	—	—	−70 to −85
	Melting point (T_m) (ASTM D1238)	°C	—	—	58–112
	Softening point (Vicat) (ASTM D1525)	°C	—	—	36–86
OPTICAL	**Clarity/transparency**	—	—	—	—
	Transmittance (ASTM D1003)	%	—	—	—
	Refractive index (23°C) (ASTM D542)	—	—	—	1.467–1.498

Data sources:
1947: *Modern Plastics Encyclopedia*, vol. 3, *Charts* (New York: Plastics Catalogue Corporation).
1967: Shell Chemicals, *Properties of Plastics*, rev. metric ed. (London: Shell International Chemical Company and Shell Chemicals UK).
2016: George Wypych, *Handbook of Polymers*, 2nd ed. (Toronto: ChemTec).

15.7 | Gutta-percha rubber

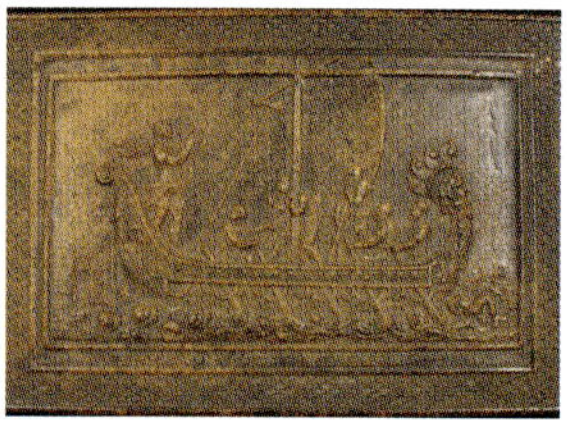

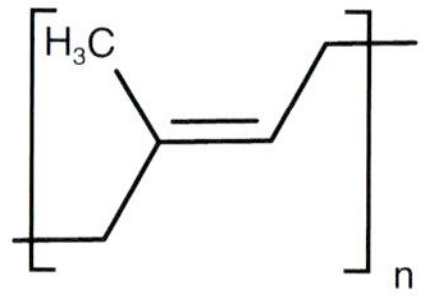

CHEMICAL STRUCTURE

FTIR SPECTRUM

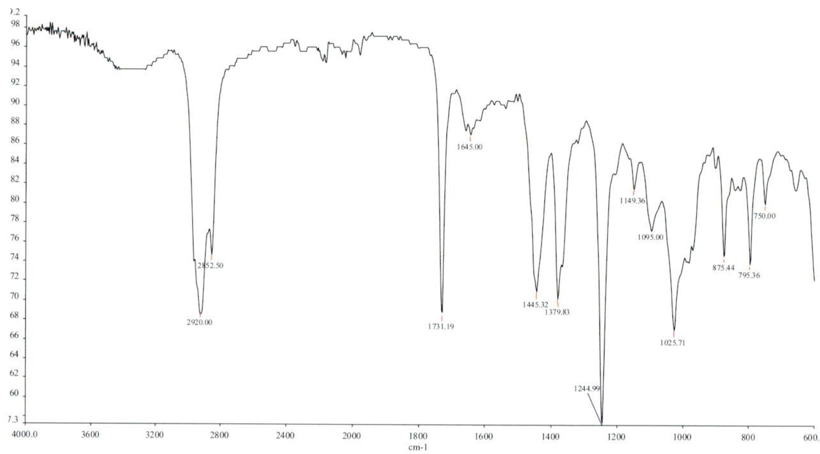

Date of first manufacture
1850s.

Alternate/trade name examples
N/A

Processing
Gutta-percha is the evaporated milky fluid from the family Sapotaceae; it is mainly obtained from *Palaquium* trees from Southeast Asia. Natural gutta-percha contains 97% trans-1,4-polyisoprene. Synthetic gutta-percha is 92%–99% of the trans form. It becomes moldable when heated to a temperature of 50°–60°C and toughens again when cooled. Processing involves calendering, impregnating, and molding.

Selected applications
During the second half of the nineteenth century, gutta-percha was used for domestic and industrial purposes, as insulation for underwater telegraph cables, as filling in root canal treatments (dental sticks), and for golf balls, picture frames, inkstands, and tourist souvenirs.

Feel/look/sound/smell
Rigid material that looks like wood. No smell.

Main degradation agents
Oxygen, ozone, light, UV radiation, moisture, temperature, stress, chemicals.

Main signs of degradation and damage
Yellowing, hardening (losing elasticity), cracking, crumbling.

15.7 | Gutta-percha rubber

	PROPERTY	UNITS (SI)	DATA 1947	DATA 1967	DATA 2016
MECHANICAL	**Compression set** (ASTM D395)	%	—	—	—
	Elongation (ASTM D638)	%	—	—	250–400 (vulcanized)
	Flexural strength (ASTM D790)	MPa	—	—	—
	Hardness (ASTM D785)	Shore A	—	—	A90–A95 (vulcanized)
	Impact strength (Izod notched) (ASTM D256)	J/cm	—	—	—
	Flexural modulus (stiffness) (ASTM D790)	GPa	—	—	—
	Tear strength (ASTM D624)	kNm^{-1}	—	—	—
	Tensile strength (ASTM D638)	MPa	—	—	19.6–36.7 (vulcanized)
PHYSICAL	**Density** (ASTM D792)	g/cm^3	—	—	0.90–0.95
	Hildebrand solubility parameter (δ)	$(MPa)^{1/2}$	—	—	—
	Permeability: H_2O vapor O_2 CO_2 (ASTM D1434)	$cm^3 \cdot cm \cdot cm^{-2} \cdot s^{-1} \cdot Pa^{-1}$	— — —	— — —	172 — —
	Resistance to solvents p = poor, f = fair, g = good	Aliphatic hydrocarbons (p), esters (p), fats/oils (p), ketones (f), alcohols (g)			
	Surface tension	mNm^{-1}	—	—	31
	Water absorption (equilibrium 24 hours) ⅛-inch (3.175 mm) thickness (ASTM D570)	%	—	—	—
THERMAL	**Coefficient of linear thermal expansion** (ASTM D696)	$°C^{-1}$ ($\times 10^{-5}$)	—	—	—
	Glass transition temperature (T_g)	°C	—	—	−63 to −68
	Low-temperature performance / Brittleness temperature (ASTM D746)	°C	—	—	—
	Melting point (T_m) (ASTM D1238)	°C	—	—	58–67
	Softening point (Vicat) (ASTM D1525)	°C	—	—	—
OPTICAL	**Clarity/transparency**	—	—	—	—
	Transmittance (ASTM D1003)	%	—	—	—
	Refractive index (23°C) (ASTM D542)	—	—	—	—

Data sources:
1947: *Modern Plastics Encyclopedia*, vol. 3, *Charts* (New York: Plastics Catalogue Corporation).
1967: Shell Chemicals, *Properties of Plastics*, rev. metric ed. (London: Shell International Chemical Company and Shell Chemicals UK).
2016: George Wypych, *Handbook of Polymers*, 2nd ed. (Toronto: ChemTec).

15.8 | Hard rubber / Vulcanite

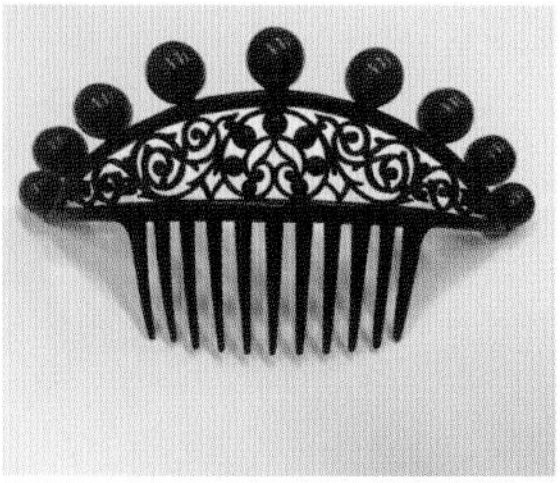

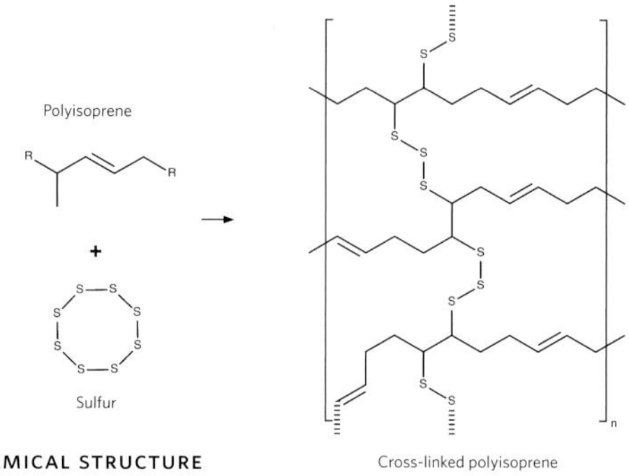

CHEMICAL STRUCTURE

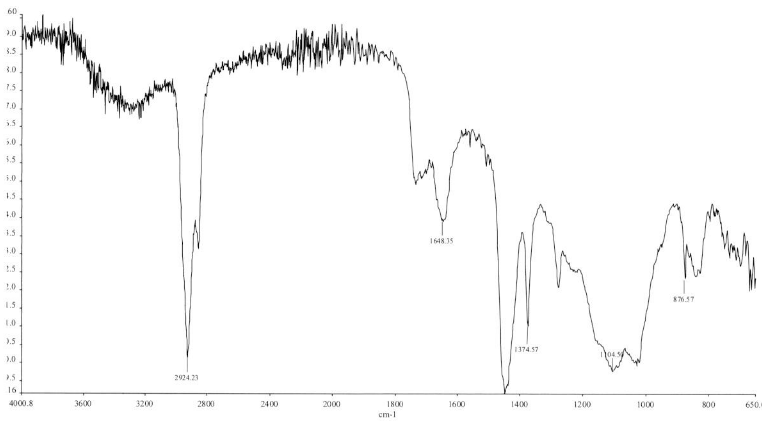

FTIR SPECTRUM

Date of first manufacture
1839 (USA), 1843 (UK). In production until the 1930s.

Alternate/trade name examples
Ebonite.

Processing
The term *hard rubber* is generally used when the amount of sulfur used in the process of vulcanizing natural rubber is more than 5% (it can be as high as 40%). The product becomes hard and tough when a high percentage of sulfur is used, resulting in a thermosetting material. This hard rubber is also called Vulcanite, for the Roman god of fire, and Ebonite, for its resemblance to ebony wood.

Selected applications
Battery housings, combs, dentures, fountain pens, knobs, matchboxes, medical applications, pipestems, Victorian jewelry.

Feel/look/sound/smell
Typically black or red.

Main degradation agents
Light, UV radiation, oxygen, ozone, moisture, temperature, stress, chemicals.

Main signs of degradation and damage
Surface becomes dull, fading, brownish discoloration, sulfur bloom.

15.8 | Hard rubber / Vulcanite

	PROPERTY	UNITS (SI)	DATA 1947		DATA 1967	DATA 2016
			NO FILLER	MINERAL FILLER		
MECHANICAL	**Compressive strength** (ASTM D695)	MPa	55.2–82.7	—	—	—
	Elongation (ASTM D638)	%	7.5	1.25	—	—
	Flexural strength (ASTM D790)	MPa	103.4	51.7	—	—
	Hardness (ASTM D785)	Rockwell	95	95	—	—
	Impact strength (Izod notched) (ASTM D256)	J/cm	0.27	—	—	—
	Tensile modulus (rigidity) (ASTM D638)	GPa	2.07	3.44	—	—
	Tensile strength (ASTM D638)	MPa	62.1	27.6	—	—
PHYSICAL	**Density** (ASTM D792)	g/cm^3	1.14	1.82	—	—
	Hildebrand solubility parameter (δ)	$(MPa)^{1/2}$	—	—	—	—
	Permeability: H_2O vapor O_2 CO_2 (ASTM D1434)	$cm^3 \cdot cm \cdot cm^{-2} \cdot s^{-1} \cdot Pa^{-1}$	— — —	— — —	— — —	— — —
	Resistance to solvents p = poor, f = fair, g = good	Aliphatic hydrocarbons (p), esters (p), fats/oils (p), ketones (f), alcohols (g)				
	Surface tension	mNm^{-1}	—	—	—	—
	Water absorption (equilibrium 24 hours) ⅛-inch (3.175 mm) thickness (ASTM D570)	%	0.02	0.02	—	—
THERMAL	**Coefficient of linear thermal expansion** (ASTM D696)	$°C^{-1}$ ($\times 10^{-5}$)	7.7	5.4	—	—
	Glass transition temperature (T_g)	°C	—	—	—	—
	Low-temperature performance / Brittleness temperature (ASTM D746)	°C	—	—	—	—
	Melting point (T_m) (ASTM D1238)	°C	—	—	—	—
	Softening point (Vicat) (ASTM D1525)	°C	—	—	—	—
OPTICAL	**Clarity/transparency**	—	Opaque	Opaque	—	—
	Transmittance (ASTM D1003)	%	—	—	—	—
	Refractive index (23°C) (ASTM D542)	—	—	—	—	—

Data sources:
1947: *Modern Plastics Encyclopedia*, vol. 3, *Charts* (New York: Plastics Catalogue Corporation).
1967: Shell Chemicals, *Properties of Plastics*, rev. metric ed. (London: Shell International Chemical Company and Shell Chemicals UK).
2016: George Wypych, *Handbook of Polymers*, 2nd ed. (Toronto: ChemTec).

15.9 | Nitrile-butadiene rubber (NBR)

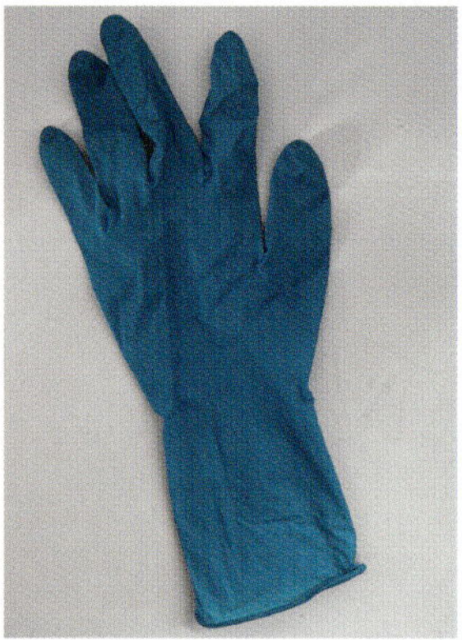

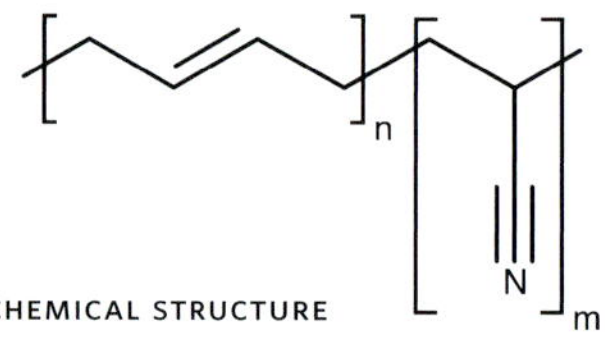

CHEMICAL STRUCTURE

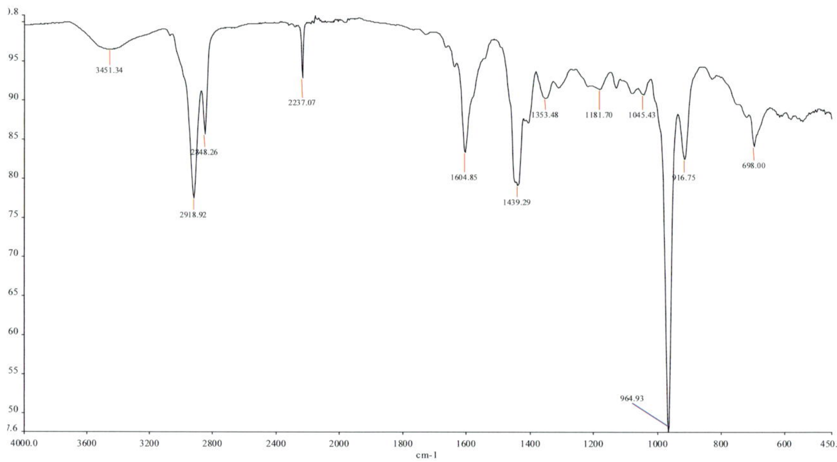

FTIR SPECTRUM

Date of first manufacture
1931 (Buna-N, Germany).

Alternate/trade name examples
Acrylonitrile-butadiene rubber, nitrile rubber, Nipol, Krynac, Perbunan NT, Butacril, Buna-N.

Polymerization
Nitrile-butadiene rubber is a copolymer of acrylonitrile and butadiene and synthesized by addition polymerization.

Processing
Molding, dipping, vulcanization.

Selected applications
Adhesives, disposable gloves, hoses, seals and gaskets, shoe soles.

Feel/look/sound/smell
Faint rubber smell.

Main degradation agents
Oxygen, ozone, temperature, light, UV radiation, moisture, stress, chemicals.

Main signs of degradation and damage
Hardening (losing elasticity).

15.9 | Nitrile-butadiene rubber (NBR)

	PROPERTY	UNITS (SI)	DATA 1947	DATA 1967	DATA 2016
MECHANICAL	**Compression set** (ASTM D395)	%	—	—	9
	Elongation (ASTM D638)	%	—	—	300-600
	Flexural strength (ASTM D790)	MPa	—	—	—
	Hardness (ASTM D785)	Shore A	—	—	A25-A95
	Impact strength (Izod notched) (ASTM D256)	J/cm	—	—	—
	Tensile modulus (rigidity) (ASTM D638)	MPa	—	—	2-5
	Tear strength (ASTM D624)	kNm^{-1}	—	—	42-65
	Tensile strength (ASTM D638)	MPa	—	—	3.1-6.5
PHYSICAL	**Density** (ASTM D792)	g/cm^3	—	—	0.92-1.01
	Hildebrand solubility parameter (δ)	$(MPa)^{1/2}$	—	—	17.90-21.30
	Permeability: H_2O vapor O_2 CO_2 (ASTM D1434)	$cm^3 \cdot cm \cdot cm^{-2} \cdot s^{-1} \cdot Pa^{-1}$ $\times 10^{12}$	— — —	— — —	— 0.0721-1.44 —
	Resistance to solvents p = poor, f = fair, g = good	Ketones (p), esters (p), alcohols (g), aliphatic hydrocarbons (g), oils/fats (g)			
	Surface tension	mNm^{-1}	—	—	—
	Water absorption (equilibrium 24 hours) ⅛-inch (3.175 mm) thickness (ASTM D570)	%	—	—	—
THERMAL	**Coefficient of linear thermal expansion** (ASTM D696)	$°C^{-1}$ ($\times 10^{-5}$)	—	—	—
	Glass transition temperature (T_g)	°C	—	—	−60 to −10
	Low-temperature performance / Brittleness temperature (ASTM D746)	°C	—	—	−28 to −55
	Melting point (T_m) (ASTM D1238)	°C	—	—	—
	Softening point (Vicat) (ASTM D1525)	°C	—	—	—
OPTICAL	**Clarity/transparency**	—	—	—	—
	Transmittance (ASTM D1003)	%	—	—	—
	Refractive index (23°C) (ASTM D542)	—	—	—	—

Data sources:
1947: *Modern Plastics Encyclopedia*, vol. 3, *Charts* (New York: Plastics Catalogue Corporation).
1967: Shell Chemicals, *Properties of Plastics*, rev. metric ed. (London: Shell International Chemical Company and Shell Chemicals UK).
2016: George Wypych, *Handbook of Polymers*, 2nd ed. (Toronto: ChemTec).

15.10 | Natural rubber (NR)

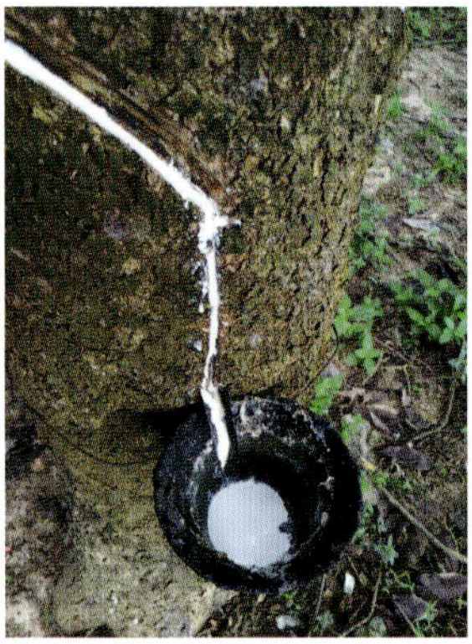

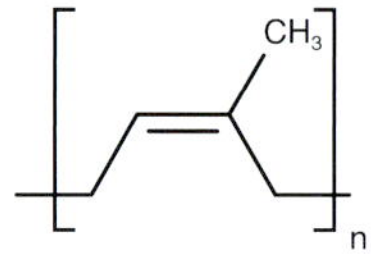

CHEMICAL STRUCTURE

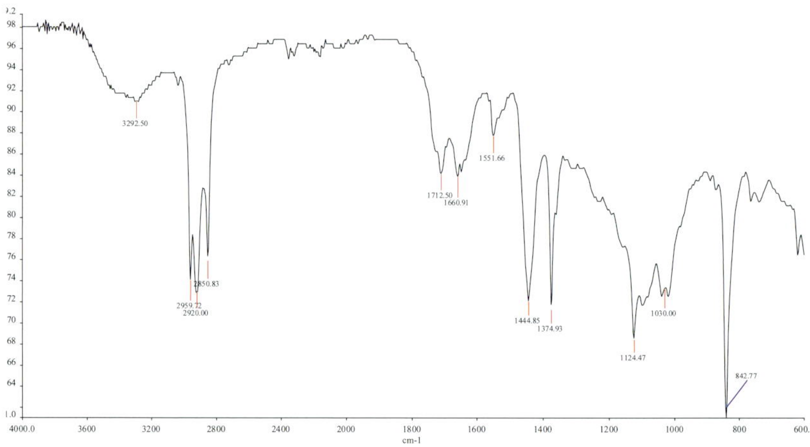

FTIR SPECTRUM

Date of first manufacture
Used by the Aztec and Maya people since the sixth century CE; 1839 (USA), 1843 (patented UK), 1954 (synthetic).

Alternate/trade name examples
Latex, pale crepe, smoked sheets, India rubber, caoutchouc.

Polymerization
Natural rubber consists of polymers of cis-1,4-isoprene, with minor impurities of other organic compounds plus water. Synthetic polyisoprene was developed in 1954 from the polymerization of 2-methyl-1,3-butadiene, as homopolymer. The cis-1,4 content in synthetic polyisoprene is 90%–98%; in natural rubber the cis-1,4 content is 100%.

Processing
Mastication, mixing, molding, vulcanization (cross-linking reaction with sulfur). Typical formulation for black rubber is rubber up to 40%, carbon black up to 60%, stearic acid 2%, sulfur 1.5%, zinc oxide 3%, and accelerator 0.9%.

Selected applications
Bathing caps, carpets, condoms, elastic bands, gas masks, gloves, hot-water bottles, inflatable boats, insulation for electric wiring, pressure hoses, safety goggles, sealant strips, shoe soles, tires, toys.

Feel/look/sound/smell
Natural rubber ranges in color from white to black and across the spectrum. Characteristic rubber smell.

Main degradation agents
Oxygen, ozone, light, UV radiation, moisture, temperature, stress, chemicals.

Main signs of degradation and damage
Softening, stickiness, hardening (losing elasticity), crumbling, discoloration, losing shape.

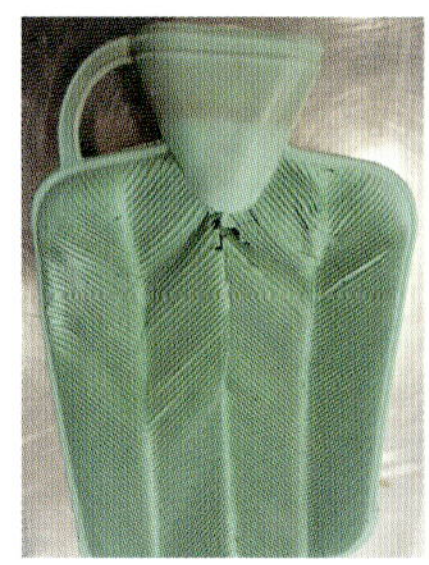

Crumbling, distorted hot-water bottle.

15.10 | Natural rubber (NR)

	PROPERTY	UNITS (SI)	DATA 1947	DATA 1967	DATA 2016
MECHANICAL	**Compression set** (ASTM D395)	%	—	—	—
	Elongation (ASTM D638)	%	—	—	100–800
	Flexural strength (ASTM D790)	MPa	—	—	—
	Hardness (ASTM D785)	Shore A Shore D	—	—	A30–A90 (unvulcanized) D30–D45 (vulcanized)
	Impact strength (Izod notched) (ASTM D256)	J/cm	—	—	—
	Flexural modulus (stiffness) (ASTM D790)	GPa	—	—	—
	Tear strength (ASTM D624)	kNm^{-1}	—	—	—
	Tensile strength (ASTM D638)	MPa	—	—	28 (vulcanized)
PHYSICAL	**Density** (ASTM D792)	g/cm^3	—	—	0.91–0.93 0.95 (vulcanized)
	Hildebrand solubility parameter (δ)	$(MPa)^{1/2}$	—	—	17–18.4
	Permeability: H_2O vapor O_2 CO_2 (ASTM D1434)	$cm^3 \cdot cm \cdot cm^{-2} \cdot s^{-1} \cdot Pa^{-1}$ $\times 10^{12}$	— — —	— — —	— 1.76 —
	Resistance to solvents p = poor, f = fair, g = good	Aliphatic hydrocarbons (p), esters (p), fats/oils (p), ketones (f), alcohols (g)			
	Surface tension	mNm^{-1}	—	—	32
	Water absorption (equilibrium 24 hours) ⅛-inch (3.175 mm) thickness (ASTM D570)	%	—	—	—
THERMAL	**Coefficient of linear thermal expansion** (ASTM D696)	$°C^{-1}$ ($\times 10^{-5}$)	—	—	0.67
	Glass transition temperature (T_g)	°C		—	−75 (natural rubber) −70 to −72 (synthetic isoprene)
	Low-temperature performance / Brittleness temperature (ASTM D746)	°C	—	—	—
	Melting point (T_m) (ASTM D1238)	°C	—	—	—
	Softening point (Vicat) (ASTM D1525)	°C	—	—	—
OPTICAL	**Clarity/transparency**	—	—	—	—
	Transmittance (ASTM D1003)	%	—	—	—
	Refractive index (23°C) (ASTM D542)	—	—	—	1.5191–1.52

Data sources:
1947: *Modern Plastics Encyclopedia*, vol. 3, *Charts* (New York: Plastics Catalogue Corporation).
1967: Shell Chemicals, *Properties of Plastics*, rev. metric ed. (London: Shell International Chemical Company and Shell Chemicals UK).
2016: George Wypych, *Handbook of Polymers*, 2nd ed. (Toronto: ChemTec).

15.11 | Thermoplastic polyurethane elastomer (TPU)

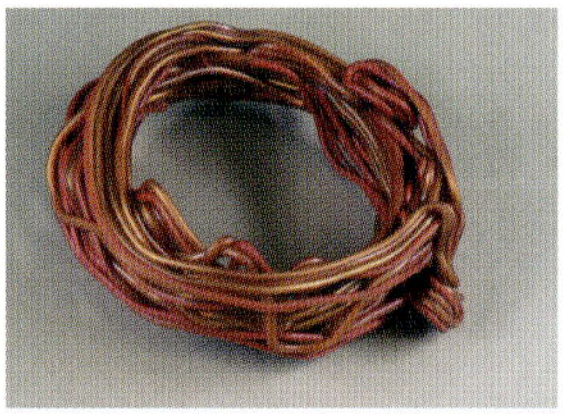

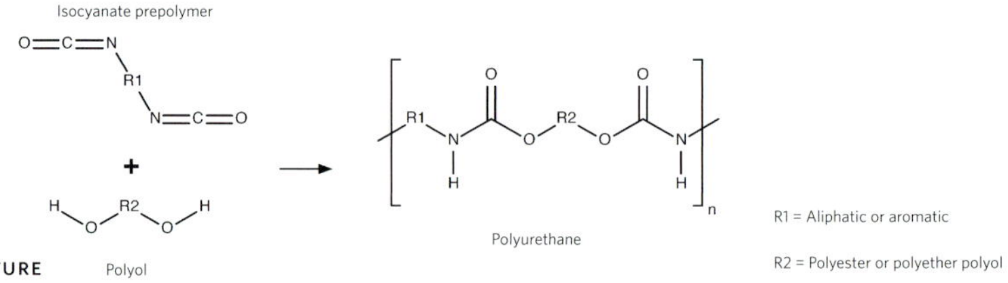

CHEMICAL STRUCTURE

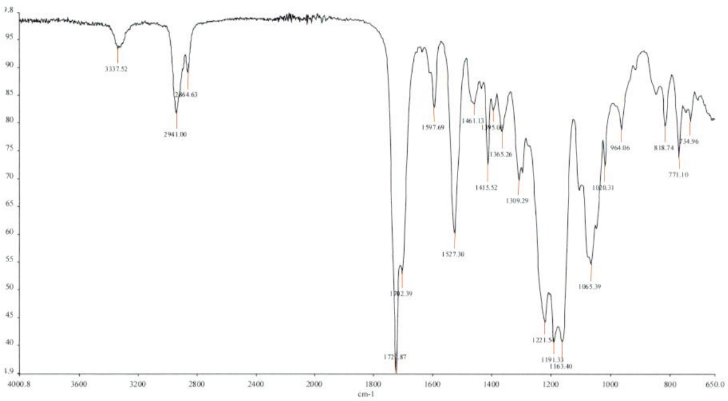

PUR-ester

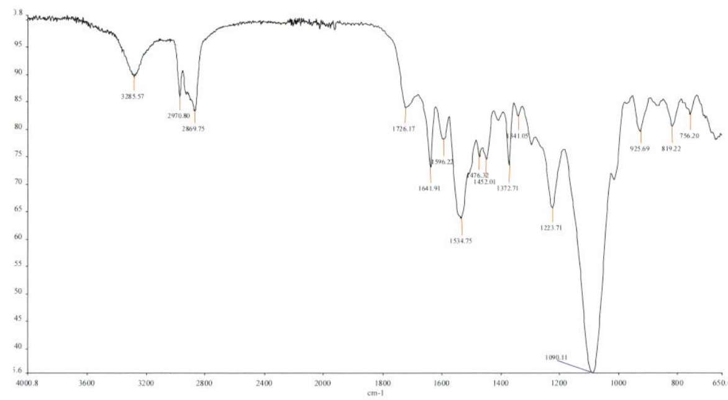

PUR-ether

FTIR SPECTRUM

Date of first manufacture
1930s (Germany).

Alternate/trade name examples
Adiprene, Texin, Vulkollan, Vulcaprene, Estane, Elastocoat, Desmophen, Urepan 600, Desmopan, Millathane, Pellethane.

Polymerization
TPUs are block copolymers consisting of alternating sequences of hard and soft segments formed by the reaction of diisocyanates with long-chain polyether diols or polyester diols and with short-chain diols (chain extenders). A very wide range of TPUs (from very soft and flexible to very hard) can be made by varying the ratio, structure, and/or molecular weight of the reaction compounds (see chap. 3).

Processing
Extrusion, injection molding, calendering, coating, additive manufacturing (3D printing, rapid prototyping).

Selected applications
Automobile parts and other products, coated fabrics, design/art objects, films, footwear/soles, hoses, rollers, seals, tires, wheels.

Feel/look/sound/smell
Off-white to yellow without colorants, all other colors with colorants.

Main degradation agents
Light, UV radiation, oxygen, ozone, moisture (changes in relative humidity), temperature, stress, chemicals. Polyurethane (PUR) ether elastomers mostly oxidize (light, UV radiation, oxygen, temperature). PUR ester elastomers mostly hydrolyze (moisture, temperature).

Main signs of degradation and damage
Yellowing, cracking, crumbling, softening, liquid exudate, solid white material efflorescence. PUR ether elastomers show yellowing and surface crumbling. PUR ester elastomers show softening, stickiness, and formation of liquid or white solid crystals on the surface or the peeling of coatings.

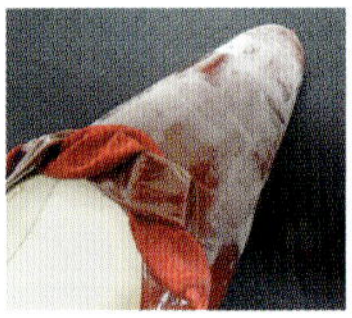

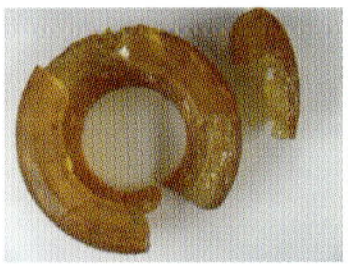

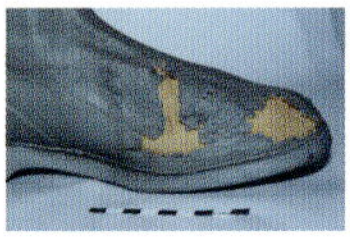

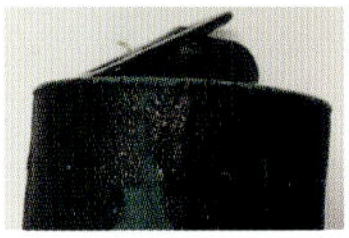

Top to bottom: white exudate (bloom) on shoe surface; cracked and crumbling roller skate wheel; imitation leather peeling from boot surface; sticky surface of a belt.

15.11 | Thermoplastic polyurethane elastomer (TPU)

	PROPERTY	UNITS (SI)	DATA 2016	DATA 2017	
				ETHER TYPE	ESTER TYPE
MECHANICAL	**Compression set** (ASTM D395)	%	11-87	1-70	10-58
	Elongation (ASTM D638)	%	300-1500	77-1000	10-1000
	Flexural strength (ASTM D790)	MPa	5.5-75.2	1.55-75.2	—
	Hardness (ASTM D785)	Shore A Shore D	A62-A98 D28-D73	A25-A101 D22-D83	A30-A101 D15-D78
	Impact strength (Izod notched) (ASTM D256)	J/cm	—	—	0.640-0.850
	Flexural modulus (stiffness) (ASTM D790)	GPa	0.017-1.99	0.0162-1.99	0.004-0.830
	Tear strength (ASTM D624)	kNm^{-1}	33-256	12.4-257	5.43-255
	Tensile strength (ASTM D638)	MPa	17-66	0.800-68.9	0.8-80
PHYSICAL	**Density** (ASTM D792)	g/cm^3	1.02-1.12	1.04-1.27	0.65-1.44
	Hildebrand solubility parameter (δ)	$(MPa)^{1/2}$	22.8	—	—
	Permeability: H_2O vapor O_2 CO_2 (ASTM D1434)	$cm^3 \cdot cm \cdot cm^{-2} \cdot s^{-1} \cdot Pa^{-1}$ $\times 10^{10}$	— 7.0 —	— — —	— — —
	Resistance to solvents p = poor, f= fair, g = good	Alcohols (g), aliphatic hydrocarbons (g), aromatic hydrocarbons (p), esters (g), ketones (g)			
	Surface tension	mNm^{-1}	36.3-39.0	—	—
	Water absorption (equilibrium 24 hours) ⅛-inch (3.175 mm) thickness (ASTM D570)	%	—	1.08-1.50	—
THERMAL	**Coefficient of linear thermal expansion** (ASTM D696)	$°C^{-1}$ ($\times 10^{-5}$)	8.6	8.64-179	15-210
	Glass transition temperature (T_g)	°C	−44 to −66	−76 to −10	−53.9 to −13.9
	Low-temperature performance / Brittleness temperature (ASTM D746)	°C	—	−70 to −35	—
	Melting point (T_m) (ASTM D1238)	°C	170-220	118-218	71-225
	Softening point (Vicat) (ASTM D1525)	°C	60.6-71.1	58-145	38-163
OPTICAL	**Clarity/transparency**	—	—	—	—
	Transmittance (ASTM D1003)	%	—	85-95	90
	Refractive index (23°C) (ASTM D542)	—	—	—	—

Data sources:
1947: *Modern Plastics Encyclopedia*, vol. 3, *Charts* (New York: Plastics Catalogue Corporation).
1967: Shell Chemicals, *Properties of Plastics*, rev. metric ed. (London: Shell International Chemical Company and Shell Chemicals UK).
2016: George Wypych, *Handbook of Polymers*, 2nd ed. (Toronto: ChemTec).
2017: MatWeb, www.matweb.com.

15.12 | Polyurethane (thermosetting) (PUR)

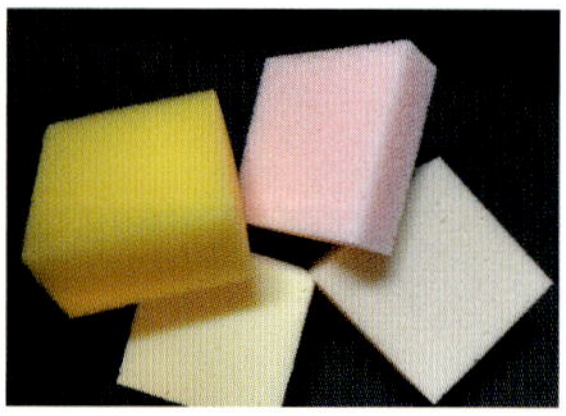

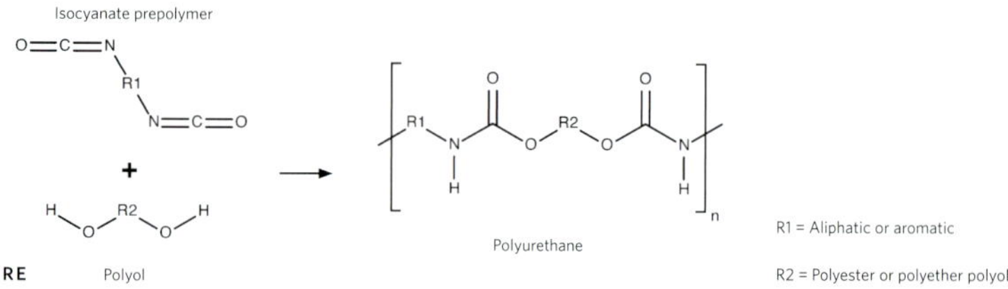

CHEMICAL STRUCTURE

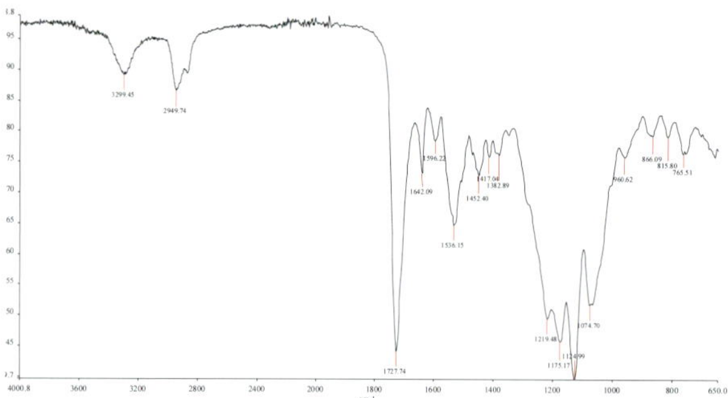

Polyester urethane

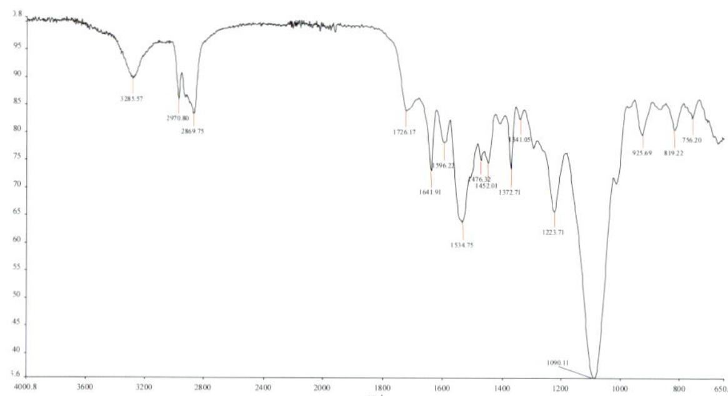

Polyether urethane

FTIR SPECTRUM

Date of first manufacture
1937 (PUR ester), 1957 (PUR ether).

Alternate/trade name examples
Desmopan, Durethan U, Moltopren.

Polymerization
Polyurethane foams can be either flexible or rigid foams:

- Thermosetting flexible polyurethane ether foam can be formed from a polyether polyol (the product made by adding propylene oxide to a hydroxyl such as diethylene glycol) with a diisocyanate using the trifunctional trimethylolpropane as the cross-linking agent.
- Thermosetting rigid polyurethane can be formed when three different diacids react with trimethylolpropane, creating a polyester polyol that reacts with a diisocyanate, resulting in a polyester polyurethane.

Processing
Slab stock foaming is a one-step process in which reactants are mixed directly while a blowing agent, catalysts, and other additives are added simultaneously. Depending on the formulations of the reactants, cold or hot molding can be applied.

Selected applications
Flexible foams are used for car seats, furniture cushioning, mattresses, packaging, and sound- and shock-absorbing materials. Rigid foams are used mostly as thermal insulating material.

Feel/look/sound/smell
Without colorants, PUR looks off-white to yellow. Freshly made PUR foam has the smell of a new car.

Main degradation agents
Light, UV radiation, oxygen, ozone, moisture, temperature, stress, chemicals. PUR ether foams mostly oxidize (light, UV radiation, oxygen, temperature). PUR ester foams mostly hydrolyze (moisture, temperature).

Main signs of degradation and damage
PUR ether foams show yellowing and surface crumbling. PUR ester foams show loss of resilience and formation of white solid crystals on the surface.

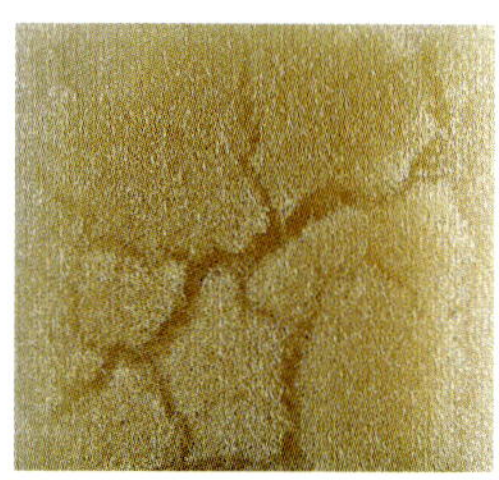

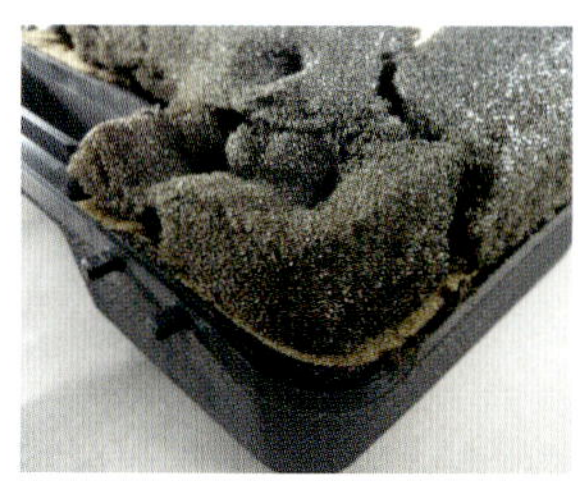

Left to right: crumbled, yellowing PUR ether foam with an intact layer under its damaged surface; loss of resilience and the formation of white crystals on the surface of PUR ester foam.

15.12 | Polyurethane (thermosetting) (PUR)

	PROPERTY	UNITS (SI)	DATA 2016			DATA 2021
			ESTER-TYPE RIGID FOAM	ETHER-TYPE FLEXIBLE FOAM	ESTER-TYPE FLEXIBLE FOAM	THERMOSET ELASTOMERS
MECHANICAL	**Compression set** (ASTM D395)	%	—	—	—	3.0-93
	Compressive strength (ASTM D695)	MPa	—	0.06	0.24	0.015-1.31
	Elongation (ASTM D638)	%	760	360	525	15-1100
	Hardness (ASTM D785)	Shore A Shore D Shore OO	A55	A15	A30	A10-A105 D40-D80 OO30-OO70
	Impact strength (Izod notched) (ASTM D256)	J/cm	—	—	—	2.40-6.70
	Tensile modulus (rigidity) (ASTM D638)	MPa	1.10	0.303	1.00	0.004-0.248
	Tear strength (ASTM D624)	kNm^{-1}	130	—	—	4.30-550
	Tensile strength (ASTM D638)	MPa	72.4	7.6	48.6	0.138-63.0
PHYSICAL	**Density** (ASTM D792)	g/cm^3	0.880	0.400	0.480	0.210-1.50
	Hildebrand solubility parameter (δ)	$(MPa)^{1/2}$	—	—	—	—
	Permeability: H_2O vapor O_2 CO_2 (ASTM D1434)	$cm^3{\cdot}cm{\cdot}cm^{-2}{\cdot}s^{-1}{\cdot}Pa^{-1}$	— — —	— — —	— — —	— — —
	Resistance to solvents p = poor, f = fair, g = good	Alcohols (g), aliphatic hydrocarbons (g), esters (p), ketones (p)				
	Surface tension	mNm^{-1}	—	—	—	—
	Water absorption (equilibrium 24 hours) ⅛-inch (3.175 mm) thickness (ASTM D570)	%	—	—	—	2.8-7.4
THERMAL	**Coefficient of linear thermal expansion** (ASTM D696)	$°C^{-1}$ (× 10^{-5})	—	—	—	95.0-153
	Glass transition temperature (T_g)	°C	—	—	—	−46.0 to −40.0
	Low-temperature performance / Brittleness temperature (ASTM D746)	°C	—	—	—	−90.0 to −27.0
	Melting point (T_m) (ASTM D1238)	°C	—	—	—	80.0
	Softening point (Vicat) (ASTM D1525)	°C	—	—	—	—
OPTICAL	**Clarity/transparency**	—	—	—	—	—
	Transmittance (ASTM D1003)	%	—	—	—	90
	Refractive index (23°C) (ASTM D542)	—	—	—	—	—

Data sources:
2016: MatWeb, www.matweb.com.
2021: MatWeb, data from overviews of various thermosetting polyurethanes.

15.13 | Silicone rubber (Si)

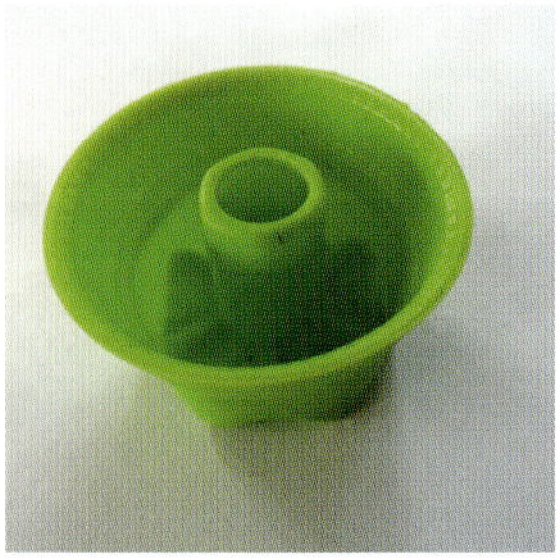

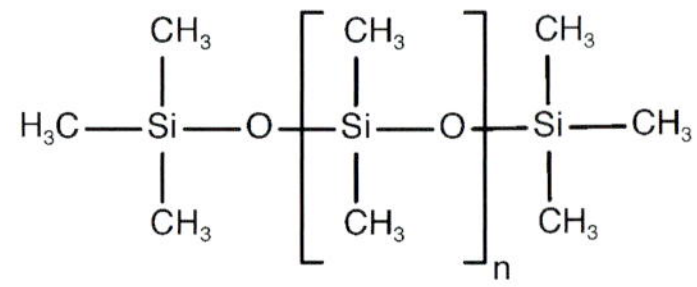

CHEMICAL STRUCTURE

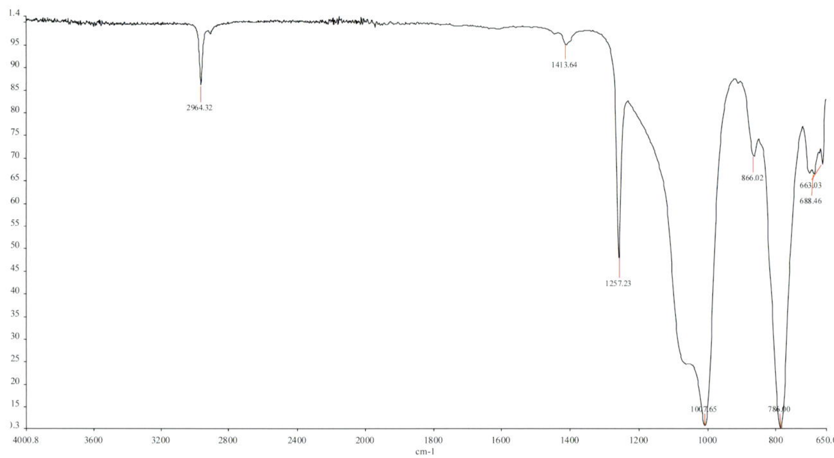

FTIR SPECTRUM

Date of first manufacture
Early 1940s, 1942 (UK).

Alternate/trade name examples
Silastic, Elastoseal, Rhodorsil, Silopren.

Polymerization
Silicones, also known as polysiloxanes, are polymerized by the reaction of silicon and methyl chloride (chloromethane). The formed dimethyl dichlorosilane is reacted with water to form polydimethyl siloxane.

Processing
Extrusion, compression molding, injection molding, calendering, casting.

Selected applications
Breast implants, electronics, food storage, medical devices, molds, sheets, shoes, sealants, sportswear. Due to its resistance to a wide range of temperatures, it is also used in kitchenware for cooking and baking.

Feel/look/sound/smell
Flexible, soft, bouncy, water repellent. No smell.

Main degradation agents
Light, UV radiation, oxygen, ozone, moisture, temperature, stress, chemicals.

Main signs of degradation and damage
Liquid exudate on surface, tearing.

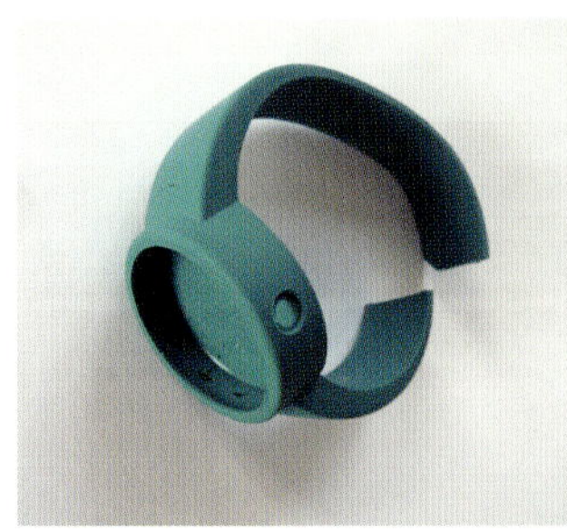

Torn watch strap.

15.13 | Silicone rubber (Si)

	PROPERTY	UNITS (SI)	DATA 1947	DATA 1988	DATA 2016
MECHANICAL	**Compression set** (ASTM D395)	%	—	—	—
	Elongation (ASTM D638)	%	—	—	220-1600
	Flexural strength (ASTM D790)	MPa	—	—	—
	Hardness (ASTM D785)	Shore A	—	—	A15-A70
	Impact strength (Izod notched) (ASTM D256)	J/cm	—	—	—
	Tensile modulus (rigidity) (ASTM D638)	MPa	—	6-12	0.7-3.45
	Tear strength (ASTM D624)	kNm^{-1}	—	—	5-40
	Tensile strength (ASTM D638)	MPa	—	28-46	0.5-9.7
PHYSICAL	**Density** (ASTM D792)	g/cm^3	—	1.8-1.9	0.97
	Hildebrand solubility parameter (δ)	$(MPa)^{1/2}$	—	—	14.9-15.6
	Permeability: H_2O vapor, O_2, CO_2 (ASTM D1434)	$cm^3 \cdot cm \cdot cm^{-2} \cdot s^{-1} \cdot Pa^{-1}$	—	—	—
	Resistance to solvents p = poor, f = fair, g = good	Alcohols (p), aliphatic hydrocarbons (g), aromatic hydrocarbons (g), oils (g), ketones (p)			
	Surface tension	mNm^{-1}	—	—	20.3-21.5
	Water absorption (equilibrium 24 hours) ⅛-inch (3.175 mm) thickness (ASTM D570)	%	—	0.2	—
THERMAL	**Coefficient of linear thermal expansion** (ASTM D696)	$°C^{-1}$ ($\times 10^{-5}$)	—	—	0.9-0.96
	Glass transition temperature (T_g)	°C	—	—	−123 to −127
	Low-temperature performance / Brittleness temperature (ASTM D746)	°C	—	—	—
	Melting point (T_m) (ASTM D1238)	°C	—	-50	−35 to −55
	Softening point (Vicat) (ASTM D1525)	°C	—	—	—
OPTICAL	**Clarity/transparency**	—	—	—	—
	Transmittance (ASTM D1003)	%	—	—	—
	Refractive index (23°C) (ASTM D542)	—	—	—	1.375-1.404

Data sources:
1947: *Modern Plastics Encyclopedia*, vol. 3, *Charts* (New York: Plastics Catalogue Corporation).
1988: Hans Domininghaus, "Guide Chart of Physical Properties," in *Die Kunststoffe und ihre Eigenschaften* (Heidelberg, Berlin: Springer-Verlag).
2016: George Wypych, *Handbook of Polymers*, 2nd ed. (Toronto: ChemTec).

15.14 | Styrene-butadiene rubber (SBR)

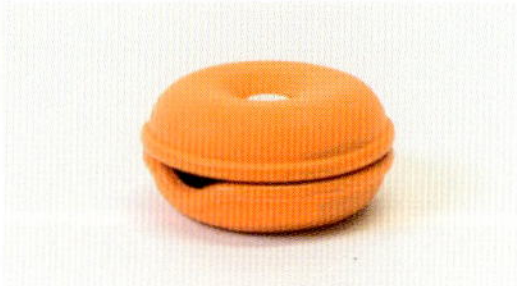
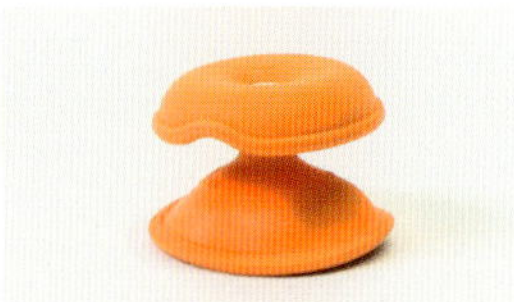
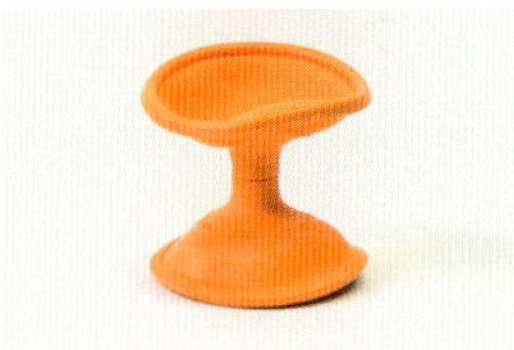

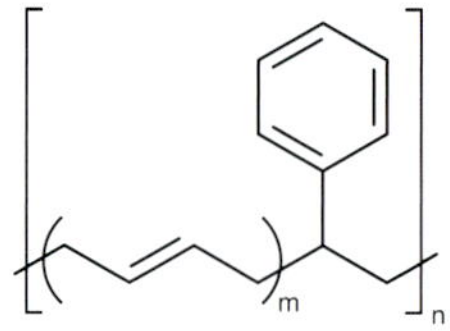

CHEMICAL STRUCTURE

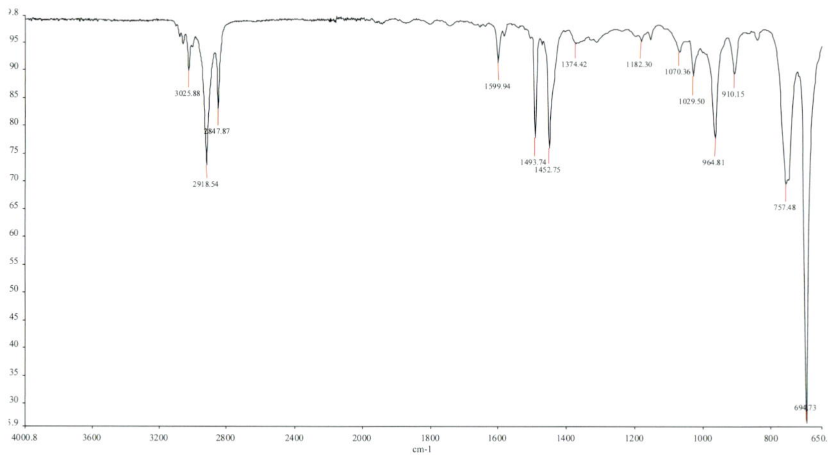

FTIR SPECTRUM

Date of first manufacture
1929 (Buna-S, Germany), 1940s (SBR, UK), 1940 (GR-S rubber, USA).

Alternate/trade name examples
Buna-SE, Stereon, Plioflex.

Polymerization
Styrene-butadiene rubber is synthesized from the polymerization of styrene and butadiene using peroxide as the vulcanization agent. The amounts of styrene and butadiene determine the hardness of the rubber.

Processing
Mixing, compression molding, foaming, calendering, vulcanization.

Selected applications
Adhesives, belts, cables, carpet, caulking, coated fabrics, flooring, gaskets, lamination, sealants, sheets, shoe products, sponges, sporting goods, tires, toys, tubing.

Feel/look/sound/smell
Characteristic rubber smell.

Main degradation agents
Oxygen, ozone, light, UV radiation, moisture, temperature, stress, chemicals.

Main signs of degradation and damage
No degradation phenomena. SBR has good weathering stability and is resistant to abrasion. It has greater durability than natural rubber but is less resistant to high temperatures.

15.14 | Styrene-butadiene rubber (SBR)

	PROPERTY	UNITS (SI)	DATA 1947	DATA 1967	DATA 2016
MECHANICAL	**Compression set** (ASTM D395)	%	—	—	34
	Elongation (ASTM D638)	%	≤700	—	380-750
	Flexural strength (ASTM D790)	MPa	—	—	—
	Hardness (ASTM D785)	Shore A	—	—	A30-A90
	Impact strength (Izod notched) (ASTM D256)	J/cm	—	—	—
	Tensile modulus (rigidity) (ASTM D638)	MPa	—	—	2-2.3
	Tear strength (ASTM D624)	kNm^{-1}	—	—	20.4-43
	Tensile strength (ASTM D638)	MPa	≤20.7	—	13-28.5
PHYSICAL	**Density** (ASTM D792)	g/cm^3	0.98	—	0.91-0.96
	Hildebrand solubility parameter (δ)	$(MPa)^{1/2}$	—	—	17.4
	Permeability: H_2O vapor O_2 CO_2 (ASTM D1434)	$cm^3 \cdot cm \cdot cm^{-2} \cdot s^{-1} \cdot Pa^{-1}$	— — —	— — —	— — —
	Resistance to solvents p = poor, f = fair, g = good	Alcohols (f/g), aliphatic hydrocarbons (p), esters (p), aromatic hydrocarbons (p), fats/oils (f), ketones (p)			
	Surface tension	mNm^{-1}	—	—	—
	Water absorption (equilibrium 24 hours) ⅛-inch (3.175 mm) thickness (ASTM D570)	%	—	—	5
THERMAL	**Coefficient of linear thermal expansion** (ASTM D696)	$°C^{-1}$ ($\times 10^{-5}$)	—	—	—
	Glass transition temperature (T_g)	°C	—	—	−25 to −55
	Low-temperature performance / Brittleness temperature (ASTM D746)	°C	—	—	—
	Melting point (T_m) (ASTM D1238)	°C	—	—	—
	Softening point (Vicat) (ASTM D1525)	°C	—	—	92
OPTICAL	**Clarity/transparency**	—	—	—	—
	Transmittance (ASTM D1003)	%	—	—	—
	Refractive index (23°C) (ASTM D542)	—	—	—	1.53-1.56

Data sources:
1947: *Modern Plastics Encyclopedia*, vol. 3, *Charts* (New York: Plastics Catalogue Corporation).
1967: Shell Chemicals, *Properties of Plastics*, rev. metric ed. (London: Shell International Chemical Company and Shell Chemicals UK).
2016: George Wypych, *Handbook of Polymers*, 2nd ed. (Toronto: ChemTec).

15.15 | Styrene-ethylene-butadiene-styrene rubber (SEBS)

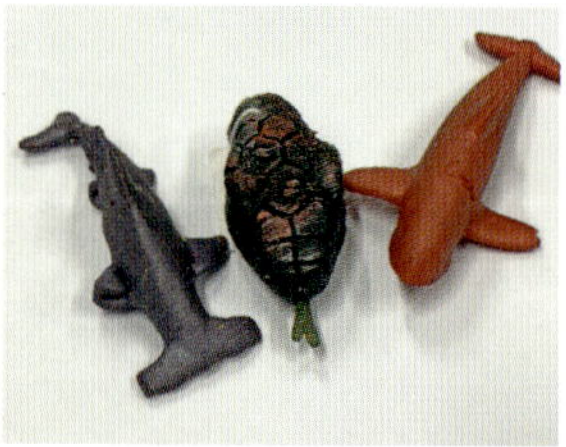

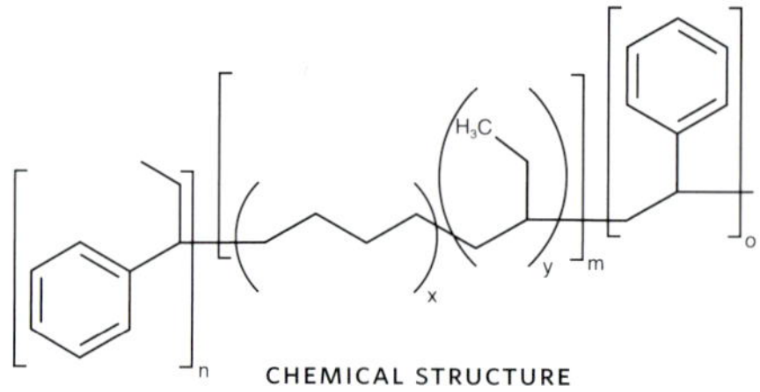

CHEMICAL STRUCTURE

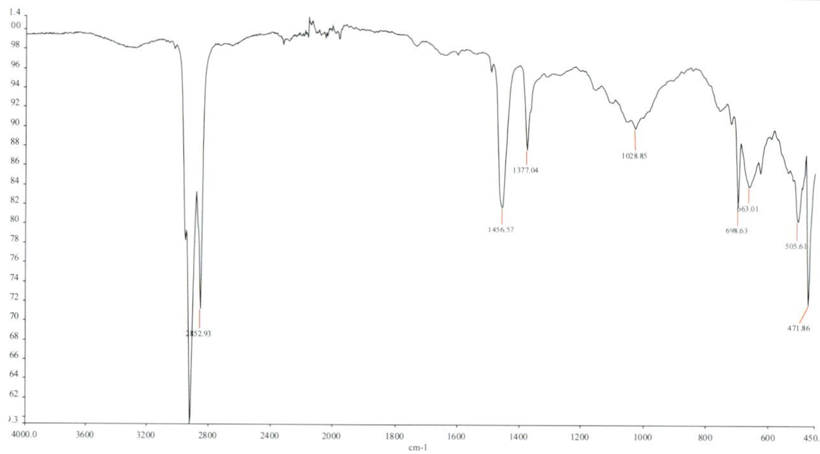

FTIR SPECTRUM

Date of first manufacture
1950s (USA).

Alternate/trade name examples
Kraton G, Evoprene G, Dryflex, Thermolast K.

Polymerization
SEBS is a styrenic triblock polymer with a hydrogenated midblock of ethylene-butylene and is synthesized by addition polymerization.

Processing
Injection molding.

Selected applications
Used in a wide variety of household goods and toys, as well as in personal care and baby products such as toothbrushes and teethers.

Feel/look/sound/smell
Sticky. No rubber smell.

Main degradation agents
Light, UV radiation, oxygen, ozone, moisture, temperature, stress, chemicals.

Main signs of degradation and damage
Easily becomes dirty due to its stickiness. No signs of degradation.

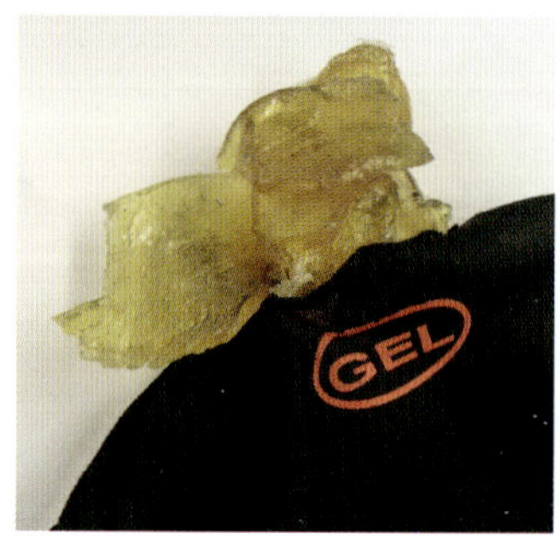

Sticky filling of a bicycle seat.

15.15 | Styrene-ethylene-butadiene-styrene rubber (SEBS)

	PROPERTY	UNITS (SI)	DATA 1947	DATA 1967	DATA 2016
MECHANICAL	**Compression set** (ASTM D395)	%	—	—	16-20
	Elongation (ASTM D638)	%	—	—	470-880
	Flexural strength (ASTM D790)	MPa	—	—	—
	Hardness (ASTM D785)	Shore A	—	—	A30-A95
	Impact strength (Izod notched) (ASTM D256)	J/cm	—	—	—
	Flexural modulus (stiffness) (ASTM D790)	MPa	—	—	100-700
	Tear strength (ASTM D624)	kNm^{-1}	—	—	21-55
	Tensile strength (ASTM D638)	MPa	—	—	3-40
PHYSICAL	**Density** (ASTM D792)	g/cm^3	—	—	0.88-0.95
	Hildebrand solubility parameter (δ)	$(MPa)^{1/2}$	—	—	—
	Permeability: H_2O vapor O_2 CO_2 (ASTM D1434)	$cm^3 \cdot cm \cdot cm^{-2} \cdot s^{-1} \cdot Pa^{-1}$	— — —	— — —	— — —
	Resistance to solvents p = poor, f = fair, g = good	Alcohol (g), solvents overall (f)			
	Surface tension	mNm^{-1}	—	—	—
	Water absorption (equilibrium 24 hours) ⅛-inch (3.175 mm) thickness (ASTM D570)	%	—	—	—
THERMAL	**Coefficient of linear thermal expansion** (ASTM D696)	$°C^{-1}$ ($\times 10^{-5}$)	—	—	—
	Glass transition temperature (T_g)	°C	—	—	−53 to −60
	Low-temperature performance / Brittleness temperature (ASTM D746)	°C	—	—	−60 to −21
	Melting point (T_m) (ASTM D1238)	°C	—	—	180-220
	Softening point (Vicat) (ASTM D1525)	°C	—	—	—
OPTICAL	**Clarity/transparency**	—	—	—	—
	Transmittance (ASTM D1003)	%	—	—	—
	Refractive index (23°C) (ASTM D542)	—	—	—	—

Data sources:
1947: *Modern Plastics Encyclopedia*, vol. 3, *Charts* (New York: Plastics Catalogue Corporation).
1967: Shell Chemicals, *Properties of Plastics*, rev. metric ed. (London: Shell International Chemical Company and Shell Chemicals UK).
2016: George Wypych, *Handbook of Polymers*, 2nd ed. (Toronto: ChemTec).

Appendix A. Types of Plastics

Polymers

Plastics

Thermoplastics

Semicrystalline Thermoplastics

- High-density polyethylene (PE-HD)
- Low-density polyethylene (PE-LD)
- Polyamide (PA)
- Polyamide-6 / Nylon-6 (PA-6)
- Polyamide-6,6 / Nylon-6,6 (PA-6,6)
- Polyamide-12 / Nylon-12 (PA-12)
- Polyethylene (PE)
- Polyoxymethylene (POM)
- Polypropylene (PP)
- Ultra-high molecular weight polyethylene (PE-UHMW)

Amorphous Thermoplastics

- Cellulose acetate (CA)
- Cellulose acetate-butyrate (CAB)
- Cellulose acetate-propionate (CAP)
- Cellulose nitrate (CN)
- Ethyl cellulose (EC)
- Expanded polystyrene (EPS)
- High-impact polystyrene (HIPS)
- Polycarbonate (PC)
- Polyethylene terephthalate (PET)
- Polymethyl methacrylate (PMMA)
- Polystyrene (PS)
- Polyvinyl chloride, plasticized (PVC-P)
- Polyvinyl chloride, rigid, unplasticized (PVC-U)
- Polyvinylidene chloride (PVDC)

Biodegradable Thermoplastics

- Cellophane (regenerated cellulose)
- Polycaprolactone (PCL)
- Polyhydroxybutyrate (PHB)
- Polylactic acid (PLA)
- Thermoplastic starch (TPS)

Thermosets

- Casein-formaldehyde (CS)
- Epoxy (EP)
- Melamine-formaldehyde (MF)
- Phenol-formaldehyde (PF)
- Unsaturated polyester (UP)
- Urea-formaldehyde (UF)

Natural polymers

- Horn
- Shellac
- Tortoiseshell
- Vulcanized fiber

Elastomers

- Balata rubber
- Butadiene rubber (BR)
- Butyl rubber / Isobutylene-isoprene rubber (IIR)
- Chloroprene rubber (CR)
- Ethylene-propylene-diene-monomer rubber (EPDM)
- Ethylene-vinyl acetate elastomer (EVAC)
- Gutta-percha rubber
- Hard rubber / Vulcanite
- Natural rubber (NR)
- Nitrile-butadiene rubber (NBR)
- Polyurethane (thermosetting) (PUR)
- Silicone rubber (Si)
- Styrene-butadiene rubber (SBR)
- Styrene-ethylene-butadiene-styrene rubber (SEBS)
- Thermoplastic polyurethane (TPU)

Compounds

Copolymers

- Acrylonitrile-butadiene-styrene (ABS)
- Acrylonitrile-styrene-acrylate (ASA)
- Polyvinyl chloride-polyvinyl acetate copolymer (PVC/PVAC)
- Styrene-acrylonitrile (SAN)

Composites

- with particle reinforcement
- with fiber reinforcement
- with fiber layer reinforcement

Polymer blends

- Acrylonitrile-butadiene-styrene + polycarbonate (ABS + PC)
- Polyethylene + polyamide-6 (PE + PA-6)
- Polypropylene + polyethylene (PP + PE)

Appendix B. Conferences, Research Projects, and Related Publications

The specialty of conserving artwork made of plastic materials has recently seen rapid progress. The following lists of notable conservation organizations, conferences and meetings, and research projects—and some of the publications that grew out of them—trace the history of activities that have advanced the field. Please note that although every attempt was made to be inclusive, some meetings and projects may have been overlooked.

B.1 CHRONOLOGY OF PLASTICS CONFERENCES AND MEETINGS

For publications related to the events listed below, see sections B.2 and B.3, "Chronology of Research Projects" and "Conference Proceedings and Preprints," respectively.

1980s

The meeting "Modern Organic Materials," held at the University of Edinburgh, April 14–15, 1988, was the first time the conservation issues of modern materials were addressed.

1990s

In the early 1990s, Anita Quye, at that time a conservation scientist at the National Museum of Scotland in Edinburgh, organized the Historic Plastics Research Scientist Working Group—a group of conservation scientists and conservators that met twice a year from 1991 to 1996 to discuss the most pressing needs in the field of plastics conservation.

"Saving the Twentieth Century," a conference at the Canadian Conservation Institute in Ottawa in 1991, was among the first to acknowledge the complex problems involved in the conservation of modern materials, such as plastics, plywood, modern metals, and textiles.

"Polymers in Conservation," an international conference organized by Manchester Polytechnic and Manchester Museum, held in July of 1991, was devoted to the techniques and methodology employed in the storage and preservation of polymers.

The Modern Materials Working Group of ICOM-CC (International Council of Museums, Committee for Conservation) was founded in 1993. Since then, a meeting of this working group has been held every three years.

"Resins—Ancient and Modern," a conference held at the Department of Zoology, University of Aberdeen, in September 1995, included the new materials called modern resins.

A Tate Gallery conference titled "From Marble to Chocolate," in September 1995, highlighted problematic sculptures made from modern materials.

The symposium "Modern Art: Who Cares?," held in September 1997 in Amsterdam, presented the results of the research project of the same name. General conservation issues regarding modern art were raised and our collective understanding refined.

"Mortality Immortality? The Legacy of 20th-Century Art," a conference at the Getty Conservation Institute in Los Angeles, held in March 1998, was devoted to issues regarding the preservation of contemporary art.

INCCA, the International Network for the Conservation of Contemporary Art, was established in 1999 to collect, share, and preserve knowledge for the conservation of contemporary art. The group started with 23 members and now boasts 1,500 members from 780 organizations across 70 countries.

2000s

In 2001, the Modern Materials Working Group of ICOM-CC merged with the Modern and Contemporary Paintings Working Group to form the Modern Materials and Contemporary Art Working Group. Research into the problems of plastics and cultural heritage has increased tremendously as a result. In the spring of 2001, the interim meeting of the ICOM-CC Modern Materials working group was held at the University of Applied Sciences, Department of the Conservation of Cultural Property, in Cologne.

"Modern Art, New Museums," organized by the IIC Congress of the International Institute for Conservation of Historic and Artistic Works (IIC) and held in Bilbao, Spain, in September 2004, tackled problems with plastic, among other topics.

Inside Installations was a three-year project (2004–7) in which museums and other institutions in Europe joined hands in a large-scale collaborative project to investigate the care and administration of installation artworks.

"The Future of the 20th Century: Collecting, Interpreting and Conserving Modern Materials," held in July 2005, was the second annual conference of the AHRC Research Centre for Textile Conservation and Textile Studies. The use of plastics in textiles was among the topics covered.

In 2007, London's Victoria and Albert Museum organized "Plastics: Looking at the Future and Learning from the Past" to look into the issues of instability of plastic materials in collections.

The first of a series of biennial conferences called Future Talks was held in 2009. It was organized by the Conservation Department of Munich's Die Neue Sammlung—The Design Museum.

Plastics Heritage Forum is a series of conferences devoted to plastics. The first was held in 2009, with follow-ups every two to three years.

2010s

"Contemporary Art: Who Cares?"—held in June 2010—was the sequel to 1997's successful "Modern Art: Who Cares?" The outcomes of research projects in contemporary art were discussed, with special attention given to installation art.

In March of 2012, at the international conference "POPART, Preservation of Plastic Artefacts in Museum Collections," the results of a three-year (2008–11) European Research Plastics project were presented in Paris.

A conference called "The Age of Plastic: Ingenuity and Responsibility" was held at the Smithsonian American Art Museum in Washington, DC, in June 2012.

"Soft Materials and Hard Plastics in Art: Characterization, Degradation and Conservation," part of the Joint Seminar Series on Conservation Science, organized by the Art Institute of Chicago and Northwestern University, was held November 15, 2012.

The ninth North American Textile Conservation Conference, "Conserving Modernity: The Articulation of Innovation," was held in San Francisco, November 12–15, 2013. Many papers delivered at this conference dealt with plastics in textiles.

"Saving the Now: Crossing Boundaries to Conserve Contemporary Works"—a collaboration of the International Institute for Conservation of Historic and Artistic Works (IIC), the International Network for the Conservation of Contemporary Art (INCCA), and the Getty Conservation Institute—took place in Los Angeles in September 2016.

"Acting in Contemporary Art Conservation," a summit on (inter)national collaboration, was held by the Dutch Foundation for the Conservation of Contemporary Art (SBMK), November 15–16, 2018, in the Netherlands.

Under the auspices of the Plastics Heritage European Association (PHEA), the Portuguese Center for the History of Science and Technology (CIUHCT) hosted the fifth international Plastics Heritage Congress, "History, Limits and Possibilities," May 29–31, 2019, in Lisbon.

An international conference titled "Plastics Forever? Researching and Preserving Transparent Figures and Other Historical Exhibition Icons," took place September 19–20, 2019, in Dresden, at the Deutsches Hygiene-Museum.

The Icon Modern Materials Network organized "Challenges of the Modern Object," a one-day symposium held October 17, 2019, at the Science Museum, London (Dana Research Centre and Library).

2020s

The Boston Society for Architecture's Historic Resources Committee organized a conference titled "Architectural Plastics and Polymer Composites in the 21st Century: Design and Preservation of Contemporary and Traditional Architecture." It was scheduled for March 28–29, 2020, but was canceled due to COVID-19. "Conservation of Plastic Works of Art" and "Plastics Then and Now" were among the many topics that would have been addressed.[1]

The online conference "Plastics in Peril: Focus on Conservation of Polymeric Materials in Cultural Heritage," held November 16–19, 2020, was jointly organized by the University of Cambridge Museums, the Leibniz Research Museums, Deutsches Museum, Deutsches Bergbau Museum, and Museum für Naturkunde Berlin.

The American Institute for Conservation's Electronic Media Group organized a virtual symposium titled "TechFocus IV: Caring for 3D-Printed Art," presented October 25, 27, and 29, 2021.

B.2 CHRONOLOGY OF RESEARCH PROJECTS

Modern Art: Who Cares? and Contemporary Art: Who Cares?

In 1997, the INCCA project Modern Art: Who Cares? outlined the problems and concerns of conserving unstable plastic materials. As a spin-off of this project, the Cultural Heritage Agency of the Netherlands (RCE) began major research into the consolidation of PUR ether foams and the adhesion of transparent plastics.

> *Modern Art: Who Cares? An Interdisciplinary Research Project and International Symposium on the Conservation of Modern and Contemporary Art*, edited by IJsbrand Hummelen, Dionne Sillé, and Marjan Zijlmans (Amsterdam: Foundation for the Conservation of Modern Art; Netherlands Institute for Cultural Heritage, 1999).

1. The papers are available for purchase at https://www.architects.org/knowledge-communities/historic-resources-committee.

The follow-up symposium, "Contemporary Art: Who Cares?," was held in 2010. Conference proceedings were not published, but videos of the presentation can be found and viewed on the INCCA website, and many of the presenters contributed articles based on their lectures to *Inside Installations* (see below).

Project PUR-Foams

This research initiative by the Cultural Heritage Agency of the Netherlands (RCE), undertaken from 2002 to 2008, investigated the degradation of polyurethane foams and considered potential conservation treatments. This study provided new insights on a versatile and popular type of plastics often found in museum collections.

> *PUR Facts: Conservation of Polyurethane Foam in Art and Design*, by Thea B. van Oosten (Amsterdam: Amsterdam University Press, 2011).

Inside Installations

Inside Installations was a three-year project (2004–7) in which museums and other institutions in Europe joined hands in a large-scale collaborative project to investigate the care and administration of art installations.

> *Inside Installations: Theory and Practice in the Care of Complex Artworks*, edited by Tatja Scholte and Glenn Wharton (Amsterdam: Amsterdam University Press, 2011), http://library.oapen.org/handle/20.500.12657/33489.

Plastic Art

Funded by AXA Art Insurance in close collaboration with the Vitra Design Museum in Germany, the Plastic Art project helped advance the study of plastics conservation.

> *Plastic Art—A Precarious Success Story*, by Stefan Albus, Christian Bonten, Kathrin Kessler, Gabriela Rossi, and Thomas Wessel (Cologne: AXA Art Insurance, 2007).

MoDiP: Curator's Guide

In 2007, a project involving four national workshops and over a hundred plastics experts, academics, curators, and conservators was carried out to develop an online plastics resource for curators to help identify and take care of objects made of plastics. It also demonstrated how plastics have increasingly become significant and suggested themes for their interpretation and presentation. Funded by the Museums, Libraries and Archives Council, the guide was developed for MoDiP, the Museum of Design in Plastics. See https://www.modip.ac.uk/projects/curators-guide.

POPART

The Preservation of Plastic Artefacts in Museum Collections (POPART) project (2008–11) was sponsored by the European Union and addressed the state of plastics conservation issues for cultural heritage objects.

> *Preservation of Plastic Artefacts in Museum Collections*, edited by Bertrand Lavédrine, Alban Fournier, and Graham Martin (Paris: Comité des travaux historiques et scientifiques, 2012).

Project Polymers: The Degradation and Conservation of Plastics

Project Polymers is one of many research projects the Cultural Heritage Agency of the Netherlands (RCE) carried out from 2008 to 2012. It focused on identifying plastic objects; studying and monitoring the degradation of polymers; and cleaning, consolidating, and applying coatings to plastics. See http://www.kennisvoorcollecties.nl/en/projects/objects-in-context/polymers. The website R_2R—From Research to Restoration (https://www.incca.org/r2r) is one of several subprojects.

MoDiP: Identifying Plastics Toolkit

This 2013 research project was intended to help curators develop a fuller understanding of various plastic materials and their needs. Funded by Arts Council England, the resulting resource helps curators begin to identify the plastics in their collections. It includes an online identification route map and a traveling toolkit that can be borrowed. See https://www.modip.ac.uk/projects/toolkit.

MoDiP: Confronting Plastics Preservation

This 2014 research project aimed to build workforce skills and share knowledge about preservation of plastics in collections. Materials produced include a film, papers from the seminar, standard practice guidelines for the care of plastic objects, and a selection of case studies. See https://www.modip.ac.uk/research/confronting-plastics-preservation.

Collection in Plastic at the Pinacoteca

The 2014–15 exhibition at the Pinacoteca do Estabo de São Paulo in Brazil, "Acervo em plástico da Pinacoteca: Problemáticas de conservação e restauro" (Collection in Plastic at the Pinacoteca: Issues of Conservation and Restoration), was the first exhibition devoted entirely to the conservation and restoration of a museum's plastics collection. It showcased plastic artworks in the museum and covered the conservation and restoration of plastic objects, as well as the history of plastics, their growing importance, and how they have transformed modes of artistic production.

> *Acervo em plástico da Pinacoteca: Problemáticas de conservação e restauro* [Collection in plastic at the Pinacoteca: Issues of conservation and restoration], edited by Fernando Pitta, Valeria de Menodca, Patricia Schossler et al. (São Paulo: Pinacoteca do Estabo de São Paulo, 2014).

The Triumph of Bakelite: Contributions for a History of Plastics in Portugal

This 2016–18 project investigated the history of plastics in Portugal. The interdisciplinary project involved Nova ID FCT, in the NOVA School of Science and Technology, and the Faculty of Science at the University of Lisbon.

Plastics Project

The Plastics Project (2017–19) was coordinated by the Foundation for the Conservation of Contemporary Art (SBMK) and the Netherlands Institute for Conservation + Art + Science (NICAS) and was a cooperative effort of the Cultural Heritage Agency of the Netherlands (RCE), the Rijksmuseum, and the University of Amsterdam's Conservation and Restoration Department. Together they developed the Plastics Identification Tool (PIT), a do-it-yourself method for identifying, monitoring, and preserving plastics. See https://plastic-en.tool .cultureelerfgoed.nl.

PolyMuse: Conserving Polymers in Museum Collections

The PolyMuse research project (2017–20), funded by the Australian Research Council, was a collaborative project of scientists and museum professionals. Together they developed methods for extending the life span of at-risk plastic and polymer-based materials in museum collections. See https://www.polymuse.net.au.

Know, Name, and Assess Your Plastics

This international project (2018–21) focused on the identification of and care for the collections of Antwerp's Design Museum and the Stedelijk Museum voor Actuele Kunst (SMAK) in Ghent. It was a collaboration of three international partners: the Cultural Heritage Agency of the Netherlands (RCE); the Royal Institute for Cultural Heritage (KIK-IRPA), a Belgian federal scientific institution; and the Cologne Institute of Conservation Sciences (CICS) of TH Köln.

Cold Storage of Museum Objects Made of Cellulose Nitrate—Mechanical Stress versus Chemical Degradation

A 2019–21 project of the Conservation Science Department of Deutsches Museum, in Munich, this project aimed at optimizing the storage of cellulose nitrate objects and establishing sustainability and feasibility guidelines for museums.

COMPLEX: The Degradation of Complex Modern Polymeric Objects in Heritage Collections

COMPLEX is an interdisciplinary collaborative research project (2017–22) between Tate and University College London's Institute for Sustainable Heritage. The project aims to explore a new approach to polymer degradation and to preserve the plastic objects in collections.

Plastics—A Modern Material in Its Cultural Historical Context

This interdisciplinary research project (2018–22) is tackling the task of learning to read the language of plastic objects and preserve them for the future. Project partners are the Industriemuseum Landschaftsverband Rheinland (LVR), the Institut für Kunststofftechnik (IKT) at the University of Stuttgart, and the Cologne Institute of Conservation Sciences (CICS) of TH Köln. The Kunststoff-Museums-Verein provided access to its collection of around 20,000 objects.

Terminology for Conservation of Plastics

This research project (2020–24), funded by the German Federal Ministry of Education and Research (BMBF), focuses on "The Language of Objects." Project partners are the LVR Industrial Museum in Oberhausen (IMus), the Institut für Kunststofftechnik (IKT) at the University of Stuttgart, and the Cologne Institute of Conservation Sciences (CICS) of TH Köln. This project also works in close collaboration with the Know, Name, and Assess Your Plastics project (see above), the Getty Conservation Institute, and the Institute of Translation and Multilingual Communication of TH Köln.

German Democratic Plastics in Design: A Research Project on Material, Technology, and Conservation

This research project (2021–24) deals specifically with the identification of production processes and technologies for plastic materials used in the German Democratic Republic between 1949 and 1990 for the production of industrial design, as well as the objects' degradation and possible preventive and active measures. German Democratic Plastics in Design is a collaborative project of several institutes: the Getty Conservation Institute and the Wende Museum, in Los Angeles; Die Neue Sammlung—The Design Museum, in Munich; and the Cologne Institute of Conservation Sciences (CICS) of TH Köln.

Glossy Surfaces

Funded by the Flemish government, Glossy Surfaces (2021–24) is aimed at establishing conservation solutions for aromatic/aliphatic thermoplastic polyurethane ester-based coatings in fashion collections. This project is coordinated by ModeMuseum (MoMu), in Belgium, and is a cooperative effort of Portugal's Museu do Design e da Moda (MUDE), the Metropolitan Museum of Art (New York), the Department of Conservation and Restoration at the NOVA School of Science and Technology (Portugal), and Centexbel (Belgium).

Preservation of Plastics

An ongoing long-term research project (2008–) of the Getty Conservation Institute, Preservation of Plastics is dedicated to advancing the conservation of plastics. It is a project of the Modern and Contemporary Art Research Initiative and contains research components, including those exploring preservation of plastics, studies on Disney animation cels, repair studies on transparent plastics, treatment options for degraded acetate and cellulose nitrate, investigations of plastics in design collections, analysis of plastics, and mechanical characterization of plastics.

Heritage of the 20th Century—Plastics

The Cultural Heritage Agency of the Netherlands (RCE) is funding research from 2021 onward to develop tools for the management, conservation, and restoration of plastics in modern and contemporary art and design collections. One of the current research projects is Smelly Plastics (2021–22), which will identify the odors given off by new and aging plastics and investigate the risks of off-gassing to other objects and people in the vicinity.

B.3 CONFERENCE PROCEEDINGS AND PREPRINTS

The Age of Plastic: Ingenuity and Responsibility: Proceedings of the 2012 MCI Symposium, edited by Odile Madden, A. Elena Charola, Kim Cullen Cobb, Paula T. DePriest, and Robert J. Koestler (Washington, DC: Smithsonian Institution Scholarly Press, 2017). https://doi.org/10.5479/si.19492367.7.

From Marble to Chocolate: The Conservation of Modern Sculpture: Tate Gallery Conference, 18–20 September 1995, edited by Jackie Heuman (London: Archetype Publications, 1995).

Future Talks. Since 2009, the Conservation Department of Munich's Die Neue Sammlung—The Design Museum has held a conference every two years. The papers from these conferences provide a wealth of information on practical conservation issues and solutions for the conservation of works of art and objects with plastics. Proceedings are available in print, and conference programs and abstracts may be viewed online at https://dnstdm.de/en/future-talks.

ICOM-CC. The International Council of Museums, Committee for Conservation's triennial conference as well as its Modern Materials and Contemporary Art Working Group proceedings are available in print and in searchable format (going back to 1972) at https://www.icom-cc-publications-online.org.

Modern Organic Materials: Preprints of Contributions to the Modern Organic Materials Meeting Held at the University of Edinburgh, 14 & 15 April 1988, edited by the Scottish Society for Conservation and Restoration (Edinburgh: SSCR Publications, 1988).

Mortality Immortality? The Legacy of 20th-Century Art, edited by Miguel Angel Corzo (Los Angeles: Getty Conservation Institute, 1999).

Plastics in Art: History, Technology, Preservation. Kölner Beiträge zur Restaurierung und Konservierung von Kunst- und Kulturgut, Bd. 15, edited by Thea B. van Oosten, Yvonne Shashoua, and Friederike Waentig (Munich: Siegl, 2002).

Plastics in Peril: Focus on Conservation of Polymeric Materials in Cultural Heritage Virtual Conference, November 16th–19th, 2020. Book of Abstracts, 48 (Munich: Deutsches Museum). A playlist of conference addresses and papers is available on YouTube.

Plastics: Looking at the Future and Learning from the Past: Papers from the Conference Held at the Victoria and Albert Museum, London, 23–25 May 2007, edited by Brenda Keneghan and Louise Egan (London: Archetype Publications, 2008).

Polymers in Conservation: Proceedings of an International Conference Organized by Manchester Polytechnic and Manchester Museum, Manchester, 17th–19th July 1991, edited by N. S. Allen, M. Edge, and C. V. Horie (Cambridge: Royal Society of Chemistry, 1992).

Resins, Ancient and Modern: Preprints of the SSCR's 2nd Resins Conference Held at the Department of Zoology, University of Aberdeen, 13 and 14 September 1995, edited by Margot M. Wright and Joyce H. Townsend (Edinburgh: Scottish Society for Conservation and Restoration, 1995).

Saving the Twentieth Century: The Conservation of Modern Materials: Proceedings of a Conference, Symposium '91: Saving the Twentieth Century, Ottawa, Canada, 15 to 20 September 1991, edited by David W. Grattan (Ottawa: Canadian Conservation Institute, 1993).

Appendix C. Terms in Translation

English	SI unit	Dutch	French
Birefringence		Dubbele breking	Biréfringence
Brittle		Bros	Fragile
Compression set	%	Compressie set	Ensemble de compression
Compressive strength	MPa	Druk sterkte	Résistance à la compression
Creep		Kruip	Fluage
Density	g/cm^3	Dichtheid	Densité
Ductile		Ductiel	Ductile
Elongation	%	Rek	Allongement
Flexible		Flexibel/Buigzaam	Souple
Flexural modulus	MPa	Buigmodulus	Module de flexion
Flexural strength	MPa	Buigsterkte	Résistance à la flexion
Glass transition temperature	°C	Glas overgangs temperatuur	Température de transition vitreuse
Hardness		Hardheid	Dureté
Impact strength	J/cm	Slagsterkte	La résistance aux chocs
Permeability		Doorlaatbaarheid	Perméabilité
Refractive index		Brekingsindex	Indice de réfraction
Rigid/stiff		Stijf	Rigide
Shear strength	Pa	Afschuifsterkte	Résistance au cisaillement
Shrinkage		Krimp	Retrait/rétractation
Solubility		Oplosbaarheid	Solubilité
Stress/strain		Stress/spanning	Contrainte/déformation
Strong/strength		Sterk/sterkte	Fort/force
Surface tension/ Surface energy	Nm^{-1} Jm^{-2}	Oppervlakte spanning/ Oppervlakte energie	Tension superficielle Energie de surface

German	Italian	Spanish	Portuguese
Doppelbrechung	Birifrangenza	Birrefringencia	Birrefringência
Spröde	Fragile	Frágil	Frágil
Kompressionsset	Set di compressione	Conjunto de compresión	Conjunto de compressão
Druckfestigkeit	Resistenza alla compressione	Fuerza de compresión	Resistência à compressão
Kriechen	Scorrimento	Deslizamiento	Deslizamento
Rohdichte	Densità	Densidad	Densidade
Dehnbar	Duttile	Dúctil	Dúctil
Verlängerung	Allungamento	Alargamiento	Alongamento
Flexibel	Flessibile	Flexible	Flexível
Biegemodul	Modulo di flessione	Moduló de flexión	Módulo de flexão
Biegefestigkeit	Resistenza alla flessione	Resistencia a la flexión	Resistência à flexão
Glasübergangstemperatur	Temperatura di transizione del vetro	Temperatura de transición del vidrio	Temperatura de transição vítrea
Härte	Durezza	Dureza	Dureza
Schlagfestigkeit	Resistenza d'impatto	Resistencia al impacto	Resistência ao impacto
Permeabilität	Permeabilitá	Permeabilidad	Permeabilidade
Brechungsindex	Indice di rifrazione	Indice de refracción	Indice de refração
Starr/Steif	Rigido	Rigido	Rígido
Schiere Stärke	Resistenza al taglio	Resistencia a la cizalladura	Resistência ao cisalhamento
Schwindung	Restringimento	Encogimiento	Encolhimento
Löslichkeit	Solubilità	Solubilidad	Solubilidade
Stress/Belastung	Sforzo/deformazione	Tensión-deformación	Tensão/deformação
Stark/Stärke	Forte/forza	Fuerte	Forte/resistente
Oberflächen Spannung Oberflächen Energie	Tensione superficiale Energia di superficie	Tensión superficial	Tensão superficial Energia superficial

English	**SI unit**	**Dutch**	**French**
Tensile modulus	GPa	Trekmodulus	Module de traction
Tensile strength	MPa	Trekkracht	Traction
Tear strength	kNm^{-1}	Treksterkte	Résistance à la traction
Torsional strength	Pa	Torsie sterkte	Résistance à la torsion
Tough/hard		Sterk/hard	Fort/dur
Vicat softening point	°C	Vicat-verwekingspunt	Point de ramollissement Vicat
Viscoelasticity		Visco-elasticiteit	Viscoélasticité
Water absorption	%	Waterabsorptie	Absorption de l'eau
Yield point		Rekgrens	Seuil de rentabilité
Young's modulus	GPa	Young's modulus	Module de Young

German	Italian	Spanish	Portuguese
Zugmodul	Modulo di tensione	Módulo de tracción	Módulo de tração
Reissfestigkeit	Resistenza alla trazione	Tracción	Resistência à Tração
Reissfestigkeit	Resistenza alla lacerazione	Resistencia al desgarre	Resistência ao rasgo
Torsionsfestigkeit	Resistenza alla torsione	Resistencia de torsión	Resistência à torçào
Kräftig/stark	Duro	Dureza	Resistente
Vicat-Erweichungspunkt	Punto di rammollimento Vicat	Punto de reblandecimiento	Ponto de amolecimento Vicat
Viskoelastizität	Viscoelasticità	Viscoelasticidad	Viscoelasticidade
Wasseraufnahme	Assorbimento dell'acqua	Absorción de agua	Absorção de água
Fliessgrenze	Punto di snervamento	Limite de elasticidad	Ponto de cedência
Youngscher Modul	Modulo di Young	Módulo de Young	Módulo de Young

Appendix D. Plastics Nomenclature: ISO and ASTM Standards

Appendix D lists the standardized names and abbreviations for the plastics and elastomers discussed in this book according to the International Organization for Standardization (ISO) (ISO 1043-1, ISO 1629, ISO 18064) and ASTM International (formerly the American Society for Testing and Materials) (ASTM D1600-18 and ASTM D 1566-21). While the standardization of names and abbreviations for plastics and elastomers is useful and convenient, slight variances exist between ISO and ASTM nomenclature. ISO and ASTM issue regular updates to the standardization of chemical names to account for evolution and change over time. The information provided below summarizes the standards current at the time of publication. When materials featured in the fact sheets were not present in ISO or ASTM standards, the names chosen reflect common, everyday usage.

Appendix D. Standardized Names and Abbreviations for Plastics and Elastomers

Fact sheet number	Fact sheet name and abbreviation	ISO name and abbreviation	ASTM name and abbreviation
12.1	Horn	—	—
12.2	Shellac	—	—
12.3	Tortoiseshell	—	—
12.4	Vulcanized fiber	—	—
13.1	Cellulose acetate (CA)	Cellulose acetate (CA)	Cellulose acetate (CA)
13.2	Cellulose acetate-butyrate (CAB)	Cellulose acetate butyrate (CAB)	Cellulose acetate-butyrate (CAB)
13.3	Cellulose acetate-propionate (CAP)	Cellulose acetate propionate (CAP)	Cellulose acetate-propionate (CAP)
13.4	Cellulose nitrate (CN)	Cellulose nitrate (CN)	Cellulose nitrate (CN)
13.5	Ethyl cellulose (EC)	Ethyl cellulose (EC)	Ethyl cellulose (EC)
13.6	Polyamide-6 / Nylon-6 (PA-6)	—	Polyamide 6 (PA6)
13.7	Polyamide-6,6 / Nylon-6,6 (PA-6,6)	—	Polyamide 66 (PA66)
13.8	Polyamide-12 / Nylon-12 (PA-12)	—	Polyamide 12 (PA12)
13.9	Polycarbonate (PC)	Polycarbonate (PC)	Polycarbonate (PC)
13.10	High-density polyethylene (PE-HD)	Polyethylene, high density (PE-HD)	High density polyethylene plastics (HDPE)

Fact sheet number	Fact sheet name and abbreviation	ISO name and abbreviation	ASTM name and abbreviation
13.11	Low-density polyethylene (PE-LD)	Polyethylene, low density (PE-LD)	Low density polyethylene plastics (LDPE)
13.12	Ultra-high molecular weight polyethylene (PE-UHMW)	Polyethylene, ultra high molecular weight (PE-UHMW)	Ultra-high molecular weight polyethylene (UHMWPE)
13.13	Polyethylene terephthalate / Saturated polyester (PET, PETE, PETP)	Poly(ethylene terephthalate) (PET)	Poly(ethylene terephthalate) (PET) / Saturated polyester plastic (SP)
13.14	Polymethyl methacrylate (PMMA)	Poly(methyl methacrylate) (PMMA)	Poly(methyl methacrylate) (PMMA)
13.15	Polyoxymethylene (POM) / Polyacetal / Acetal	Polyoxymethylene / Polyacetal / Polyformaldehyde (POM)	Polyoxymethylene, polyacetal (POM)
13.16	Polypropylene (PP)	Polypropylene (PP)	Polypropylene (PP)
13.17	Polystyrene (PS)	Polystyrene (PS)	Polystyrene (PS)
13.18	Expanded polystyrene (EPS)	Polystyrene, expandable (PS-E)	Expanded polystyrene (EPS)
13.19	High-impact polystyrene (HIPS)	Polystyrene, high impact (PS-HI)	High impact-resistant polystyrene (HIPS)
13.20	Acrylonitrile-butadiene-styrene (ABS)	Acrylonitrile-butadiene-styrene plastic (ABS)	Acrylonitrile-butadiene-styrene plastics (ABS)
13.21	Acrylonitrile-styrene-acrylate (ASA)	Acrylonitrile-styrene-acrylate plastic (ASA)	Acrylonitrile-styrene-acrylate plastics (ASA)
13.22	Styrene-acrylonitrile (SAN)	Styrene-acrylonitrile plastic (SAN)	Styrene-acrylonitrile plastic (SAN)
13.23	Polyvinyl chloride, plasticized (PVC-P)	Poly(vinyl chloride) (PVC)	Poly(vinyl chloride) PVC
13.24	Polyvinyl chloride, rigid, unplasticized (PVC-U)	Poly(vinyl chloride), unplasticized (PVC-U)	—
13.25	Polyvinyl chloride–polyvinyl acetate copolymer (PVC/PVAC)	(vinyl chloride)-(vinyl acetate) plastic (VCVAC	Poly(vinylchloride-acetate (PVCA)
13.26	Polyvinylidene chloride (PVDC)	Poly(vinylidene chloride) (PVDC)	Poly(vinylidene chloride) (PVDC)
13.27	Cellophane (regenerated cellulose)	—	—
13.28	Polycaprolactone (PCL)	Polycaprolactone (PCL)	Poly(ε-caprolactone) (PCL)

Fact sheet number	Fact sheet name and abbreviation	ISO name and abbreviation	ASTM name and abbreviation
13.29	Polyhydroxybutyrate (PHB)	Poly(3-hydroxybutyrate) (PHB)	Polyhydroxy butyrate (PHB)
13.30	Polylactic acid (PLA)	Poly(lactic acid) (PLA)	Poly(lactic acid) (PLA)
13.31	Thermoplastic starch (TPS)	—	Thermoplastic starch (TPS)
14.1	Casein-formaldehyde (CS)	—	Casein (CS) / Caseine-formaldehyde resin (CSF)
14.2	Epoxy (EP)	Epoxide, epoxy resin (EP)	Epoxy, epoxide (EP)
14.3	Melamine-formaldehyde (MF)	Melamine-formaldehyde (MF)	Melamine-formaldehyde resin (MF)
14.4	Phenol-formaldehyde / Cast phenolic (PF)	Phenol-formaldehyde resin (PF)	Phenol-formaldehyde resin (PF)
14.5	Unsaturated polyester (UP) Glass fiber-reinforced unsaturated polyester (GRP)	Unsaturated polyester resin (UP)	Unsaturated polyester (UP)
14.6	Urea-formaldehyde (UF)	Urea-formaldehyde resin (UF)	Urea-formaldehyde resin (UF)
15.1	Balata rubber	—	—
15.2	Butadiene rubber (BR)	Butadiene rubber (BR)	Butadiene rubber
15.3	Butyl rubber / Isobutylene-isoprene rubber (IIR)	Isobutene-isoprene rubber, butyl rubber (IIR)	Isobutene-isoprene rubber (IIR)
15.4	Chloroprene rubber (CR)	Chloroprene rubber (CR)	Chloroprene rubber (CR)
15.5	Ethylene-propylene-diene-monomer rubber (EPDM)	Terpolymer of ethylene, propylene and a diene (EPDM)	Ethylene-propylene-diene plastics (EPD)
15.6	Ethylene-vinyl acetate elastomer (EVAC)	Ethylene-(vinyl acetate) plastic (EVAC)	Ethylene-vinyl acetate plastics (EVA)
15.7	Gutta-percha rubber	—	—
15.8	Hard rubber / Vulcanite	—	—
15.9	Nitrile-butadiene rubber (NBR)	Acrylonitrile-butadiene rubber/ nitrile rubber (NBR)	Acrylonitrile-butadiene rubber (NBR)
15.10	Natural rubber (NR)	Natural rubber (NR)	Natural rubber (NR)
15.11	Thermoplastic polyurethane elastomer (TPU)	Thermoplastic polyurethane (TPU)	Thermoplastic polyurethane (TPU)

Fact sheet number	Fact sheet name and abbreviation	ISO name and abbreviation	ASTM name and abbreviation
15.12	Polyurethane (thermosetting) (PUR)	Polyurethane (PUR)	Thermoset polyurethane (TSPU)
15.13	Silicone rubber (Si)	Silicone plastic (Si)	Silicone plastics (SI)
15.14	Styrene-butadiene rubber (SBR)	Styrene-butadiene rubber (SBR)	Styrene-butadiene rubber (SBR)
15.15	Styrene-ethylene-butadiene-styrene rubber (SEBS)	Styrene ethylene/butylene styrene block copolymer (TPS-SEBS)	Styrene-ethylene/butylene-styrene block copolymer (SEBS)

Glossary

Terms describing macromolecules, polymers, polymerization, polymer reactions, and polymer degradation are defined by various standards organizations, most notably the International Union for Pure and Applied Chemistry (IUPAC). IUPAC definitions are used by, or derived from, other standards organizations, including ASTM International (formerly the American Society for Testing and Materials) and the International Organization for Standardization (ISO). IUPAC, ASTM, and ISO definitions are precise, specific, and often very technical. Many of the definitions listed in this glossary are based on those definitions but have been simplified in keeping with the book overall. These and the other source documents listed below are largely available online.

Sources

ASTM, "D883-20b: Standard Terminology Relating to Plastics" (West Conshohocken, PA: ASTM International, 2020).

———, "ASTM D1600-18: Standard Terminology for Abbreviated Terms Relating to Plastics" (West Conshohocken, PA: ASTM International, 2018).

———, "F2792: Standard Terminology for Additive Manufacturing Technologies" (West Conshohocken, PA: ASTM International, 2012).

ISO, "ISO/ASTM 52900:2015: Additive manufacturing—General principles—Terminology" (Geneva: ISO, 2015).

IUPAC, *Compendium of Chemical Terminology*, 2nd ed. (the "Gold Book"), compiled by A. D. McNaught and A. Wilkinson (Oxford: Blackwell Scientific Publications, 2014).

Sylvia Katz, *Early Plastics*, 2nd ed. (Princes Risborough: Shire Publications, 1994).

———, *Plastics. Common Objects, Classic Designs. With a Collector's Guide* (New York: H. N. Abrams, 1984).

Morris Kaufman, *Giant Molecules: The Technology of Plastics, Fibers and Rubber*, pp. 74–76 (London: Aldus, 1968).

Plenco, "Glossary of Databook Tests," n.d., https://plenco.com/data/glossary.pdf.

Paul I. Smith, *Practical Plastics Illustrated* (London: Odhams Press, 1947).

Technische Hochschule Köln, "Terminology for Conservation of Plastics," 2020, https://www.th-koeln.de/en/terminology-for-conservation-of-plastics_73939.php.

United States Plastics Corporation, "Common Plastic Terminology," n.d., https://www.usplastic.com/knowledgebase/article.aspx.

Wohlers Associates, "Appendix A. Glossary of Terms," 2013, https://wohlersassociates.com/2013glossary.pdf.

The World of Plastics (London: British Plastics Federation in cooperation with the Plastics Industry in the UK, 1986).

accelerator. Chemical compound used to increase the rate of a plastic's or elastomer's cure or hardening.

addition polymer. Polymer made by addition polymerization.

addition polymerization. Polymerization by a repeated addition process in which monomers, oligomers, or macromolecules are linked together without the splitting off of water or other simple molecules. Also called *polyaddition.*

additive. Substance added to a polymer to alter one of its characteristics or to provide a specific functionality. Examples include antioxidants, antistatic agents, flame retardants, fillers, lubricants, pigments, plasticizers, slip agents, and UV absorbers.

additive manufacturing. Process of joining materials to make objects from 3D model data, usually layer upon layer. See also *3D printing.*

aging. Processes that occur in a polymeric material during a specified period of time and that usually result in changes in physical and/or chemical structure and the properties of the material.

amorphous. Without regular form, not crystalline, a highly disordered state.

annealing. Procedure for preventing or removing stresses within a plastic through the use of controlled heating and subsequent cooling of the plastic.

antioxidant. Chemical substance (an additive) that prevents or slows oxidation of plastics when they are exposed to air.

antistatic agent. Chemical substance applied to the surface of a plastic article or incorporated into the plastic from which the object is made. The antistatic agent renders the surface less susceptible to the accumulation of electrostatic charges, which attract and hold fine dirt or dust.

atactic. Polymer structure that has repeating units with no regular stereochemical configuration.

backbone. The linear chain to which all other polymer chains—long, short, or both—may be regarded as being pendant. Also called *main chain.*

binder jetting. An additive manufacturing process in which a liquid bonding agent is selectively deposited to join powder materials.

blister. Undesirable rounded elevation of the surface of a plastic whose boundaries may or may not be sharply defined. A blister may burst and flatten.

block copolymer. An essentially linear copolymer that has repeated sequences of polymeric segments with different chemical structures.

bloom. Visible exudation or efflorescence on a surface caused by plasticizers or lubricants. See *exudation.*

blowing agent. Compounding ingredient used to produce gas—by chemical or thermal action, or both—in the manufacture of cellular plastics.

blow molding. Low-cost method for mass-producing hollow objects that involves blowing air into a hollow tube of softened thermoplastic so that it expands to line the mold.

bond scission. Chemical reaction that results in the breaking of the backbone or bonds of a polymer molecule. Some bond scissions are classified according to the mechanism of the scission process: hydrolytic, mechano-chemical, thermal, photo-chemical, or oxidative scission. Also called *main-chain scission.*

branch. An oligomeric or polymeric offshoot from a macromolecular chain. An oligomeric branch may be termed a *short-chain branch*. A polymeric branch may be termed a *long-chain branch*.

branch point. Point on a chain where a branch is attached.

branched polymer. Polymer composed of molecules having a branched structure that is chainlike either between branch junctions or between each chain end and a branch junction. The branches are composed of the same monomers as the backbone chain.

brittle. A plastic that is hard and easily broken.

CAD (computer-aided design). Use of computing software in design processes. May also refer to a digital file produced by such software and used in additive manufacturing, other production techniques, or building construction.

calendering. Method of processing thermoplastics and thermoplastic elastomers between three or more hot and cold rollers rotating in opposite directions.

casting. (1) To form a plastic object by pouring a fluid monomer-polymer solution into an open mold where it finishes polymerizing. (2) Forming plastic film or sheet by pouring liquid plastic onto a moving belt or precipitating it in a chemical bath.

catalysis. Acceleration of a chemical reaction through the addition of a comparatively small amount of a foreign substance called a *catalyst*.

cell. A small cavity in plastic or elastomer surrounded partially or completely by walls.

cellular plastic. A plastic of which the density is reduced by the presence of numerous small cavities (cells), interconnected or not, that are dispersed throughout the mass. Also called *expanded plastic, foamed plastic.* A cellular plastic is often simply called a *foam.*

cellulose. The most frequently encountered carbohydrate, cellulose provides the framework of the main part of plant cell walls. Its fundamental molecules are polymers, based on two glucose esters.

chain. All or part of a macromolecule, an oligomer molecule, or a block, comprising a linear or branched sequence of constitutional units between two boundary constitutional units, each of which may be either an end group, a branch point, or an otherwise designated characteristic feature of the macromolecule. A cyclic macromolecule has no end groups but may nevertheless be regarded as a chain.

chain polymerization. Chain reaction in which the growth of a polymer chain proceeds exclusively by reactions between monomers and reactive sites on the polymer chain, with regeneration of the reactive sites at the end of each growth step. A chain polymerization consists of initiation and propagation reactions, and may also include termination and chain transfer reactions. Propagation in chain polymerization usually occurs without the formation of small molecules.

chain scission. See *bond scission*.

chemical bonds. Links between atoms in molecules arising from interactions between electrons.

chemically foamed plastic. Cellular plastic in which the cells are formed by gases generated from thermal decomposition or chemical reaction of the constituents.

chiral center. (1) An asymmetric carbon atom. (2) A carbon atom bonded with four different atoms.

cis-trans isomerism. If two groups with the same formula attached to a double bond are on the same side of a polymer chain, the isomer is called *cis*. If two groups attached to the double bond are on opposite sides of a polymer chain, the isomer is called *trans*.

closed-cell cellular plastic. Cellular plastic in which almost none of the cells are interconnected.

coefficient of linear thermal expansion (CLTE). The measurement of a material's change in dimensions with change in temperature.

coextrusion. Involves the extrusion process where two or more layers of different material are extruded together to form the wall of the tubing or item being formed.

cold flow. Change in dimension or shape in certain plastics when subjected to continuous stress at room temperature.

commodity plastics. Plastics that are used in high volume and in a wide range of applications, such as film for packaging, photographic and magnetic tape, clothing, beverage and trash containers, and a variety of household products where mechanical properties and service environments are not critical. Such plastics are low cost and have relatively low mechanical properties. Also called *consumer plastic.*

composite. (1) A structure or an entity made up of distinct components. (2) A complex material, such as fiberglass, in which two or more distinct, structurally complementary substances combine to produce structural or functional properties not present in any individual components. (3) Reinforced laminates (i.e., canvas phenolic, glass epoxy, etc.).

compound. A combination of ingredients before being processed or made into a finished product. Sometimes used as a synonym for *material, formulation*.

compression molding. Commonly used method for thermosetting plastics. Thermosetting powder is placed in the lower part of a two-part mold, the top part is closed, heat and pressure are applied, the plastic flows around the mold, and the final molded object is ejected.

compression set. Unrecoverable deformation (strain) that remains in a material after compressive loading has been removed.

compressive strength. Ability to resist a crushing force, calculated from the load at failure and the original cross-sectional area of the specimen.

condensation polymerization. Polymerization in which the growth of polymer chains proceeds by repeated condensation reactions in which monomers, oligomers, and macromolecules are linked together with the splitting off of water or other simple molecules of low mass. Also called *polycondensation.*

configuration. Structural arrangement of atoms in a molecule.

conjugated double bonds. Double bonds separated by a single bond.

consumer plastics. See *commodity plastics*.

copolymer. Polymer derived from more than one species of monomer.

copolymerization. Polymerization in which a copolymer is formed.

covalent bonding. Chemical bond with electron pairs shared between two atoms.

crazing. Fine cracks at or just under the surface of a plastic.

creep. Plastic deformation under continuous stress. The rate of deformation is governed by time.

cross-link. A small region in a macromolecule from which at least four chains emanate. Cross-links are formed by reactions involving sites or groups on existing macromolecules or by interactions between existing macromolecules. The small region may be an atom, a group of atoms, or a number of branch points connected by bonds, groups of atoms, or oligomeric chains. In the majority of cases, a cross-link is a covalent structure, but the term is also used to describe sites of weaker chemical interactions, portions of crystallites, and even physical interactions and entanglements.

cross-linking. Formation of chemical bonds to create a three-dimensional molecular network polymer resulting in changes in physical properties.

crystallinity. Molecular structure resulting from the formation of solid crystals with a geometric pattern.

curing. Creation of a polymer through polymerization and/or cross-linking.

degradation. Process leading to a significant change in the structure of a material, typically characterized by a change of properties (e.g., integrity, molecular mass or structure, mechanical strength) and/or by fragmentation, affected by environmental conditions, proceeding over a period of time and comprising one or more steps.

degree of polymerization. The number of monomeric units in a macromolecule, an oligomer molecule, a block, or a chain.

delamination. Separation of a laminate along the planes of its layers.

density. Mass per unit volume of a substance. For solids and liquids, typically, temperature would also be provided with density.

depolymerization. Process of converting a polymer or copolymer into a monomer or a mixture of monomers.

diene. Unsaturated hydrocarbon with two double bonds between carbons.

dimensional stability. Ability of a part to retain its size and proportions over time.

directed energy deposition. An additive manufacturing process in which focused thermal energy (e.g., from a laser, electron beam, or plasma arc) is used to fuse materials by melting them as they are being deposited.

ductility. A plastic's ability to deform under tensile stress. See also *malleability*.

durometer. A numerical scale for measuring the hardness of rubber or plastic based on the depth of penetration of an indenter point on the surface of a test specimen.

elasticity. Tendency for a plastic or elastomer to return to its original shape after removal of the force responsible for a change in form.

elastomer. (1) Material that if stretched will return to its original size and shape when released. (2) A rubbery polymer (*elastomer* is a shortening of *elastic polymer*). Elastomers are generally cross-linked but have a low cross-link density.

elongation. Fractional increase in length of a material stressed in tension.

embossing. Techniques used to create depressions of a specific pattern in plastic film and sheeting.

engineering plastics. Those plastics and polymeric compositions for which well-defined property measurements are available, such that engineering rather than empirical methods can be used for the design and manufacture of products that require definite and predictable performance in structural applications over a substantial temperature range.

environmental stress cracking. Crazing or cracking due to the combination of residual or applied stress in the material and chemical, thermal, or electromagnetic elements in an object's environment.

ester. A compound formed by the elimination of water during the reaction between an alcohol and an acid. Frequently used as plasticizers in rubber and plastic compounds.

expanded plastic. See *cellular plastic*.

extrusion. Process for molding continuous lengths of tubing, rod, film, and structural elements for building construction.

exudation. Migration of liquid constituents to the surface. Also called *bloom*.

feedstock. Unprocessed material used as a supply for a manufacturing process.

filler. Inert material added to a polymer to improve its properties, usually in powder or fiber form, such as wood pulp, cotton flock, and talc.

flexural modulus. Ratio of stress within the elastic limit of stress to corresponding strain.

flexural strength. Ability of a material to withstand a bending force calculated from the load required to break a standard test specimen when tested in a three-point loading situation.

flow lines. Patterns visible in moldings indicating the direction of the flow of the molten plastic in the mold.

foam, foamed plastic. See *cellular plastic*.

functional groups. Groups of atoms that confer particular characteristics in compounds containing them: hydroxyl groups of alcohols, carboxyl groups of organic acids, amino groups of amines.

functionality. Number of covalent bonds that a monomer can form with other monomers or with a polymer.

gauge. Thickness, size, or capacity of something as a standard measure.

glass transition. Reversible change in an amorphous polymer (or in amorphous regions of a partially crystalline polymer) from a viscous or rubbery condition to a hard and relatively brittle one, or vice versa.

glass transition temperature (T_g). Temperature range where a thermoplastic changes from a glassy state to a rubbery one.
gloss. The shine or luster of the surface of a material.

hardness. A plastic's resistance to permanent deformation. If it is very difficult to deform a plastic, it is called hard; when it is easy to deform, it is called soft. Hardness should not be confused with strength.
haze. Degree of cloudiness in a plastic.
heat deflection temperature. Temperature at which a loaded material deforms.
heat sealing. Method of joining plastic films by simultaneous application of heat and pressure to areas in contact. Heat may be supplied conductively or dielectrically.
homopolymer. Polymer derived from one species of monomer.
hydrolysis. Reaction of macromolecule with water involving the rupture of one or more bonds in the macromolecule.

impact resistance. Relative susceptibility of material to fracture by stress at high speeds.
impact strength. Resistance of a material to sudden impact or shock. Impact strength is determined by Izod or Charpy test.
impermeability. Characteristic that prevents the passage of other substances into or through a material.
impregnate. (1) To fill the voids and interstices of material with a compound. (2) To thoroughly soak a material that has an open or porous structure with a compound.
inhibitor. Substance used in low concentration that suppresses a chemical reaction. Inhibitors, unlike catalysts, are consumed during the reaction.
injection molding. Process whereby softened plastic is forced along inside a metal cylinder by a screw. The screw then stops and plunges the molten plastic into a cold mold. The remains of the sprue can usually be seen on the finished object.
isomers. Polymers made up of the same elements but in different arrangements. The differing arrangements provide different property values.
isotactic. Describes substituents positioned on the same side of the polymer chain.
Izod test. Method of determining the impact strength of a plastic material.

lamination. Method of making strong sheets of plastic by compressing thermoset-impregnated paper or fabric between steel plates.
latex. Milky fluid that comes from certain plants, primarily the rubber tree. Used to produce natural rubber.
linear polymer. (1) Polymer whose molecules consist of chains without cross-links. (2) A thermoplastic.
luminous transmittance. Amount of light that will pass through a plastic.

macromolecule. Molecule of high relative molecular mass (molecular weight), the structure of which essentially comprises the multiple repetition of units derived from molecules of low relative molecular mass. Macromolecules have at least 1000 atoms. Also called *polymer molecule.*
main chain. See *backbone.*
main-chain scission. See *bond scission.*
malleability. (1) Ability of a plastic or rubber to deform under compressive stress. (2) Ability to be shaped.
material extrusion. Manufacturing process in which material is selectively dispensed through a nozzle or orifice.
material jetting. Additive manufacturing process in which droplets of build material are selectively deposited.
melting temperature. Temperature at which the crystallinity in a semicrystalline polymer disappears and the polymer converts into a viscous liquid.
mer. See *monomer.*
migration. Transfer, usually undesirable, of a constituent of a plastic material to another contacting material.
modulus of elasticity. See *tensile modulus.*
mold. (1) To shape a substance in a cavity or on a frame. (2) The form in which the shaping takes place.
mold seam. Vertical line formed at the point of contact of mold halves. The prominence of the line depends on the tightness with which the mating mold sections fit together.
mold shrinkage. Automatic shrinkage that occurs when still-warm moldings are taken from the mold and allowed to cool.
molecular mass (MW). Sum of the masses of the atoms making up a molecule. Also called *molecular weight.*
monomer. Chemical compound, usually of low molecular mass, that can be converted into a polymer by combining it with itself or with other chemical compounds.

nondiene. Describes polymer chains with no double bonds.
number average molecular weight (Mn). Total number of molecules in a unit mass of polymer independent of their shape or size.

olefin. An unsaturated simple hydrocarbon with one double bond per molecule, such as ethylene or butylene. Olefins are chained into polymers, for example, polyethylene.
oligomer. A molecular complex of chemicals that consists of a few repeating units, in contrast with a polymer, where the number of monomers is in principle infinite. Dimers, trimers, and tetramers are oligomers composed of two, three, and four monomers, respectively.
opaque. Describes a material or substance that does not permit light to pass through.
open-cell cellular plastic. Cellular plastic in which almost all the cells are interconnecting.

parting line. See *mold seam.*
pendant group. A small group of atoms that hangs off the backbone chain of a polymer. Also called a *side chain.*

permeability. Ability of a material to transmit gases and liquids by passage through one surface and out at another surface by diffusion and sorption processes. Not to be confused with *porosity*.

plastic. (1) Capable of being deformed in shape and then remaining in the new shape when the deforming force has been removed. (2) A material that contains as an essential ingredient one or more organic polymeric substances of large molecular weight, is solid in its finished state, and, at some stage in its manufacture or processing into finished articles, can be shaped by flow. Rubber, textiles, adhesives, and paint, which may in some cases meet this definition, are not considered plastics. See also *elastomer*.

plastic memory. Ability of a plastic to return to its original shape when hot.

plasticizer. Chemical additive that makes a polymer more flexible and moldable.

poly. Prefix meaning "having many parts"; for example, polystyrene is made from many units of styrene.

polyaddition. See *addition polymerization*.

polycondensation. See *condensation polymerization*.

polymer. A product resulting from a chemical change involving the successive addition of a large number of relatively small molecules (monomers) to form the polymer.

polymer blend. A mixture of two or more polymers or copolymers to create a new polymer with different properties.

polymer molecule. See *macromolecule*.

polymer network. A three-dimensional structure formed by interchained covalent bonds.

polymerization. Process by which monomers are made to react together to form linear polymer chains or a three-dimensional network of polymer chains.

polyol. Polyfunctional alcohol with at least two hydroxyl groups.

polyolefin. Polymer with the general formula $(CH_2CH_R)_n$.

porosity. Property of a material that contains very fine continuous holes that allow the passage of gases, liquids, and solids in through one surface and out at another surface. Not to be confused with *permeability*.

powder bed fusion. Additive manufacturing process in which thermal energy selectively fuses regions of a powder layer.

rapid prototyping. Additive manufacturing of a design for form, fit, or functional testing, often iterative.

release agent. A lubricant, often wax, used to coat a mold cavity to prevent the molded piece from sticking to it, and thus to facilitate its removal from the mold.

resin. Originally this meant one of a group of organic solids or semisolids such as shellac or rosin secreted by certain plants and insects. Since the early years of the plastics industry, it has meant a natural or synthetic organic compound consisting of a noncrystalline or viscous liquid substance. Synthetic resins compose a large class of synthetic products that have some of the physical properties of natural resins but are different chemically. Resins are glassy and amorphous, and soften by heat.

ring-opening polymerization. Addition polymerization of a reactive end group of a polymer chain with a cyclic monomer. Reaction takes place at the broken bond of the cyclic monomer.

Rockwell hardness. Measure of a material's surface hardness. Rockwell hardness is measured using different scales corresponding to different indenter diameters and loads.

rotational molding. Process used for the manufacture of hollow plastics, including large storage tanks. It involves placing a thermoplastic powder into a mold and heating the mold in an oven while the mold rotates on perpendicular axes sequentially. Also known as *roto-molding* or *rotational casting*.

saturated compound. A compound without double bonds where all the carbon atoms are connected by single bonds.

semisynthetic plastics. Plastics made from natural materials treated with chemicals.

set. Strain remaining after complete release of the force producing a deformation.

shear strength. (1) Ability of a material to withstand shear stress or stress at which a material fails in shear. (2) Maximum shear stress that a material is capable of sustaining.

sheet lamination. An additive manufacturing process in which sheets of material are bonded to form an object.

Shore hardness. Measure of a material's surface hardness. Shore hardness is measured using different scales corresponding to different indenter diameters and loads.

side chain. See *pendant group*.

softening point. Temperature at which a plastic softens. Measurement involves a standard specimen deforming to a predetermined extent when subjected to a particular type of mechanical stress while being heated.

specific gravity. Density of a material in the molded form compared to the density of water.

sprue. (1) In injection molding, the channel through which the molten plastic is injected into the mold cavity. (2) The piece of hardened plastic left behind after unmolding, which is broken off.

stabilizers. Additives that are mixed with other substances (e.g., polymers) to inhibit reactions that result in undesirable chemical degradation.

stereoregular polymers. Tactic-, isotactic-, or syndiotactic-ordered arrangement of pendant groups along the polymer chains.

stereospecific polymerization. Polymerization with repeating units that have the same configuration.

stiff. A plastic is stiff when it is difficult to bend or fold. Stiffness is the opposite of flexible. A plastic is flexible when it is capable of being bent or flexed.

strain. Change, due to the application of a force, in the original size of an object.

strength. Capable of withstanding stress without breaking. Also referred to as *toughness* or *durability*.

stress crack. External or internal cracks in a plastic caused by tensile stresses of less than its short-time mechanical strength.

surface tension. Molecular force that tends to contract any given volume of liquid into a form with the minimum surface area.

surface treatment. Any method of treating a hydrophobic polymer so as to alter the surface and render it receptive to inks, paints, lacquers, and adhesives. Includes chemical, flame, and electronic treatments.

surfactant. A compound that affects interfacial tensions between two liquids. It usually reduces surface tension.

syndiotactic. Substituents positioned so that they alternate on the sides of the polymer chains.

synthesis. Building of chemical compounds from their elements or from simpler compounds.

tactic polymer. Regular polymer in which all the repeating units are identical.

tear strength. (1) Force required to initiate or continue a tear in a material under specified conditions. (2) Force acting substantially parallel to the major axis of the test specimen.

tensile elongation. Measure of a material's ductility. It represents the amount that a standard specimen stretches within a specific gauge length before it fractures and is expressed as the percentage of change in gauge length.

tensile modulus. Represents a material's ability to deform elastically when a force is applied to it. The tensile modulus of an object is defined as the slope of its stress-strain curve in the elastic deformation zone. Also called *modulus of elasticity* or *Young's modulus.*

tensile strength. Maximum tensile stress or pull required to break a standard test specimen.

terpolymer. Polymer composed of three different monomers.

thermal oxidation. Degradation induced by the combined action of heat and oxygen.

thermoforming. Shaping of heat-softened thermoplastic sheet through heat and/or vacuum.

thermoplastic. Plastic material which, when softened in a mold under heat and pressure, forms a shape that can be resoftened and remolded.

thermoplastic elastomer. Polymer with both thermoplastic and elastic properties.

thermoset. Plastic material that, under heat and pressure, polymerizes into a form that, due to cross-linking of the molecules, cannot be resoftened and remolded.

3D printing. Fabrication of objects through material deposition using a nozzle, print head, or other printer technology. See also *additive manufacturing.*

torsion. Strain created by twisting.

toughness. A measure of the energy a sample can absorb before it breaks. Plastic is said to be tough when it is strong, durable, and not easy to cut. Elastic plastics cut much more easily. See also *elasticity, strength.*

UV stabilizer. Any chemical compound that, when admixed to a material, selectively absorbs UV rays and minimizes chemical and/or physical changes that may be caused by ultraviolet radiation.

vacuum forming. See *thermoforming.*

van der Waals forces. Attraction of intermolecular forces between molecules.

vat photopolymerization. Additive manufacturing process in which liquid photopolymer in a vat is selectively cured by light-activated polymerization.

virgin grade material. Plastic material that has not undergone any processing other than that required to prepare it for manufacturing.

viscosity. Degree of internal friction of a liquid. Degree of mobility or fluidity.

vulcanization. Chemical process that makes rubber moldings elastic and rubbery. For example, sulfur combines with the rubber molecule under pressure at elevated temperatures (125°C–180°C) to cross-link the rubber chains, and thus changes its properties.

warping. Distortion in a plastic object's dimensions after molding or at degradation.

water absorption. Amount of water absorbed by a standard test specimen after immersion for twenty-four hours at 23°C, calculated as a percent gain in weight.

weather resistance. Ability of plastic to retain its original physical properties and appearance upon prolonged exposure to outdoor weather.

weight average molecular weight (Mw). Average mass of the individual chains that make up the overall molecular weight of the polymer.

yield point. Point at which a material will continue to elongate at no substantial increase in load during a short test period.

yield stress. Force that must be applied to a plastic to initiate flow.

Young's modulus. See *tensile modulus.*

References

Note: Names beginning with "van" are alphabetized by the name following the particle, such as Oosten, Thea B. van.

Agrawal, S. L., S. Mandot, S. Bandyopadhyay, R. Mukhopadhyay, and A. S. Deuri. 2005. "The Effect of Waxes on Rubber Vulcanisates." *Progress in Rubber, Plastics and Recycling Technology* 21 (2): 139–53. https://doi.org/10.1177/147776060502100204.

Allen, Norman S., Alfonso Chirinis-Padron, and Terence J. Henman. 1985. "The Photo-stabilisation of Polypropylene: A Review." *Polymer Degradation and Stability* 13 (1): 31–76. https://doi.org/10.1016/0141-3910(85)90133-8.

Al-Malaika, S. 1999. "Photostabilizers." In *Polypropylene: An A–Z Reference*, edited by J. Karger-Kocsis, 581–90. Dordrecht: Kluwer. https://doi.org/10.1007/978-94-011-4421-6_79.

Althöfer, Heinz. 1980. *Moderne Kunst: Handbuch der Konservierung*. Düsseldorf: Schwann.

Beerkens, Lydia. 2007. "Nothing but the Real Thing: Considerations on Copies, Remakes and Replicas in Modern Art." *Tate Papers*, 8 (Autumn). https://www.tate.org.uk/research/publications/tate-papers/08/nothing-but-the-real-thing-considerations-on-copies-remakes-and-replicas-in-modern-art.

———. 2012. "Artist Interview with Henk Peeters." In *The Artist Interview: For Conservation and Presentation of Contemporary Art, Guidelines and Practice*, edited by L. Beerkens, P. 't Hoen, IJ. Hummelen, V. van Saaze, T. Scholte, and S. Stigter, 112–19. Heyningen: JAP SAM Books.

———. 2014. "Upson—Installation" (unpublished condition report). The Hague: Gemeentemuseum Den Haag.

Birley, A. W. 1989. "Review of Thermoplastic Elastomers: A Comprehensive Review." *British Polymer Journal* 21 (2): 181. https://doi.org/10.1002/pi.4980210214.

British Plastics Federation. n.d. "Polymer: Bio-Based/Degradables." Plastipedia. Accessed May 27, 2021. https://www.bpf.co.uk/plastipedia/polymers/polymer-bio-based-degradables.aspx.

Brown, Roger, ed. 1999. *Handbook of Polymer Testing: Physical Methods*. Plastics Engineering 50. New York: Marcel Dekker.

Brydson, J. A. 1999. *Plastics Materials*. 7th ed. Oxford: Butterworth-Heinemann.

Campos, Cristian. 2007. *Plastic*. New York: Collins Design.

Cecchini, Cecilia, and Marco Petroni, eds. 2015. *Plastic Days: Materials and Design*. Cinisello Balsamo, Milan: Silvana Editoriale.

Charvat, Robert A., ed. 2003. *Coloring of Plastics: Fundamentals*. 2nd ed. New York: Wiley-Interscience.

Chiantore, Oscar, and Antonio Rava. 2005. *Conservare l'arte contemporanea: Problemi, metodi, materiali, ricerche*. Milano: Electa.

Cook, J. Gordon. 1993. *Handbook of Textile Fibres, II: Man-Made Fibres*. 5th ed. Durham: Merrow Publishing.

Crawford, R. J., and G. Stephens. 1985. "Microhardness Testing of Plastics Using Ball and Conical Indenters." *Polymer Testing* 5 (2): 113–36. https://doi.org/10.1016/0142-9418(85)90015-7.

Cudell, Ana, Rita Veiga, Thea van Oosten, Anna Laganà, Henk van Keulen, and Inês Venade. 2011. "Strategies for the Conservation of Cellulose Acetate Artworks—A Case Study of Two Plastic Books." In *ICOM-CC 16th Triennial Conference Lisbon 19–23 September 2011: Preprints*, edited by Janet Bridgland, 1–8. Lisbon: Critério-Produção Grafica. https://www.icom-cc-publications-online.org/1247/Strategies-for-the-conservation-of-cellulose-acetate-artworks--case-study-of-two-plastic-books.

Davis, Anthony, and David Sims. 1983. *Weathering of Polymers*. London: Elsevier Applied Science Publishers.

Deckard, Carl R. 1989. "Method and Apparatus for Producing Parts by Selective Sintering." US Patent 4,863,538, filed October 17, 1986; issued September 5, 1989. https://patents.google.com/patent/US4863538A/en.

Derrick, Michele, Dusan Stulik, and Eugena Ordonez. 1993. "Deterioration of Cellulose Nitrate Sculptures Made by Gabo and Pevsner." In *Saving the Twentieth Century: The Conservation of Modern Materials: Proceedings of a Conference, Symposium '91: Saving the Twentieth Century, Ottawa, Canada, 15 to 20 September, 1991*, edited by David W. Grattan, 169–82. Ottawa: Canadian Conservation Institute.

Domininghaus, Hans. 1988. "Guide Chart of Physical Properties of Plastics." In *Die Kunststoffe und ihre Eigenschaften,* 3rd ed. Berlin, Heidelberg: Springer-Verlag.

Duce, C., L. Bernazzani, E. Bramanti, A. Spepi, M. P. Colombini, and M. R. Tiné. 2014. "Alkyd Artists' Paints: Do Pigments Affect the Stability of the Resin? A TG and DSC Study on Fast-Drying Oil Colours." *Polymer Degradation and Stability* 105 (July): 48–58. https://doi.org/10.1016/j.polymdegradstab.2014.03.035.

Elias, Hans-Georg. 2003. *An Introduction to Plastics*. 2nd ed. Weinheim: Wiley-VCH.

Freinkel, Susan. 2011. *Plastic: A Toxic Love Story*. Boston: Houghton Mifflin Harcourt.

Gilbert, Marianne, ed. 2017. *Brydson's Plastics Materials*. 8th ed. Oxford: Butterworth-Heinemann.

Glenz, Wolfgang. 1984. *The Plastics Industry in Western Europe: Facts and Figures*. Munich: Hanser Publishers.

Grattan, David W., ed. 1993. *Saving the Twentieth Century: The Conservation of Modern Materials: Proceedings of a Conference, Symposium '91: Saving the Twentieth Century, Ottawa, Canada, 15 to 20 September, 1991*. Ottawa: Canadian Conservation Institute.

Gross, Richard A., and Bhanu Kalra. 2002. "Biodegradable Polymers for the Environment." *Science* 297 (5582): 803–7. https://doi.org/10.1126/science.297.5582.803.

Hancock, Thomas. 1920. *Personal Narrative of the Origin and Progress of the Caoutchouc or India-Rubber Manufacture in England*, centenary ed. London: Reprinted by James Lyne Hancock. First published 1857.

Harmsen, Paulien F. H., and Martijn Hackmann. 2012. *Groene bouwstenen voor biobased plastics: Biobased routes en marktontwikkeling*. Groene grondstoffen. Wageningen: Wageningen UR Food & Biobased Research. http://edepot.wur.nl/241016.

Hicks, Edward. 1961. *Shellac: Its Origin and Applications*. New York: Chemical Publishing Company.

Horie, C. V. 2010. *Materials for Conservation: Organic Consolidants, Adhesives and Coatings*. 2nd ed. Amsterdam: Butterworth-Heinemann.

Huke, D. W. 1961. *Introduction to Natural and Synthetic Rubbers*. London: Hutchinson Scientific & Technical.

Hummelen, IJsbrand, Dionne Sillé, and Marjan Zijlmans, eds. 1999. *Modern Art: Who Cares? An Interdisciplinary Research Project and International Symposium on the Conservation of Modern and Contemporary Art*. Amsterdam: Foundation for the Conservation of Modern Art; Netherlands Institute for Cultural Heritage.

Izzo, Francesca Caterina, Elisabetta Zendri, Guido Biscontin, and Eleonora Balliana. 2011. "TG-DSC Analysis Applied to Contemporary Oil Paints." *Journal of Thermal Analysis and Calorimetry* 104 (2): 541–46. https://doi.org/10.1007/s10973-011-1468-y.

Kaminitz, Marian. 1988. "Amazonian Ethnographic Rubber Artifacts." In *Preprints of Contributions to the Modern Organic Materials Meeting. Edinburgh, 14–15 April, 1988*, 143–50. Edinburgh: Scottish Society for Conservation and Restoration.

Katz, Sylvia. 1984. *Classic Plastics: From Bakelite to High-Tech: With a Collector's Guide*. London: Thames and Hudson.

———. 1994. *Early Plastics*. 2nd ed. Shire Album 168. Princes Risborough: Shire.

Kaufman, Morris. 1968. *Giant Molecules: The Technology of Plastics, Fibers, and Rubber*. London: Aldus.

Keneghan, Brenda. 1996. "Plastics? Not in My Collection." *V & A Conservation Journal* (21): 4–6.

Kessler, Kathrin, Thea van Oosten, and Henk van Keulen. 2004. "The AXA Art Conservation Project in Cooperation with the Vitra Design Museum: Research into Glassfibre-Reinforced Polyester." In "Modern Art, New Museums: Contributions to the Bilbao Congress, 13–17 September 2004," edited by Ashok Roy and Perry Smith. Supplement. *Studies in Conservation* 49, suppl. 2: 86–90. https://doi.org/10.1179/sic.2004.49.s2.019.

Kinloch, A. J. 1987. *Adhesion and Adhesives: Science and Technology*. London: Chapman and Hall.

Knapczyk, Jerome K., and Robert H. M. Simon. 1992. "Synthetic Resins and Plastics." In *Riegel's Handbook of Industrial Chemistry*, 9th ed., edited by James A. Kent, 623–707. New York: Van Nostrand Reinhold.

Krevelen, D. W. van. 1990. *Properties of Polymers: Their Correlation with Chemical Structure, Their Numerical Estimation and Prediction from Additive Group Contributions*. 3rd compl. rev. ed. Amsterdam: Elsevier.

Labet, Marianne, and Wim Thielemans. 2009. "Synthesis of Polycaprolactone: A Review." *Chemical Society Reviews* 38 (12): 3484–504. https://doi.org/10.1039/B820162P.

Laganà, Anna. 2012. "Cleaning Cellulose Acetate (CA): A Case Study." In *Preservation of Plastic Artefacts in Museum Collections*, edited by Bertrand Lavédrine, Alban Fournier, and Graham Martin, 258–61. Paris: Comité des travaux historiques et scientifiques.

Laganà, Anna, Michael Doutre, Melissa David, Michal Łukomski, Ella Hendriks, and Maarten van Bommel. 2021. "In Search of a Perfect Bond: An Evaluation of Potential Adhesives to Repair Transparent Poly(Methyl Methacrylate) Objects." In *Transcending Boundaries: Integrated Approaches to Conservation. ICOM-CC 19th Triennial Conference Preprints, Beijing, 17–21 May 2021*, edited by Janet Bridgland, 1–10. Paris: International Council of Museums.

Laganà, Anna, Julia Langenbacher, Rachel Rivenc, Maëlle Caro, Vincent Dion, and Tom Learner. 2017. "The Future of Looking Younger: A New Face for PMMA: Research into Fill Materials to Repair Poly(Methyl Methacrylate) in Contemporary Objects and Photographs." In *ICOM-CC 18th Triennial Conference Preprints, Copenhagen, 4–8 September 2017*, edited by Janet Bridgland, art. 0902. Paris: ICOM Committee for Conservation. https://www.icom-cc-publications-online.org/1656/The-future-of-looking-younger--A-new-face-for-PMMA-Research-into-fill-materials-to-repair-polymethyl-methacrylate-in-contemporary-objects-and-photographs.

Laganà, Anna, and Thea B. van Oosten. 2011. "Back to Transparency, Back to Life: Research into the Restoration of Broken Transparent Unsaturated Polyester and Poly(methyl methacrylate) Works of Art." In *ICOM-CC 16th Triennial Conference, Lisbon, September 19–23, 2011: Preprints*, edited by Janet Bridgland, 1–9. Almada, Portugal: Critério Produção Gráfica, Lda. https://www.icom-cc-publications-online.org/1260/Back-to-transparency-returning-to-life-Research-and-development-of-a-restoration-method-for-transparent-artworks-made-of-Unsaturated-Polyester-UP-and-Polymethylmethacrylate-PMMA-resins.

Laganà, Anna, Rachel Rivenc, Julia Langenbacher, John Griswold, and Tom Learner. 2014. "Looking through Plastics: Investigating Options for the Treatment of Scratches, Abrasions, and Losses in Cast Unsaturated Polyester Works of Art." In *ICOM-CC 17th Triennial Conference Preprints, Melbourne, 15–19 September 2014*, edited by Janet Bridgland, art. 1005. Paris: ICOM Committee for Conservation. https://www.icom-cc-publications-online.org/1491/Looking-through-plastics-Investigating-options-for

-the-treatment-of-scratches-and-losses-in-cast-unsaturated-polyester-and-polymethylmethacrylate-works-of-art.

Lavédrine, Bertrand, Alban Fournier, and Graham Martin, eds. 2012. *Preservation of Plastic Artefacts in Museum Collections*. Paris: Comité des travaux historiques et scientifiques.

Lemaire, J., R. Arnaud, and J. Lacoste. 1988. "The Prediction of the Long-Term Photoageing of Solid Polymers." *Acta Polymerica* 39 (1–2): 27–32. https://doi.org/10.1002/actp.1988.010390106.

Lieverloo, Karin van, ed. 2009. *Heringa/Van Kalsbeek: Country Trash*. Nijmegen: BnM Uitgevers.

Loadman, John. 2014. *Tears of the Tree: The Story of Rubber—A Modern Marvel*. Oxford: Oxford University Press.

Loadman, M. J. R. 1993. "Rubber: Its History, Composition and Prospects for Conservation." In *Saving the Twentieth Century: The Conservation of Modern Materials: Proceedings of a Conference, Symposium '91: Saving the Twentieth Century, Ottawa, Canada, 15 to 20 September, 1991*, edited by David W. Grattan, 59–80. Ottawa: Canadian Conservation Institute.

———. 1995. *The Exploitation of Natural Rubber*. London: Malaysian Rubber Producers Research Association. Reprint accessed May 28, 2021. https://holdenslatex.com/content/pdf/history.pdf.

Lokensgard, Erik. 2004. *Industrial Plastics: Theory and Applications*. 4th ed. Albany, NY: Thomson Delmar Learning.

Lower, Stephen. 2017. "Polymers and Plastics: An Introduction to Their Structures and Properties." Chem1 Virtual Textbook (website). Last modified October 23, 2017. http://www.chem1.com/acad/webtext/states/polymers.html.

Lucas, Josh. 2017. "Kaari Upson." *Interview Magazine* (April 19). https://www.interviewmagazine.com/art/kaari-upson-1.

MatWeb: Material Property Data. n.d. (Online database of engineering materials). Accessed August 24, 2021. http://www.matweb.com/.

Megens, Frans. 1991. "Wil Fruytier." PhD diss., Universiteit Nijmegen.

Modern Plastics Encyclopedia. 1947. Vol. 3, *Charts*. New York: Plastics Catalogue Corporation.

Mossman, Susan, ed. 1997. *Early Plastics: Perspectives, 1850–1950*. London: Leicester University Press; Science Museum.

Mullins, Charlotte. 2004. *RW: Rachel Whiteread*. Modern Artists. London: Tate Publishing.

Mutar, Mohammed Ali, and Noor Mohammed Abdul Hassan. 2017. "Synthesis and Characterization of New Alkyd Resins (Short, Medium and Long) Based on Sunflower Oil and Linoleic Acid as Binder for Paints." *International Journal of Chemical & Petrochemical Technology* 7 (1): 1–16. http://www.tjprc.org/view_paper.php?id=7926.

Odian, George G. 1991. *Principles of Polymerization*. 3rd ed. New York: Wiley.

Oosten, Thea B. van. 1999. "Material-Analysis (FTIR) of the Synthetic Rubber." In "Analysis and Investigation Report on the Rubber Material Used by Whiteread for SLAB II, September 1999." ICN Project 99-092; document 99/72, object number 2345 (unpublished report). Instituut Collectie Nederland.

———. 2002. "A Survey of Problems with 'Early' Plastics." In *Contributions to Conservation: Research in Conservation at the Netherlands Institute for Cultural Heritage (ICN)*, edited by Jaap A. Mosk and Norman H. Tennent, 87–96. London: James & James.

———. 2011. *PUR Facts: Conservation of Polyurethane Foam in Art and Design*. Amsterdam: Amsterdam University Press.

Oosten, Thea B. van, Irina Fundeanu, Clementine Bollard, Cristina de Castro, and Anna Laganà. 2008. "Lights Out! The Conservation of Polypropylene Wall Tapestries." In *Plastics: Looking at the Future and Learning from the Past: Papers from the Conference Held at the Victoria and Albert Museum, London: 23–25 May 2007*, edited by Brenda Keneghan and Louise Egan, 97–105. London: Archetype Publications.

Oosten, Thea van, and Katharina Haider. 2007. "Condition Report of the Polypropylene Tapestry 'Fête I' by Wil Fruytier." Amsterdam: Cultural Heritage Agency of the Netherlands. https://www.incca.org/sites/default/files/field_attachments/condition_report_fete_1.pdf/condition_report_fete_1.pdf.

Oosten, Thea B. van, Ineke Joosten, and Luc Megens. 2006. "Man-made Fibres from Polypropylene to Works of Art." In *The Future of the 20th Century: Collecting, Interpreting and Conserving Modern Materials: Postprints*, edited by Cordelia Rogerson and Paul Garside, 61–66. London: Archetype Publications.

Oosten, Thea van, and Anna Laganà. 2011. "Mending Broken Pieces: Research into Methods and Materials for Adhering Broken Unsaturated Polyester Artworks." In *Future Talks 009: The Conservation of Modern Materials in Applied Arts and Design: Papers from the Conference Held at the Pinakothek der Moderne, Munich, 22–23 October 2009*, edited by Tim Bechthold, 70–77. Munich: Die Neue Sammlung.

Oosten, Thea B. van, Anna Laganà, and Marleen Wagenaar. 2012. "Collection Survey: Survey at the Stedelijk Museum Amsterdam." In *Preservation of Plastic Artefacts in Museum Collections*, edited by Bertrand Lavédrine, Alban Fournier, and Graham Martin, 115–21. Paris: Comité des travaux historiques et scientifiques.

Oosten, Thea B. van, Yvonne, Shashoua, and Friederike Waentig, eds. 2002. *Plastics in Art: History, Technology, Preservation*. Kölner Beiträge zur Restaurierung und Konservierung von Kunst- und Kulturgut, Bd. 15. Munich: Siegl.

Pauly, S. 1999. "Permeability and Diffusion Data." In *Polymer Handbook*, vol. 2., 4th ed., edited by J. Brandrup, E. H. Immergut, and E. A. Grulke, VI/543–69. New York: John Wiley & Sons.

Pörschmann, Dirk, and Margriet Schavemaker, eds. 2015. *ZERO*. Amsterdam: Stedelijk Museum.

Pratt, Ken. 2009. "Liet Heringa & Maarten van Kalsbeek: Mixed Doubles." *Wound Magazine* (Fall): 1–6. https://heringavankalsbeek.nl/?portfolio=mixed-doubles.

Pullen, Derek, and Jackie Heuman. 1988. "Cellulose Acetate Deterioration in the Sculptures of Naum Gabo." In *Preprints of Contributions to the Modern Organic Materials Meeting: Held at the University of Edinburgh, 14 and 15 April 1988*, 56–66. Edinburgh: Scottish Society for Conservation and Restoration.

Quye, Anita. 1993. "Examining the Plastic Collections of the National Museums of Scotland." In *Conservation Science in the UK: Preprints of the Meeting Held in Glasgow, May 1993,* edited by Norman H. Tennant, 48. London: James & James.

Quye, Anita, and Colin Williamson, eds. 1999. *Plastics: Collecting and Conserving*. Edinburgh: NMS Publishing.

Riara, Martin M., Abdallah S. Merenga, and Charles M. Migwi. 2013. "Creep and Recovery Behavior of Compression Molded Low Density Polyethylene/Cellulose Composites." *Journal of Polymers* 2013: art. 209529. https://doi.org/10.1155/2013/209529.

Roberts, A. D., ed. 1988. *Natural Rubber Science and Technology*. Oxford Science Publications. Oxford: Oxford University Press.

Rodrigo, Evert, and Lydia Beerkens. 2005. "For the Benefit of Science." In *Modern Art: Who Cares? An Interdisciplinary Research Project and an International Symposium on the Conservation of Modern and Contemporary Art,* edited by IJsbrand Hummelen, Dionne Sillé, and Marjan Zijlmans, 42–48. London: Archetype Publications.

Rodriguez, Ferdinand, Gordon P. Bierwagen, George B. Kauffman, Alan N. Gent, J. Preston, and Malcom P. Stevens. 2016. "Major Industrial Polymers." *Encyclopedia Britannica Online*. https://www.britannica.com/topic/industrial-polymers-468698.

Rossem, Arnold van. 1958. *Rubber: Winning, Eigenschappen, Verwerking*. The Hague: Sevire.

Shah, Vishu. 2013. "Plastics Testing." In *Handbook of Measurement in Science and Engineering*, edited by Myer Kutz, 1463–96. Hoboken, NJ: John Wiley & Sons.

Shashoua, Yvonne. 2004. "Modern Plastics: Do They Suffer from the Cold?" In "Modern Art, New Museums: Contributions to the Bilbao Congress, 13–17 September 2004," edited by Ashok Roy and Perry Smith. Supplement. *Studies in Conservation,* 49, suppl. 2: 91–95. https://doi.org/10.1179/sic.2004.49.s2.020.

———. 2008. *Conservation of Plastics: Materials Science, Degradation and Preservation*. Amsterdam: Elsevier/Butterworth-Heinemann.

Shashoua, Yvonne, and Clare Ward. 1995. "Plastics: Modern Resins with Ageing Problems." In *Resins: Ancient and Modern*, edited by Margot M. Wright and Joyce H. Townsend, 33–37. Edinburgh: Scottish Society for Conservation and Restoration.

Shell Chemicals. 1967. *Properties of Plastics*. Rev. metric ed. London: Shell International Chemical Company and Shell Chemicals UK.

Simmons, H. E. 1921. *Rubber Manufacture: The Cultivation, Chemistry, Testing, and Manufacture of Rubber, with Sections on Reclamation of Rubber and the Manufacture of Rubber Substitutes*. New York: Van Nostrand. https://archive.org/details/rubbermanufactur00simm.

Sparke, Penny, ed. 1990. *The Plastics Age: From Modernity to Post-Modernity*. London: Victoria and Albert Museum.

Staudinger, H. 1920. "Über Polymerisation." *Berichte der deutschen chemischen Gesellschaft (A and B Series)* 53 (6): 1073–85. https://doi.org/10.1002/cber.19200530627.

Tate. 2017a. "Five Things to Know: Rachel Whiteread" (website). Accessed August 24, 2021. https://www.tate.org.uk/art/artists/rachel-whiteread-2319/five-things-know-rachel-whiteread.

———. 2017b. "Tate Britain Exhibition: Rachel Whiteread, 12 September 2017–24 January 2018" (website). Accessed August 24, 2021. https://www.tate.org.uk/whats-on/tate-britain/exhibition/rachel-whiteread.

Then, Edward, and Victoria Oakley. 1993. "A Survey of Plastic Objects at the Victoria and Albert Museum." *V & A Conservation Journal* (6): 11–14.

Ullmann's Polymers and Plastics: Products and Processes, vol. 1. 2016. Weinheim: Wiley-VCH.

Upson, Kaari. 2017. "Interview for Whitney Biennial 2017" (video). Whitney Museum of American Art. https://www.youtube.com/watch?v=blrSTu1da7I.

Waentig, Friederike. 2008. *Plastics in Art: A Study from the Conservation Point of View*. Petersberg, Germany: Imhof.

Ward, Pete. 1997. *Fantastic Plastic: The Kitsch Collector's Guide*. London: New Burlington Books.

Webster, Robert. 1994. *Gems: Their Sources, Descriptions and Identification*. 5th ed., revised by Peter G. Read. Oxford: Butterworth-Heinemann.

Wegen, D. H. van. 1999. "On the Way to *Une seconde d'éternité*." In *Modern Art: Who Cares? An Interdisciplinary Research Project and an International Symposium on the Conservation of Modern and Contemporary Art,* edited by IJsbrand Hummelen, Dionne Sillé, and Marjan Zijlmans, 115–18. Amsterdam: Foundation for the Conservation of Modern Art; Netherlands Institute for Cultural Heritage.

Woodruff, Maria Ann, and Dietmar Werner Hutmacher. 2010. "The Return of a Forgotten Polymer—Polycaprolactone in the 21st Century." *Progress in Polymer Science* 35 (10): 1217–56. https://doi.org/10.1016/j.progpolymsci.2010.04.002.

Wypych, George. 2016. *Handbook of Polymers*. 2nd ed. Toronto: ChemTec.

Yarsley, V. E., W. Flavell, P. S. Adamson, and N. G. Perkins. 1964. *Cellulosic Plastics: Cellulose Acetate, Cellulose Ethers, Regenerated Cellulose, Cellulose Nitrate*. London: Published for the Plastics Institute by Iliffe Books.

Young, Robert J., and P. A. Lovell. 1991. *Introduction to Polymers*. 2nd ed. London: Chapman & Hall.

Contributors

Conservation scientist **Thea B. van Oosten** is a Fellow of the International Institute for Conservation of Historic and Artistic Works and an associate researcher at the University of Amsterdam. She began her career in 1975 as a conservation scientist at the Cultural Heritage Agency of the Netherlands (RCE), where, in 1989, she was named senior conservation scientist and initiated the agency's research program on plastics.

Specializing in the conservation of modern and contemporary art and design objects, she has contributed to several publications and books, including *Modern Art: Who Cares?* (Foundation for the Conservation of Modern Art and the Netherlands Institute for Cultural Heritage, 1999), *Plastics in Art: History, Technology, Preservation* (Siegl, 2002), *PUR Facts: Conservation of Polyurethane Foam in Art and Design* (Amsterdam University Press, 2011), *Preservation of Plastic Artefacts in Museum Collections* (Comité des travaux historiques et scientifiques, 2012), and *Plastic Days: Materials and Design* (Silvana Editoriale, 2015).

Since her retirement from the RCE in 2011, van Oosten has been a freelance conservation scientist and adviser on the conservation of plastics in modern and contemporary art and design. She continues to share her knowledge and experience by teaching courses and workshops and lecturing on this topic around the world.

With contributions by

Lydia Beerkens is a senior conservator of modern and contemporary art at the Conservation Institute (SRAL), Maastricht, Netherlands.

Ana Cudell is a freelance conservator based in Porto, Portugal.

Anna Laganà is senior research specialist at the Getty Conservation Institute, Los Angeles.

Rita Veiga is a freelance conservator based in Vila Nova de Gaia, Portugal.

Illustration Credits

Fig. 1.9: Photo: Pavel Kubarkov / Shutterstock
Fig. 1.12: © Joe Colombo / Victoria and Albert Museum, London
Fig. 2.2: Getty Conservation Institute, Plastics Reference Collection / Photo: Anna Laganà
Fig. 2.3: The Work of Naum Gabo © Nina and Graham Williams / Tate, 2020
Fig. 2.4: Photo: Pictorial Press Ltd. / Alamy Stock Photo
Fig. 2.5: Photo: Panther Media GmbH / Alamy Stock Photo
Fig. 2.7: Image courtesy of Gufram
Fig. 2.8: Digital Image © The Museum of Modern Art / Licensed by SCALA / Art Resource, NY
Fig 2.9: Digital Image © The Museum of Modern Art / Licensed by SCALA / Art Resource, NY
Fig. 2.10a: Compostable plastics logo courtesy of Europe Bioplastics
Fig. 2.10b: Courtesy of Biodegradable Products Institute
Fig. 2.11: Periodic table of thermoplastics courtesy of and © Tangram Technology, Ltd.
Fig. 2.12: ©Andreas Slominski / Image courtesy of Museum MMK für Moderne Kunst / Photo: Axel Schneider, Frankfurt am Main
Fig. 2.14: © 2021 Artists Rights Society (ARS), New York / ADAGP, Paris
Fig. 2.15: Image courtesy of Collection S.M.A.K., Ghent / Photo: Rony Heirman
Fig. 2.16: © 2018 Hattula Moholy-Nagy / VG Bild-Kunst, Bonn / Artists Rights Society (ARS), New York
Fig. 2.17: Photo courtesy of Kartell
Fig. 2.18: © Madeleine Berkhemer
Fig. 2.20: © The Museum of Modern Art / Licensed by SCALA / Art Resource, NY
Fig. 2.21: Photo: PA Images / Alamy Stock Photo
Fig. 2.22: Image courtesy of Collection Kröller-Müller, Otterlo, the Netherlands
Fig. 2.23; p. 168 (detail); back cover: © Jean Shin / Commissioned by University Art Museum, Albany, NY
Fig. 3.4; fact sheet 15.10 (top): https://commons.wikimedia.org/wiki/File:Rubber_tree_in_Habiganj_Upazila,_Bangladesh._(7).jpg. Creative Commons (CC BY-SA-4.0) / Photo: Salim_Khandoker
Fig. 3.8: Photo courtesy of Field Museum, B78625
Fig. 3.9: Photo courtesy of BDA Museum, 08-2006 39
Fig. 3.10: Photo: Qiu Ju Song / Shutterstock
Fig. 3.11: Photo: John Frost Newspapers / Alamy Stock Photo
Fig. 3.12: From Robert Shanks and Ing Kong, "Thermoplastic Elastomers," in *Thermoplastic Elastomers*, ed. Adel El-Sonbati (London: IntechOpen, 2012). 10.5772/36807. Creative Commons (CC BY 3.0)
Fig. 4.1: From D.W. Van Krevelen, *Properties of Polymers*, 3rd ed., p. 49. © 1990 Elsevier
Fig. 4.3: From George Odian, *Principles of Polymerization*, 3rd ed., p. 34. © 1991 John Wiley & Sons
Fig. 4.4: From azom.com/images/Article_Images/ImageForArticle_510_4388520220035886443.gif
Figs. 4.13a, 4.13b, 4.13c: From D.W. Van Krevelen, *Properties of Polymers*, 3rd ed., p. 413. © 1990 Elsevier
Fig. 4.15: Adapted from matweb.com/reference/compressivestrength.aspx
Fig. 4.18: Adapted from matweb.com/reference/flexuralstrength.aspx
Fig. 4.24: Adapted from midwestinnovators.com/material-properties/durometer.html
Fig. 4.25: From artmolds.com/shore-hardness
Fig. 4.32: From Bertrand Lavédrine, Alban Fournier, and Graham Martin. *Preservation of Plastic Artefacts in Museum Collections.* © 2012 Comité Des Travaux Historiques et Scientifiques
Fig. 4.36a: Image courtesy of Collection Kröller-Müller, Otterlo, the Netherlands
Fig. 4.38: © The Museum of Modern Art / Licensed by SCALA / Art Resource, NY
Fig. 4.41a: Photo: Anton Starikov / Alamy Stock Photo
Fig. 4.41b: Photo: Metta Stock / Alamy Stock Photo
Fig. 4.41c: Photo: Danny Smythe / Alamy Stock Photo
Fig. 4.42: Photo: https://commons.wikimedia.org/wiki/File:Birefringence_of_stress_areas_on_a_DVD_jewel_case.jpg. Creative Commons (CC BY-SA 3.0) / Photo: Mk2010
Fig. 4.43: Photo: Shine Mathew / Shutterstock
Fig. 5.2: Photo: Peter Sobolev / Shutterstock
Fig. 5.3b: Photo: Vikentiy Elizarov / Shutterstock
Fig. 5.5: Photo: Oleksiy Mark / Shutterstock
Fig. 5.7: From Ulrich Meier-Westhues, Karsten Danielmeier, and Peter Kruppa, *Polyurethanes: Coatings, Adhesives and Sealants.* © Vincentz Network. Used with permission.
Fig. 5.8: From injectionmoldingonline.com/Molding101/InjectionMolding.aspx
Fig. 5.9: From msicarolina.com/2016/05/02/what-is-thermostat-molding/
Fig. 5.12: Photo: Pixel B / Shutterstock
Fig. 5.13: Photo: Chinahbzyg / Shutterstock
Fig. 5.15: Photo: ClassicStock / Alamy Stock Photo
Fig. 5.17; fact sheet 13.19: Images courtesy of Hindleys
Fig. 5.18: Image courtesy of Hapco Inc.
Fig. 5.19; front cover: © Gaetano Pesce / Photo: Corsi Design, www.corsidesign.it
Fig. 5.24: From Harold F. Giles Jr. and John R. Wagner Jr., *Extrusion: The Definitive Processing Guide and Handbook*, 2nd ed., fig. 47.5. © 2013 Elsevier. Used with permission.
Fig. 5.29: Laszlo Lipot, notesandsketches.co.uk
Figs. 5.30, 5.32, 5.38a: Adapted from printspace3d.com
Fig 5.31: Photo: Shutterstock
Fig. 5.33: The Art Institute of Chicago / Art Resource, NY
Fig. 5.35: From stratasys.com. Used with permission.
Fig. 5.37: © Eric van Straaten / Image from https://ericvanstraaten.com/
Figs. 6.1, 6.2, 6.3a, 6.3b, 6.4, 6.5, 6.6a, 6.6b, 6.7: © Henk Peeters / Photo: Lydia Beerkens
Figs. 7.1, 7.2a, 7.2b, 7.2c, 7.2d, 7.2e: © Wil Fruytier / Photo: Thea B. van Oosten
Figs. 7.3, 7.4, 7.5: © Wil Fruytier / Photo: Anna Laganà
Figs. 8.1, 8.2a, 8.2b, 8.3, 8.4, 8.5, 8.6a, 8.6b, 8.7a, 8.7b: © Lourdes Castro / Artists Rights Society (ARS), New York / Photo: Ana Cudell
Figs. 9.1, 9.2: © Liet Heringa and Maarten van Kalsbeek
Figs. 9.3, 9.4, 9.5, 9.6, 9.7: © Liet Heringa and Maarten van Kalsbeek / Photo: Anna Laganà
Figs. 10.1, 10.2, 10.3, 10.4: © The Kaari Upson Trust / Photo: Lydia Beerkens
Figs. 11.1, 11.2, 11.3, 11.4, 11.5: © Rachel Whiteread / Courtesy of the artist and Luhring Augustine, New York / Photo: Kim Sluijter
Fact sheet 13.5: Image courtesy of Nebraska State Historical Society
Fact sheet 13.6: Photo: https://commons.wikimedia.org/wiki/File:Toothbrush_20050716_004.jpg. Public domain via Wikimedia Commons / Photo: Jonas Bergsten
Fact sheet 13.8: Photo: https://commons.wikimedia.org/wiki/File:Sheepshank_knot.jpg. Creative Commons (CC BY-SA-3.0) / Photo: Tom Murphy VII
Fact sheet 13.29: Image courtesy of Science and Society Picture Library
Fact sheet 13.30: Photo: Stockphotograf / Shutterstock
Fact sheet 15.1: Image courtesy of the Marian Library, Crèche Collection, University of Dayton
Fact sheet 15.2: Photo: Studiovin / Shutterstock
Fact sheet 15.4: Photo: EQRoy / Shutterstock
Fact sheet 15.5: Photo (bottom): Andy.LIU / Shutterstock
Fact sheet 15.6: Photo (bottom): PEPE / Alamy Stock Photo
Fact sheet 15.14: Images courtesy of Museum of Design in Plastics, Arts University Bournemouth

Index

Note: *Italic* page numbers indicate figures and tables.